Micronutrient Fertilizer Use in Pakistan

Micronutrient research has been an important component of the soil fertility and plant nutrition program in Pakistan since the identification of zinc deficiency in rice in 1969. Since then, considerable progress has been made on diagnosis and management of micronutrient nutrition problems in crops. However, now there is growing R&D evidence that micronutrient malnutrition in humans could be addressed through enriching staple food grains with micronutrients.

This book presents the latest R&D information on micronutrient problems in crop plants/cropping systems and their corrective measures. The current status, the constraints, and economic benefits of using micronutrient fertilizers for optimizing crop productivity and soil resource sustainability are discussed along with estimating future potential requirement of micronutrient fertilizers to optimize crop productivity, produce quality, and soil resource sustainability.

Wide-scale preventable micronutrient deficiencies in human populations originate from micronutrient-deficient soils over which staple cereals and other food crops are grown. This book summarizes R&D information on fertilizer use-based micronutrient biofortification in staple food grains to address *"hidden hunger"* in human populations. The book also presents the best management practices by which micronutrient deficiencies could be corrected in crop plants in a farmer-friendly manner.

Features

- Reviews the micronutrients R&D carried out in Pakistan over the past five decades
- Focuses on soil–plant analysis techniques for effective prognosis and diagnosis of micronutrient disorders
- Presents spatial variability maps of micronutrient deficiencies in agricultural soils and crops
- Provides value–cost ratios of using micronutrient fertilizers for major crops
- Works out current use level of micronutrient fertilizers and their potential future requirements in the country
- Discusses agronomic biofortification approach for enriching crop-based food with micronutrients to address *"hidden hunger"*
- Presents a compelling case for enhanced use of the deficient micronutrient fertilizers to optimize crop productivity, farmer income, and national economy
- Presents micronutrient fertilizer use recommendations for salient crops and discusses fertilizer use for micronutrients in the context of 4R nutrient stewardship
- Recommends future R&D needed for optimizing micronutrient nutrition of crops

Micronutrient Fertilizer Use in Pakistan

Historical Perspective and 4R Nutrient Stewardship

Abdul Rashid
Munir Zia
Waqar Ahmad

CRC Press
Taylor & Francis Group
Boca Raton New York London

CRC Press is an imprint of the
Taylor & Francis Group, an **informa** business

First edition published 2023
by CRC Press
6000 Broken Sound Parkway NW, Suite 300, Boca Raton, FL 33487-2742

and by CRC Press
4 Park Square, Milton Park, Abingdon, Oxon, OX14 4RN

CRC Press is an imprint of Taylor & Francis Group, LLC

ISBN: 978-1-032-30762-6 (hbk)
ISBN: 978-1-032-32269-8 (pbk)
ISBN: 978-1-003-31422-6 (ebk)

DOI: 10.1201/9781003314226

Typeset in Times
by SPi Technologies India Pvt Ltd (Straive)

Dedicated to

The late Dr. Faqir Muhammad Chaudhry
for initiating systematic research on micronutrient nutrition
of crops in Pakistan and establishing first AAS-equipped
micronutrient research lab in the country in the early 1970s.
His pioneer research group on micronutrients established zinc
deficiency in rice, wheat, and maize crops in the country, and
elucidated the mechanism of higher susceptibility to zinc deficiency
in flooded rice compared to wheat.

Contents

List of Figures

List of Tables

List of Annexures

List of Abbreviations

AARI	Ayub Agricultural Research Institute
AAS	Atomic absorption spectrophotometer
ARI	Agricultural Research Institute
AB-DTPA	Ammonium bicarbonate-diethylenetriaminepentaacetic acid
AJK	Azad Jammu and Kashmir
BCR	Benefit:cost ratio
BMP	Best Management Practices
DALY	Disability-adjusted life years
DTPA	Diethylenetriaminepentaacetic acid
EDDHA	Ethylene-diamine di (ortho-hydroxyphenylacetic acid)
EDTA	Ethylene-diamine tetra-acetic acid
FAO-UN	Food and Agriculture Organization of the United Nations
FFC	Fauji Fertilizer Company Limited
FYM	Farmyard manure
HWS	Hot-water soluble
ICP	Inductively coupled plasma-emission spectrometry
Ha	Hectare
IFA	International Fertilizer Industry Association
IFDC	International Fertilizer Development Center
IRRI	International Rice Research Institute
IPNS	Integrated plant nutrition system
IZA	International Zinc Association
KP	Khyber Pakhtunkhwa (previously North West Frontier Province, NWFP)
Kt	Killotonne (kiloton)
Mg	Mega gram (equal to metric tonne)
NARC	National Agricultural Research Center
NFDC	National Fertilizer Development Center
NPK fertilizers	Nitrogen, phosphorus, and potassium fertilizers
NFC	National Fertilizer Corporation
Mha	Million hectare
NIA	Nuclear Institute of Agriculture
NIAB	Nuclear Institute for Agriculture and Biology
NPK	Fertilizer containing N, P, and K
PARC	Pakistan Agricultural Research Council
PKR	Pakistani Rupee
PSQCA	Pakistan Standards & Quality Control Authority

RSF&STI	Rapid Soil Fertility and Soil Testing Institute
RDA	Recommended dietary allowance
R&D	Research and Development
VCR	Value:cost ratio
WHO	World Health Organization

Foreword

Plants use minerals present in soil and water in order to grow. If they don't get enough of any nutrient it can seriously affect their health. Hence, ensuring proper plant nutrition by using organic and mineral fertilizers to supplement the nutrients already available in the soil is essential.

"A mineral plant nutrient is an element which is essential or beneficial for plant growth and development or for the quality attributes of the harvested product of a given plant species grown in its natural or cultivated environment". According to this new definition of plant nutrients, atleast 20 nutrients can benefit plant health. This includes micronutrients such as boron (B), chlorine (Cl), copper (Cu), iron (Fe), manganese (Mn), molybdenum (Mo), nickel (Ni) and zinc (Zn) as well as often overlooked nutrients such as aluminum (Al), cobalt (Co), iodine (I), selenium (Se), silicon (Si) and sodium (Na). This new definition opens the scope for innovation to enhance crop yield and nutritional quality, and optimize nutrient management and overall agronomic performance.

Micronutrients are an essential component of balanced fertilization. In Pakistan, widespread zinc deficiencies affect yield and zinc density in grain, impacting the wellbeing of farming communities, and leading to input (nitrogen, phosphorus, water, land, energy, etc.) use inefficiencies and their environmental consequences. For instance, nitrogen use efficiency in Pakistan is close to 30%, well below the global average. Better nitrogen fertilizer management is essential to improve nitrogen use efficiency, but it is not enough. If zinc, boron and iron deficiencies remain widespread, nitrogen use efficiency will remain low, whether the source, rate, time and place of nitrogen application is right or not. Low nitrogen use efficiency also means high nitrous oxide emissions and their related climate change impact, a global threat that needs to be urgently tackled.

The Scientific Panel on Responsible Plant Nutrition calls for "a new paradigm for plant nutrition", and has identified ten higher-level questions that need to be resolved within the next 20 years. Micronutrients are relevant to most of these challenges. A key question related to micronutrients is how cropping systems can deliver high quality, more nutritious food while a handful of micronutrient-poor crops dominate the global food and feed chains. Agronomic biofortification with zinc, selenium and iodine has proved to be highly successful. Now, it is time to scale up these practices to maximize their benefits.

Micronutrients, just like macronutrients, must be applied using best management practices. 4R Nutrient Stewardship also applies to micronutrients to get the best from their use. In this connection, agronomic research and the fertilizer industry must work closely together to develop suitable micronutrient fertilizer products for various application methods (soil application, fertigation, foliar spray or seed treatment), as well as decision-making support tools that will advise farmers on the best application rate, time and place. An increasing number of micronutrient-containing commercial products are now available on the market. Encouraging their adoption is the main challenge. It requires effective last-mile delivery to the farmers and, where needed, policy incentives when farmers don't see direct benefits.

Whether in Pakistan or globally, multi-stakeholder dialogue and partnerships are key to materialize these opportunities into tangible impacts. Let's work all together towards this collective goal.

The authors earn credit for a unique publication that not only covers current usage of micronutrients at the country scale but also for working out their potential requirements. The gap in between is fairly wide; hence role of the fertilizer industry is crucial in generating demand and maintaining adequate supply of the quality products. The book is also unique in the way that it also covers soil, plant, and fertilizer related analytical methodologies for micronutrients as a ready reference for all the stakeholders, i.e., academia, field agronomists, extension staff, research scientists, and the industry.

Patrick Heffer
Deputy Director General,
International Fertilizer Association (IFA)

Preface

Eight micronutrients (i.e., boron (B), chlorine (Cl), copper (Cu), iron (Fe), manganese (Mn), molybdenum (Mo), nickel (Ni), and zinc (Zn)) are required for normal growth and reproduction of higher plants. The soils of Pakistan are predominantly alkaline–calcareous and have low organic matter content; hence, they are conducive to deficiencies of certain micronutrients in plants. The first-ever micronutrient deficiency identified in the country was of Zn in rice, in 1969. In the 1970s, cotton and rice yield increases with B fertilizer were also observed. Since then, micronutrient research has been an important component of the soil fertility and plant nutrition program.

The micronutrient status report for Pakistan, compiled in 1998 by the lead author of this book, helped to crystallize information on the nature and extent of micronutrient deficiencies in various crops and geographic areas and crop yield increases and profitability with micronutrient fertilizer use. That report helped in determining that field-scale micronutrient problems were limited to deficiencies of Zn, B, and Fe. It also helped in developing recommendations for deficiency management, determining potential micronutrient fertilizer requirements, and suggesting future research and development (R&D) needs. Since then, considerable R&D progress has been made on various aspects of micronutrients. Along with other advancements, there is growing evidence that micronutrient malnutrition in humans (*hidden hunger*) can be addressed by enriching staple food grains with micronutrients. Consequently, now micronutrient R&D is receiving greater emphasis.

The first four chapters of this book deal with the prognosis and diagnosis of micronutrient deficiencies, research data quality, and historical developments regarding micronutrient R&D in Pakistan. Micronutrient deficiencies are ascertained by considering crop sensitivity, plant deficiency symptoms, soil–plant analysis, and crop yield increases with micronutrient application. An FAO global study on micronutrients, published in 1982, identified Zn and Fe deficiencies and suspected excess B in soils and crops of Pakistan. Subsequent local research verified field-scale deficiencies of Zn and Fe in several crops, but refuted B toxicity; rather, widespread B deficiency was identified and established in several crops. Salient milestones regarding micronutrient R&D in the country are the establishment of a) field-scale Zn deficiency in rice in the 1970s; b) Fe chlorosis in fruit plants and some legume crops in the 1980s; c) B and Zn deficiency in cotton in the 1990s; and d) B deficiency in rice in the 2000s.

The local research also helped in the adoption of a multi-nutrient soil test, ammonium bicarbonate-diethylene-triamine-penta-acetic acid (AB-DTPA), for assessing Zn, Cu, Fe, and Mn (along with N, P, and K), and dilute HCl method for soil B. Plant analysis diagnostic criteria were also established for several local crop genotypes. The first atomic absorption spectrometer (AAS) for agricultural research in Pakistan became functional in 1973. Now, many laboratories have

AAS and even inductively coupled plasma spectrometer (ICP) for micronutrient analyses. Notable scientific advancements also include improved understanding of residual and cumulative effects of soil-applied micronutrient fertilizers and apparent soil micronutrient balances. Furthermore, cost-effective yield increases with micronutrient fertilizers have led to a vital "*pull force*" for the use of micronutrient fertilizers in a wide variety of crops. As a result, major fertilizer companies and numerous small vendors are now marketing a variety of micronutrient fertilizer products.

Chapters 5 and 6 discuss the micronutrient status of soils and crops, and crop yield and produce quality improvement with micronutrient fertilization. The post-1998 research has re-affirmed that crop deficiencies of economic significance are limited to Zn, B, and Fe. The earlier understanding was that Zn deficiency is the most important micronutrient problem; however, now the extent and severity of Zn and B deficiencies appear to be almost equal. Undoubtedly, research information for B is still far less than for Zn. The micronutrient-deficient areas in cotton–wheat belt in Punjab and Sindh provinces, and in rainfed Pothohar plateau have been mapped. Micronutrient deficiencies are hampering crop productivity as well as impairing produce quality. For instance, soil B fertilization of rice improves grain cooking quality and wheat grains are biofortified with foliar Zn fertilization. In general, soil application of 2–5 kg Zn ha^{-1} and 0.75–1.0 kg B ha^{-1} are adequate for 2–6 crops grown in the same field. Foliar sprays (2–3) of Zn, B, and Fe are also effective in managing crop deficiencies. To address Fe chlorosis in field crops and vegetables, the use of tolerant genotypes is suggested. Micronutrient fertilizer use is highly cost-effective, more so by foliar feeding. However, foliar Zn increases wheat yield as well as grain Zn density, which is so badly needed to address Zn malnutrition in the Pakistani population. Full adoption of Zn fertilizer use for wheat crop in Punjab province alone is estimated to provide an additional annual return of 405–1216 Million International Dollars (I\$), at a cost per disability-adjusted life years (DALYs), saved of I\$ 461–619.

Chapters 7 and 8 deal with the availability and quality aspects of micronutrient fertilizers, and their use efficiency by crops. Chapter 7 elaborates on specifications of standard micronutrient fertilizer products, available products in the country, national fertilizer policy, micronutrient fertilizer marketing, and fertilizer quality issues. Chapter 8 deals with micronutrient fertilizer use efficiency, residual and cumulative effects of micronutrient fertilization, and apparent soil micronutrient balances. Whereas the cost of foliar feeding is relatively less, soil fertilization leaves a beneficial residual effect for 2–6 subsequent crops. Considering economics and practical feasibility, growers can opt for either approach. For managing Zn deficiency in rice, the use of Zn-enriched nursery seedlings is an inexpensive and easier-to-adopt option. In-country field research revealed that even in the absence of Zn and B fertilization, the apparent soil balances remain positive in cotton–wheat as well as rice–wheat systems. Despite positive soil balances, however, these crops suffer from Zn and B deficiency – presumably because of high sorption of these micronutrients in calcareous soils. Thus, soil balances appear to be of little value to predict the micronutrient fertilizer requirement of crops. Also, contrary to general apprehension that regular B fertilization may lead to B build-up to toxic levels, field experimentation in major cropping systems of the country has revealed that 0.75–1.0 kg B ha^{-1} applied after every 2–6 crop seasons is quite safe. However, periodic soil testing is suggested to monitor the situation.

Chapters 9–11 deal with the current use and potential requirement of micronutrients in the country, strategies for improving micronutrient management, and fertilizer-based approach to address human malnutrition. Despite convincing evidences about the highly cost-effective impact of micronutrient fertilizers, their current use in the country is far less compared to the crop requirements. The potential fertilizer requirements *per annum* – estimated by using the recommended fertilizer rates approach, by considering the residual effect of soil-applied micronutrients – are about 12,537 tonnes of elemental Zn, 2396 tonnes of elemental B, and around 1298 tonne of FeSO$_4$. The gap between the current use and their potential requirement in the country is quite vast. For instance, the potential fertilizer Zn requirement is five times higher than its current use level. The situation is much worse in the case of fertilizer B, because its potential requirement is about 24-times higher than its current use level. The anticipated enhanced adoption of Zn and B fertilizer use over the next 10 years has been worked out. The constraints to micronutrient fertilizer use appear to be growers' ignorance about the need and benefits of fertilizing, inadequate purchasing power, poor-quality micronutrient fertilizer products, and application problems. The suggested strategies for improving the situation include awareness enhancement, subsidy on micronutrient fertilizers, provision of micronutrient-fortified fertilizers, and enhanced availability of quality micronutrient products. Multi-year multi-country field R&D, including Pakistan, has established that staple food grains of wheat and rice can be biofortified with Zn, iodine (I), and selenium (Se) by their simultaneous foliar feeding. Fertilizer use-based biofortification of Zn in wheat enhances crop productivity as well. In Pakistan, three high-Zn wheat cultivars, "*Zincol-2016*", "*Akabr-2019*", and "*Nawab-2021*", have been introduced, which can be further biofortified to the desired level by foliar feeding of Zn. Recent pilot trials of feeding "*Zincol-2016*"-based flour in Pakistan have confirmed significant improvement in human plasma Zn concentration, i.e., from 680 µg L^{-1} to 794 µg L^{-1}. In the follow-up second trial, 50 households were fed with Zn-biofortified and control wheat flour for a period of eight weeks. A significant increase in plasma Zn concentration was noted after four weeks (mean difference 41.5 µg L^{-1}); however, the difference was not present at the end of the intervention period, i.e., eight weeks. Further evidence is being collected at a larger scale under the BiZiFED2 project. Chapter 11 also discusses human I deficiency in the context of inadequate I in dietary sources because of its low content in soils, cereal grains, and irrigation waters in the country.

Chapter 12 deals with some of the misconceptions about micronutrients, i.e., B toxicity apprehension, the belief that Zn and P fertilizers should not be applied together, and the assertion that chelated micronutrient products are necessarily more effective than inorganic fertilizer products.

Chapter 13 pertains to R&D needs regarding micronutrient nutrition management. The final chapters (Chapters 14 and 15) of the book are devoted to strategies for managing micronutrient deficiencies in crop plants in the context of 4R nutrient stewardship and recommendations regarding R&D, technology transfer, and policy improvement needs. Fertilizer use for managing micronutrient deficiencies is discussed in the context of 4R nutrient stewardship, i.e., use of the right source, right rate, right time, and right place of fertilizer use. This important topic has been elaborated by suggesting 4R packages of micronutrient fertilizer use in major crops of the country, i.e., wheat, rice, cotton, maize, and sugarcane.

In short, wise use of micronutrient fertilizers, even on a fraction of the total deficient cultivated area in the country, can result in greater crop productivity, higher farm income, improved national economy, and soil resource sustainability. The book also aims at enhancing stakeholder awareness about the sound use of micronutrient fertilizers in the context of 4Rs, i.e., in which product form, at which dosage, at which crop growth stage, and by which method, the deficient micronutrients must be applied. The book culminates with key messages for the farming community aiming at effective micronutrient deficiency diagnosis and management.

Abdul Rashid
Munir Zia
Waqar Ahmad

Acknowledgements

The idea to write this book was conceived during the execution of FAO-Pakistan project GCP/PAK/130/USA "*Soil Fertility Management for Sustainable Intensification in Pakistan: Baseline Input Atlas and Promotion of Soil Fertility with Private Sector*", where IPNI's 4R Nutrient Stewardship was intensively promoted for the first time in Pakistan. To source relevant information for the write-up of this publication, all the relevant institutions/departments across Pakistan were requested to share summarized information about the relevant research work carried out within the domain of their institutes. Accordingly, focal persons (Associate Professor Dr. Abdul Wakeel – University of Agriculture Faisalabad; Professor Dr. Mohammad Jamal Khan – University of Agriculture Peshawar; Dr. Abid Subhani – Barani Agricultural Research Institute Chakwal; Dr. Khalid Mahmood – Nuclear Institute for Agriculture and Biology; Dr. Masood A. Shakir – Ayub Agricultural Research Institute, Faisalabad; Dr. Muhammad Aslam – International Center for Agricultural Research in the Dry Areas) and others were nominated by the respective institutes for sharing of relevant information.

Thanks are also extended to the staff of National Fertilizer Development Center (NFDC), Islamabad (Mr. Abdul Jalil, Mr. Sultan Muhammad, Dr. Ahmad Ali, and Dr. Muhammad Islam), and the representatives of national fertilizer companies, especially Fauji Fertilizer Company Limited, Engro Fertilizers, and Jaffer Brothers Pvt. Ltd. for sharing unpublished data about applied aspects of micronutrient fertilizers on important crops and their marketed volume(s). Additionally, the authors are grateful to Dr. Nisar Ahmad (Ex-Chief, NFDC, Islamabad); Dr. Ejaz Rafique (Land Resources Research Institute, NARC); Dr. Shahzada Munawar Mehdi (Ex-Director, Rapid Soil Fertility Survey and Soil Testing Institute, Lahore); Mr. Nasser Jaffer and Mr. Nadir Jaffer (Jaffer Brothers Pvt. Ltd.); Mr. Rao Muhammad Tariq (Fauji Fertilizer Company Limited); Dr. Tariq Aziz (Principal/Associate Professor Sub-Campus of the University of Agriculture Faisalabad, Depalpur; Dr. Qadir Baloch (late), and Dr. M. Yaqub Mujahid (HarvestPlus Pakistan); Dr. Muhammad Younis Arain (Southern Zone Agricultural Research Center of PARC); and Dr. Sagheer Ahmad (National Sugar Crops Research Institute of PARC, Thatta) for thought-provoking discussions and sharing useful information.

Last but not the least, the authors would also like to mention the technical support extended by Mr. Mike Kucera (Agronomist USDA, Natural Resources Conservation Service) and Mr. Khalil Hamid (Program Specialist, USDA/Foreign Agriculture Service at Washington, DC) for their guidance in finalizing contents of this book.

Authors Biographies

Abdul Rashid, PhD, is an accomplished soil fertility and crop nutrition expert with a special interest in micronutrients. He earned his PhD from the University of Hawaii, Honolulu, USA. Dr. Rashid spent a major part of his career at Pakistan's National Agricultural Research Center, where, finally, he served as the Director General. Later, he served as a Member (Bio-Sciences) of the Pakistan Atomic Energy Commission. He led a well-conceived research program for effective prognosis/diagnosis and cost-effective management of micronutrient disorders in field crops. His salient R&D contributions are identification, establishment, and mapping spatial variability of micronutrients; management of boron deficiency in cotton and rice grown in irrigated calcareous soils; and determination of plant analysis diagnostic norms for boron, zinc, and ferrous iron for many crops. He has published his research effectively, and his farmer-friendly micronutrient use technologies are adopted in the country. His consistent leadership and effective advocacy led to a *"pull force"* for micronutrient fertilizer use in the country. Currently, he is pursuing agronomic biofortification R&D for enriching staple cereal grains with micronutrients to combat *"hidden hunger"*. His monumental research contributions and persistent technology transfer to stakeholders have earned him many prestigious awards and recognitions, like *IFA Norman Borlaug Award, IPNI Science Award, East West Center Distinguished Alumni Award*, and *Fellowship of Pakistan Academy of Sciences*. Currently, he is Co-Chair of the Scientific Review Panel (Agricultural Sciences) of Pakistan's Higher Education Commission.

Munir Zia, PhD, is R&D Coordinator at Fauji Fertilizer Company Limited, Pakistan, since 2001, and also holds adjunct positions of Associate Professor and Research Scientist at the University of Nottingham, UK, and Center for Environmental Geochemistry, UK. Dr. Zia is an expert in soil fertility, plant nutrition, and environmental geochemistry with research experience across Pakistan, Australia, the US, and the UK. His particular interests are in the development and testing of enhanced efficiency fertilizers, spatial variability of essential plant nutrients especially micronutrients, crop nutrition, environmental risk assessment, GIS-based soil fertility mapping, and agriculture-linked human nutrition. He earned his doctorate degree in Soil Science in

2006 and has since worked as a Research Scientist at the University of South Australia (Endeavour Research Fellow, 2008/09); USDA – Agricultural Research Service-USA (Fulbright Research Scientist, 2009/10); UoN-BGS Center for Environmental Geochemistry-UK (Commonwealth Professional Fellow, 2012); Visiting Scientist at CEG –Keyworth (the Royal Society of Chemistry Fellowship, 2014); Executive Fellow at the University of Adelaide (Endeavour Fellowship, 2016); and Professional Fellow at BGS-UoN (Commonwealth Fellowship, 2018). He has published 50 research papers/book chapters to his credit in peer-reviewed journals and books with >2000 citations (h-index 20). He also holds FAO (UN) Chair of Asian Soil Partnership – Pillar 2 for Asia Region.

Waqar Ahmad, PhD, is an expert in soil and environmental chemistry, with an interest in carbon accounting and management for constrained ecologies, soil-linked human security, digital agriculture, and teaching in these domains for creating future leaders. He earned his PhD from the University of Sydney, NSW, Australia. He has been the recipient of the Endeavour Award 2015 – Commonwealth Scientific and Industrial Research Organization (ÇSIRO), in Perth, Western Australia. Dr. Ahmad has been coordinating for Research, Development, and Extension (RD&E) organizations since 2003. As a consultant with FAO-UN, he led two projects on "4R Nutrient Stewardship" – (*IPNI's philosophy for nutrients management*) funded by the United States Agency for International Development (USAID) and USDA-Foreign Agricultural Services in Pakistan. These projects were implemented in collaboration with the public and private sectors. He led the process of developing Soil Fertility Atlases for four provinces of Pakistan. The Atlases provide a comprehensive account of soil fertility status, classification, dominant soil series, parent material, native best management practices, fertilizer use trends, and respective yields at farm-gate, with a management strategy for resource-based improvement. While working with the FAO, he demonstrated expertise in training, designing curriculum, resource mobilization, and engaging communities. He has published significant research outcomes (~40) in highly reputed journals. Currently, he is associated as a Fellow with the University of Queensland, Australia, an Adjunct Associate Professor at the Centre of Excellence in Geology, at the University of Peshawar, Pakistan, and is a focal person of FAO-UN Asian Soil Partnership.

1 Micronutrients and Their Deficiencies

1.1 INTRODUCTION: MICRONUTRIENTS AND THEIR DEFICIENCIES

Seventeen elements are essentially needed by all higher plants for their normal growth and reproduction. Carbon (C), hydrogen (H), and oxygen (O) are obtained from CO_2 (in the atmosphere) and water. The remaining 14 elements are obtained by plants from the soil. Six of these, i.e., nitrogen (N), phosphorus (P), potassium (K), calcium (Ca), magnesium (Mg), and sulfur (S), act as "building blocks" in plants and, hence, are needed in large quantities. These elements required in large quantities are known as major or macronutrients. The other eight essential elements are needed by plants in minuscule quantities and, hence, are known as micronutrients. These are boron (B), chlorine (Cl), copper (Cu), iron (Fe), manganese (Mn), molybdenum (Mo), nickel (Ni), and zinc (Zn). Micronutrients are as essential for plants as macronutrients; the only difference is that micronutrients are required in much lesser quantities than macronutrients (Table 1.1). An inadequate supply of any micronutrient results in abnormal growth and/or reproduction of plants.

In the past, micronutrients were also termed "trace elements"; this term is inappropriate because it lacks the requisite message of essentiality. The newest element added to the list of micronutrients for plants is Ni as its essentiality was recognized in 1987 (Brown et al., 1987; Table 1.1). Some soil scientists and plant nutritionists like to call Ni "the forgotten essential trace element". Nickel is taken up from soil as Ni^{2+} and is readily mobile in plants. Some plant species translocate Ni to developing seeds. The concentration of Ni ≥ 10 mg kg^{-1} could be toxic in some plant species and ≥ 50 mg kg^{-1} becomes toxic even in moderately tolerant species. Like other heavy metal micronutrients, as soil pH increases, Ni availability to plants decreases. High divalent concentrations of Zn, Cu, or Fe can inhibit Ni uptake. As trace amounts of Ni are found in most fertilizers as a contaminant, it is unlikely to become deficient in plants. In case Ni deficiency is observed, foliar sprays of Ni salts or organic Ni ligands are suggested. Ni-lignosulfonate is preferred due to potential safety concerns.

Recently, Brown et al. (2021) have initiated an open scientific debate to reconsider the definition of "plant nutrient". In this context, they have stated that aluminum (Al), cobalt (Co), sodium (Na), selenium (Se), and silicon (Si) are known to beneficially impact plant growth; hence, they have designated these five elements as "essential for some plants", which otherwise were known to be "beneficial elements in plant nutrition" (Kirkby, 2012). Additionally, vanadium (V)

DOI: 10.1201/9781003314226-1

1

TABLE 1.1
Salient Characteristics of Essential Micronutrients for Higher Plants, Compared with Some Macronutrients

Nutrient	Year Essentiality Established; Source	Average Concentration in Plants*	Typical Concentration in Plants	Form Absorbed	Mobility in Plant
Micronutrients		$(mg\ kg^{-1})$	$(mg\ kg^{-1})$		
Molybdenum (Mo)	Arnon and Stout, 1939	0.1	0.1	MoO_4^{2-}	Moderately mobile
Nickel (Ni)	Brown et al., 1987	0.1	0.05	Ni^{2+}	Mobile
Copper (Cu)	Sommer, 1931	6	6	Cu^{2+}, Cu^+	Relatively immobile, but mobile under sufficient conditions
Zinc (Zn)	Sommer and Lipman, 1926	20	20	Zn^{2+}	Variably mobile
Boron (B)	Warington, 1923	20	20	$H_3BO_3^0$	Relatively immobile
Manganese (Mn)	McHargue, 1922	50	50	Mn^{2+}	Relatively immobile
Iron (Fe)	Sachs, 1860	100	100	Fe^{2+}	Relatively immobile
Chlorine (Cl)	Broyer et al., 1954	100	100	Cl^-	Mobile
Macronutrients					
Sulfur (S)		1000		SO_4^{2-}, SO_2	Relatively immobile
Phosphorus (P)		2000		H_2PO_4, HPO_4^{2-}	Mobile
Magnesium (Mg)		2000		Mg^{2+}	Mobile
Calcium (Ca)		5000		Ca^{2+}	Relatively immobile
Potassium (K)		10000		K^+	Mobile
Nitrogen (N)		15000		NO_3^-, NH_4^+	Mobile

Data from Kirkby, 2012; Tandon 1995, 2013; Havlin et al., 2017; Bell and Dell, 2008.
* Concentration expressed on a dry matter weight basis.

and iodine (I) are also considered beneficial to some plants but not necessary for the completion of the plant life cycle (Broadley et al., 2012; Havlin et al., 2017). Brown et al. (2021) believe that reconsidering the definition of "plant nutrient" has the potential to revitalize innovation and discovery in the realm of plant

TABLE 1.2

Salient Reviews in International Literature Documenting Micronutrient Deficiencies in Pakistan

Reference	Micronutrient(s)	Geographic Domain	Remarks
Sillanpää (1982, 1990)	All micronutrients	Global, including Pakistan	Sillanpää (1982) suspected B toxicity in salt-affected soils of Pakistan, particularly in the cotton belt. However, post-1982 extensive local research has no evidence of B toxicity
Mortvedt (1991)	All micronutrients	Global	A monograph
Rashid and Ryan (2004)	All micronutrients	Mediterranean-type soils, including Pakistan	Review of micronutrient R&D
Rashid and Ryan (2008)	All micronutrients	Near-East region, including Pakistan	Book chapter in *Micronutrient Deficiencies in Global Crop Production*
Alloway (2008)	All micronutrients	Global, including Pakistan	A monograph
Bell and Dell (2008)	All micronutrients	Global, including Pakistan	Book for practitioners and stakeholders in the fertilizer industry and for concerned policy makers
Imtiaz et al. (2010)	All micronutrients	Global, including Pakistan	Review on the role of micronutrients in crop production and human health
Ahmad et al. (2012)	Zinc deficiency in soils, crops, and humans	Global, including Pakistan	Global review on Zn deficiency in soils, crops, and humans
Ahmad et al. (2012)	Boron deficiency in soils and crops	Global, including Pakistan	Review on B deficiency in soils and crops
Rehman et al. (2012)	Zinc nutrition of rice production systems	Global, including Pakistan	Review on Zn nutrition in rice production systems
Ryan et al. (2013)	All micronutrients	Middle East-West Asia region, including Pakistan	Review on micronutrient problems and their management in the Middle East–West Asia region
Rehman et al. (2018)	Boron nutrition of rice	Global, including Pakistan	Review on B deficiency in rice and impacts of B fertilization on rice productivity and grain quality
Rehman et al. (2018)	Zinc nutrition in wheat-based cropping systems	Global, including Pakistan	Review of R&D concerning Zn nutrition in wheat-based cropping systems
Nadeem et al. (2019)	Micronutrients fertilizer use in rice–wheat cropping systems	South Asia, including Pakistan	Review on micronutrient fertilizer use in rice–wheat cropping system

nutrition, by way of leading to a better understanding of the role of "beneficial elements" in plant, animal, and human nutrition, following the holistic concept of "one nutrition".

In recent times, micronutrient nutrition of crops has gained more importance because of the following:

- Increased incidence of micronutrient deficiencies due to crop intensification, cultivation of high yielding crop cultivars, and higher susceptibility of new crop varieties compared with landraces and low-yielding old cultivars.
- Decreased use of organic manures and minimal recycling of crop residues.
- Increased use of micronutrient-free major nutrient fertilizers.
- Soil nutrient imbalances, primarily because of inadequate use of micronutrient fertilizers.
- Appreciable improvements in crop yield and produce quality with micronutrient fertilization, in micronutrient-deficient soil situations.
- The realization that low micronutrient density food grains lead to malnutrition (*hidden hunger*) in humans and consequent health problems.

Deficiencies of B, Zn, and Fe are well established in most irrigated and rainfed crops grown in 22 million hectares (Mha) cultivated soils. Some salient reviews in international literature documenting and discussing micronutrient deficiencies in Pakistan are listed in Table 1.2.

REFERENCES

Ahmad, W., M. Watts, M. Imtiaz, I. Ahmed, and M. H. Zia. 2012. Zinc deficiency in soils, crops and humans: a review. *Agrochimica* LVI(2): 65–97.

Ahmad, W., M. H. Zia, S. S. Malhi, A. Niaz, and Saifullah. 2012. Boron deficiency in soils and crops: A review. *Crop Plant*, Aakash K Goyal (Ed.), ISBN: 978-953-51-0527-5, In Tech, Available at: https://www.intechopen.com/chapters/35614

Alloway, B. J. (Ed.). 2008. *Micronutrient Deficiencies in Global Crop Production.* Springer, Dordrecht.

Arnon, D. I., and P. R. Stout. 1939. Molybdenum as an essential element for higher plants. *Plant Physiology* 14:599–602.

Bell, R. W., and B. Dell. 2008. *Micronutrients for Sustainable Food, Feed, Fibre and Bioenergy Production.* International Fertilizer Industry Association, Paris, France.

Broadley, M., P. Brown, I. Cakmak, et al. 2012. Beneficial elements. In: Marschner, P. *Marschner's Mineral Nutrition of Higher Plants*, 3rd ed. Academic Press, Amsterdam.

Brown, P. H., R. M. Welch, and E. E. Cary. 1987. Nickel: A micronutrient essential for higher plants. *Plant Physiology* 85:801–803.

Brown, P. H., F. H. Zhao, and A. Dobermann. 2021. What is a plant nutrient? Changing definitions to advance science and innovation in plant nutrition. *Plant and Soil.* doi:10.1007/s11104-021-05171-w.

Broyer, T. C., A. B. Charlton, C. M. Johnson, and P. R. Stout. 1954. Chlorine: a micronutrient element for higher plants. *Plant Physiology* 29:526–532.

Havlin, J. L., S. L. Tisdale, W. L. Nelson, and J. D. Beaton. 2017. *Soil Fertility and Fertilizers: An Introduction to Nutrient Management*, 8th ed. Pearson, Chennai, India.

Imtiaz, M., A. Rashid, P. Khan, M. Y. Memon, and M. Aslam. 2010. The role of micronutrients in crop production and human health. *Pakistan Journal of Botany* 42:2565–2578.

Kirkby, E. 2012. Introduction, definition and classification of nutrients. In: Marschner, P. (Ed.), *Marschner's Mineral Nutrition of Higher Plants*, 3rd ed., pp: 3–5. Academic Press, Amsterdam.

McHargue, J. S. 1922. The role of manganese in plants. *Journal of the American Chemical Society* 44:1592–1598.

Mortvedt, J. J.1991. Micronutrient fertilizer technology. In: Mortvedt, J. J., Cox, F. R., Shuman, L. M., Welch, R. M. (Ed.) *Micronutrients in Agriculture*. SSSA Book Series No. 4. Madison, WI. pp. 89–112.

Nadeem, F., M. Farooq, A. Nawaz, and R. Ahmad. 2019. Boron improves productivity and profitability of bread wheat under zero and plough tillage on alkaline calcareous soil. *Field Crops Research* 239:1–9.

Rashid, A., and J. Ryan. 2004. Micronutrient constraints to crop production in soils with Mediterranean-type characteristics: A review. *Journal of Plant Nutrition* 27:959–975.

Rashid, A. and J. Ryan. 2008. Micronutrient constraints to crop production in the Near East: Potential significance and management strategies. In: Alloway, B.J. (Ed.), *Micronutrient Deficiencies in Global Crop Production*, pp: 149–180. Springer, Dordrecht, Netherlands.

Rehman, H., T. Aziz, M. Farooq, A. Wakeel, and Z. Rengel. 2012. Zinc nutrition in rice production systems: a review. *Plant and Soil* 361:203–226.

Rehman, A., M. Farooq, L. Ozturk, M. Asif, and K. H. M. Siddique. 2018. Zinc nutrition in wheat-based cropping systems. *Plant and Soil* 422:283–315.

Rehman, A., M. Farooq, A. Rashid, et al.2018. Boron nutrition of rice in different production systems: a review. *Agronomy for Sustainable Development*. doi:10.1007/s13593-018-0504-8.

Ryan, J., A. Rashid, J. Torrent, et al. 2013. Micronutrient constraints to crop production in the Middle East-West Asia region: Significance, research, and management. *Advances in Agronomy* 122:1–84.

Sachs, J. 1860. Berichteuber die physiologische Tatigkeitan der Versuchsstation in Tharandt. IV. VegetationsversuchemitAusschluss des Bodensuber die NahrstoffeundsonstigenErnahrungsbedingungen von Mais, Bohnen und anderenPflanzen. *Landwirtschaft s Versuchsstation* 2:219–268.

Sillanpää, M. 1982. *Micronutrients and the Nutrient Status of Soils: A Global Study*. FAO Soils Bulletin 48, FAO, Rome.

Sillanpää, M. 1990. *Micronutrient Assessment at Country Level: An International Study*. FAO Soils Bulletin 63. FAO, Rome.

Sommer, A. L. 1931. Copper as an essential element for plant growth. *Plant Physiology* 6:339–345.

Sommer, A. L., and C. B. Lipman. 1926. Evidence on the indispensible nature of zinc and boron for higher green plants. *Plant Physiology* 1:231–249.

Tandon, H. L. S. (Ed.). 1995. *Micronutrients in Soils*, Crops and Fertilizers. Fertilizer Development and Consultation Organization, New Delhi, India. 164 pp.

Tandon, H. L. S. 2013. *Micronutrient Handbook – From Research to Application*, 2nd ed. Fertilizer Development and Consultation Organization, New Delhi, India. 234 pp.

Warington, K. 1923. The effect of boric acid and borax on the broad bean and certain other plants. *Annals of Botany* 27:629–672.

2 Prognosis and Diagnosis of Micronutrient Deficiencies

2.1 INTRODUCTION

Diagnosis refers to the identification of a disorder at the time of sampling by determining micronutrient status in the plants, whereas prognosis is the prediction of the possibility of a micronutrient deficiency impairing growth at some later crop growth stage. Micronutrient deficiency prognosis can be made from either soil or plant analysis. By contrast, usually, only plant analysis can diagnose micronutrient deficiencies (Bell, 1997). The accuracy of micronutrient deficiency predictions can be improved when the plant or soil analysis is supplemented by information on soil type, soil-crop management practices, crop yield target, and probable weather conditions between sampling and the crop harvest. Complementary use of soil testing and plant analysis is more effective for the diagnosis/prognosis of micronutrient deficiencies.

The approaches employed for ascertaining deficiencies of micronutrients in Pakistani soils and crops are as follows:

(i) Crop sensitivity to micronutrient deficiencies
(ii) Deficiency symptoms in plants
(iii) Soil testing and plant analysis
(iv) Crop responses to micronutrients

2.2 CROP SENSITIVITY TO MICRONUTRIENT DEFICIENCIES

Crop genotypes vary in their micronutrient requirement and ability to utilize soil micronutrient supplies. Various species, and even varieties of the same species, have been observed to differ in their susceptibility to micronutrient deficiencies when grown in low micronutrient soils and/or solution culture (Rashid and Ryan, 2004; Chaudry et al., 2007; Hussain et al., 2012b, 2012a, 2013; Rehman et al., 2012). For example, cotton, sunflower, legumes clover, and canola have higher B requirements than cereals (Shorrocks, 1997). Hence, non-cereals are

DOI: 10.1201/9781003314226-2

more susceptible to B deficiency (Rashid et al., 2002d; Rashid, 2006a). Despite the small amount of B uptake by cereals, these crops have a relatively high B requirement for flower fertilization and seed set growth stages (Shorrocks, 1997) and consequently suffer yield losses in low B soil situations. Contrarily, because of their low B requirement or more efficient soil B utilization, barley, maize, oat, and pea are categorized as tolerant to B deficiency (Table 2.1). Since the initiation of micronutrient research in Pakistan in the early 1970s, drastic variations in susceptibility of various crop species to micronutrient deficiencies have been observed. For example, while studying the mechanism of Zn deficiency in rice, it was observed that cv. IR-6 was highly susceptible to Zn deficiency, whereas the country's premier cultivar, Basmati-370, was quite tolerant (Chaudhry et al., 1977). In the case of non-aromatic rice, B requirement of cv. *Shandar* was less than cv. *Sarshar*. Application of B at 1.0 kg ha^{-1} was the best economical dose for cv. *Shandar*, while, 1.5 kg B ha^{-1} was the best suitable dose for the maximum harvest of cv. *Sarshar* (Shah et al., 2016). Similarly, chickpea cv. C-44 was highly sensitive to Fe chlorosis and cv. CM-72 was tolerant (Rashid and Din, 1992). Maize and onion were highly sensitive to Zn deficiency, whereas sugarcane and wheat were observed to be quite tolerant (Rashid and Fox, 1992).

Moreover, recently developed high-yielding varieties of previously considered tolerant crop species have been found more susceptible to Zn deficiency. Salient recent examples in Pakistan are substantial grain yield increases with soil Zn fertilization of wheat cvs. Faisalabad-2008, Lasani-2008, and Sehar-2006 in Punjab (Zou et al., 2012; Ram et al., 2016). Recently, Hussain et al. (2012a) categorized 40 wheat genotypes released in Punjab (Pakistan) during the last five decades for grain yield and yield components, grain phytate concentration, and concentration and bioavailability of minerals Zn, Fe, and Ca. They observed that mean grain Zn concentrations in the currently grown wheat cultivars were significantly lower (~14%) than in obsolete cultivars released during the Green Revolution (1965–1976). Compared to obsolete cultivars, the current cultivars have higher phytate:mineral ratio in grains, indicating poor bioavailability of Zn to humans. Hussain et al. (2012a) carriedout field scale screening of 65varieties of bread wheat (including 40 indigenous varieties) for bioavailable Zn in grains. The estimated bioavailable Zn in grains ranged from 1.52 to 2.15 mg Zn for 300 g of wheat flour, indicating that only 21 ± 3% of grain Zn was bioavailable. Similarly, during field trials in Pothohar belt, the wheat variety "*Inqalab-91*" yielded better response in terms of increased grain yield to applied iron fertilizer @ 10 kg Fe per hectare compared to Chakwal-97 (Chaudry et al., 2007).

Thus, micronutrient fertilizer recommendations depend on the crop species and even the variety being grown in soils prone to micronutrient deficiency. In consideration of genetic variability, locally grown field crops, vegetables, and fruits considered sensitive to micronutrient deficiencies are listed in Table 2.1. Some examples of differential susceptibility of Pakistani crop genotypes to micronutrient deficiencies are listed in Table 2.2.

TABLE 2.1

Crop Sensitivity to Micronutrient Deficiencies

Crop Sensitivity Level	
High	**Medium**

Zinc

High	Medium
Apple (*Malus domestica*)	Alfalfa (*Medicago sativa*)
Beans, field (*Vicia faba*)	Barley (*Hordeum vulgare*)
Citrus (*Citrus* spp.)	Clover (*Trifolium* spp.)
Cowpea (*Vigna sinensis*)	Cotton (*Grossypium hirsutum*)
Grape (*Vitis vinifera*)	Potato (*Solanum tuberosum*)
Maize (*Zea mays*)	Sorghum (*Sorghum bicolor*)
Millet (*Panicum miliaceum*)	Sugarcane (*Saccharum officinarum*)
Onion (*Allium cepa*)	Sugarbeet (*Beta vulgaris*)
Peach (*Prunus persica*)	Sunflower (*Helianthus annuus*)
Pear (*Pyrus communis*)	Tomato (*Lycopersicon esculentum*)
Plum (*Prunus domestica*)	Wheat (*Triticum* spp.)
Rice (*Oryza sativa*)	

Boron

High	Medium
	Cabbage (*Brassica oleracea*)
Alfalfa (*Medicago sativa*)	Cherry (*Cerasus* spp.)
Apple (*Malus domestica*)	Grape (*Vitis vinifera*)
Brassica (*Brassica* spp.)	Lettuce (*Lactuca sativa*)
Cauliflower (*Brassica oleracea*)	Maize (*Zea mays*)
Clovers (*Trifolium* spp.)	Mustard (*Brassica juncea*)
Cotton (*Grossypium hirsutum*)	Olive (*Olea europaea*)
Eucalyptus (*Eucalypus* spp.)	Pea (*Pisum sativum*)
Peanut (*Arachis hypogaea*)	Peach (*Prunus persica*)
Rapeseed (*Brassica napus*)	Pear (*Pyrus communis*)
Rice (*Oryza sativa*)	Radish (*Raphanus sativus*)
Sugarbeet (*Beta vulgaris*)	Spinach (*Spinacia oleracea*)
Sunflower (*Helianthus annuus*)	Sweet Potato (*Ipomoea batatas*)
Turnip (*Brassica rapa*)	Tobacco (*Nicotiana tabaccum*)
	Tomato (*Lycopersicon esculentum*)
	Carrot (*Daucus carota*)

Iron

High	Medium
	Alfalfa (*Medicago sativa*)
Bean, field (*Vicia faba*)	Bean, field (*Vicia faba*)
Berries (*Morus spp.*)	Cotton (*Grossypium hirsutum*)
Chickpea (*Cicer arietinum*)	Maize (*Zea mays*)
Citrus (*Citrus spp.*)	Oats (*Avena sativa*)
Grape (*Vitis vinifera*)	Pea (*Pisum sativum*)
Peanut (*Arachis hypogaea*)	Rice (*Oryza sativa*)
Soybean (*Glycine max*)	Sorghum (*Sorghum bicolor*)
Walnut (*Juglans* spp.)	Wheat (*Triticum* spp.)

Adapted from Rashid and Ryan, 2004, with permission.

TABLE 2.2
Differential Susceptibility of Crop Genotypes to Micronutrient Deficiencies

Micronutrient/ Species	Genotype/Cultivar Tolerant (Less susceptible)	Susceptible	Reference
Zinc			
Rice (*Oryza sativa*)	Basmati-370, Kashmir Basmati, KS-282	NR-1, NIAB-6, DM-25	Kausar (1998)
	Basmati-370	IR-6	Chaudhry et al. (1977)
Wheat (*Triticum aestivum*)	LU-26, Chakwal-86, Punjab-86	Faialabad-83, Pak-81, Barani-83	Kausar and Tahir (1994)
	Bakhtawar	Chakwal-86	Imtiaz et al. (2006)
Cotton (*Gossypium hirsutum*)	FH-900, VH-137	NIAB-78, CIM-446	Irshad et al. (2004)
Boron			
Rapeseed (*Brassica napus*)	CON-II, CON-III, Gannyou-5	BARD-1, Sheerali, Westar	Rashid et al. (2002a)
Wheat (*Triticum aestivum*)	Inqlab-91	Chakwal-97	Chaudry et al. (2007)
	Inqlab-91, Bakhtawar	Rahtas-90, Sindh-81	Rashid et al. (2002c)
	Punjab-85, Pasban, Rohtas, Inqlab-91	Kahinoor-83, LU-26, Punjab-81, Blue Silver	Kausar et al. (1990); NIAB (1997)
	Bhakhar 2000; Dirk; AS-2000; Vatan (v-87092)	Shafaq 2006, Sehar 2006, Inqlab 91, Sehar 06; Auqab 2000	Maqsood et al. (2009a, 2009b)
Rice (*Oryza sativa*)	Shahkar, IR 8, IR 6 (high B accumulation in grains)	DR 92, DR 83 (low B accumulation in grains)	Bhutto et al. (2013)
Rice (*Oryza sativa*)	Shandar (lower B requirement)	Sarshar (higher B requirement)	Shah et al. (2016)
	Basmati-370, IR-6	Super Basmati, Basmati-6129	Rashid et al. (2002a)
	Basmati-198, M-25, KS-282	NR-1, IR-6, Basmati-385	NIAB (1997)
Maize (*Zea mays*)	YHS-202, Sultan	Agaiti-72, DTC	NIAB (1997)
Peanut (*Arachis hypogaea*)	Golden	BARD-479, BARI-2000	Rafique et al. (2014)
Iron			
Chickpea (*Cicer arietinum*)	CM-72	C-44, CM-88	Rashid and Din (1992)

2.3 DEFICIENCY SYMPTOMS IN PLANTS

The characteristic symptoms may appear if a plant is deficient in a micronutrient. These symptoms can prove a useful indicator of micronutrient deficiencies, particularly in disorders like Zn deficiency-induced bronzing in young rice plants and Fe deficiency-induced interveinal chlorosis in young leaves of legume crops, chili, deciduous fruits, and citrus (Table 2.3). However, micronutrient deficiency diagnosis based on plant symptoms is not easy, because i) apparent visual symptoms may be caused by many factors other than specific micronutrient stress; ii) a peculiar

TABLE 2.3

Salient Physiological Disorders in Plants Caused by Micronutrient Deficiencies

Micronutrient	Crop	Disorder	Symptoms
Zinc	Maize	Leaf chlorosis	Broad chlorotic bands on both sides of the midrib
	Rice	Bronzing	Brown rusty spots on leaves, predominantly on lower leaves
	Wheat		Stunted plants; young to middle leaves develop yellow patches that extend lengthways toward the tip and base
	Cotton		Shortened internodes, upward curling of younger leaves
	Apple	Little leaf	Small, malformed leaves, with shortened internodes
	Fruit trees, like apple	Rosetting	The shoots fail to extend, and the small leaves bunch together at the tip in a "rosette" type cluster
	Citrus, Peach, Plum	Interveinal chlorosis	Bleached spots (chlorosis) on leaves, in between the veins
Boron	Rice	Delayed flowering and sterile panicles	Flowering is delayed by a few days. Panicle sterility, on lower part of ears, without exhibiting any symptoms on the foliage
	Wheat	Shoots wither; New leaves die back from the tip	Splitting of youngest leave close to the mid rib accompanied by saw tooth notches along the leaf edge; gradually, the symptoms spread to older growth
	Cotton	Flower and boll shedding	Premature flower and boll shedding that lead to less boll bearing; boll size is also reduced
	Maize		Barren cobs in maize; cobs lack grains on the tip ends
	Sorghum		Head size is reduced, without any symptoms on the foliage
	Rapeseed		Stunted growth and delayed flowering
	Peanut	Hollow heart	Malformed kernels, due to depression and dark-colored cotyledons from inside

(*Continued*)

TABLE 2.3 (CONTINUED)
Salient Physiological Disorders in Plants Caused by Micronutrient Deficiencies

Micronutrient	Crop	Disorder	Symptoms
	Beets	Heart rot	Death of center of "crown", rotting of the center of the root
	Cauliflower	Hollow stem	Rotting of the center of the stem
	Potato	Internal browning	Irregular brown tissue within normal color flesh
	Turnip	*Raan* (brown heart)	Rotting of the center of the root
	Apple	Bitter pit	Decay or corking of the flesh under the skin
	Mango	Internal necrosis	Too soft and spongy tissue; reduced shelf life
	Citrus		Abnormally shaped fruits
Iron	Peanut, Soybean, Chili, Strawberry	Interveinal chlorosis	Interveinal chlorosis on young leaves; in acute deficiency whole leaf becomes yellow with necrotic lesions and even the plant may die
	Deciduous fruits, Citrus	Interveinal chlorosis	Interveinal chlorosis on young leaves; in severe deficiency, whole leaves turn bleached and may die
Copper	Cereals	Wither tip	Chlorosis of leaves, withering of tips of leaves and inflorescences, and distortion of young leaves; in acute deficiency, wheat ears look normal but lack grains
Manganese	Oat	Gray speck	Irregular gray-brown streaks or specks on leaves
	Sugarbeet	Speckled yellows	Chlorosis between leaf veins, inward curling of leaves
	Pea	Marsh spot	Brown area in the center of the seed
Molybdenum	Cauliflower	Whiptail	Reduction or suppression of leaf blades
Chlorine	Wheat	Take-all	Root rot
Nickel	Cowpea	Leaflet tip necrosis	Necrosis initiating from the tip of the leaf, and marked yellowing
	Pecan	Mouse ear	Leaf expansion is delayed and decreased

Data from Jones, 1994; Irshad et al., 2004; Rashid, 2006b, Rashid et al., 2007; Bell and Dell, 2008; Tandon, 2003; Liu et al., 2014.

visual symptom may be caused by more than one nutrient or even by more than one factors; iii) deficiency of one micronutrient deficiency may relate to an excessive supply of another nutrient, e.g., the well-known P-induced Zn deficiency; and iv) micronutrient deficiency symptoms are difficult to be distinguished in the field, because disease, insect, or herbicide damage can resemble certain micronutrient deficiency symptoms. Therefore, only an experienced plant nutritionist might be able to identify deficiency symptoms to diagnose a micronutrient deficiency disorder. Frequently, deficiency symptoms appear late during the growth period for taking effective remedial measures. Moreover, all crops do not exhibit typical micronutrient deficiency symptoms, and, thus, may suffer yield losses because of micronutrient deficiency without showing any symptoms. For example, B-deficient cereals (like wheat and rice) suffer panicle sterility without exhibiting any other symptoms on the foliage (Rashid et al., 2004; Rerkasem and Jamjod, 2004. The same is true for Zn-deficient millet (Shar et al., 2012) because vegetative growth of its Zn-deficient plants remained unaffected but heads of the Zn-deficient plants bore strikingly fewer and shriveled grains (Rashid and Fox, 1992). Thus, micronutrient deficiency symptoms may be used only as a supplement to other diagnostic techniques like plant tissue analysis and soil testing.

2.4   SOIL TESTING FOR MICRONUTRIENTS

The availability of micronutrients is generally assessed by soil testing using chemical extractants, inorganic salts/acids, and/or chelating agents. Only a small fraction of the total micronutrient content is extracted by the extractants which is a good index of the supplying capacity of the soil to plants. The soil test values are traditionally assumed to indicate "available" contents. The success of soil testing depends largely on the collection of representative soil samples, in addition to accurate laboratory analyses. Soil sampling should sub-divide fields into appropriately sized management units based on the soil type and past management history, and problem areas should be avoided. Areas that test adequate for micronutrients should not be fertilized with micronutrients, while areas that are inadequate should have fertilizers applied uniformly at the appropriate rate.

Soil testing for micronutrients is used to predict the capacity of a soil to provide plant micronutrients to a crop, during its growing season. Ideally, a soil test should identify the existence of a micronutrient deficiency, and the degree of its deficiency in terms of anticipated yield loss. In Pakistan, soil testing for micronutrients is carried-out in almost all research laboratories. The routine soil test for metal micronutrients (Zn, Cu, Fe, and Mn) is the DTPA method of Lindsay and Norvell (1978). However, because of the economy of time and effort, many research laboratories use the "universal" soil test method for the alkaline soils (AB-DTPA method of Soltanpour and Workman, 1979), as certain macronutrients (NO_3-N, P, K) and metal micronutrients (Zn, Cu, Fe, Mn) can be measured by a single extraction. Micronutrients in soil extracts are measured by atomic absorption spectrophotometry (AAS) or by inductively coupled plasma spectrometry (ICP). Many research institutions and agricultural universities also analyze soils for B. The predominant soil test for B is the hot-water extraction (HWE) procedure (Berger and Truog, 1940). Research studies by NARC and NIAB have established that a less tedious and less error-prone method, the dilute HCl test of Ponnamperuma et al. (1981), is equally appropriate for evaluating B fertility of alkaline soils in Pakistan (Kausar et al., 1990; Rashid et al., 1994b, 1997c, 1997d; Rashid, 1996a, 1996b). Dilute HCl extracts slightly less soil B, and the relationship is: HCl B = 0.030 + 0.841 (HWE B) (Rashid et al., 1994b; Rafique et al.,

TABLE 2.4

Generalized Soil Test Interpretation Criteria for Micronutrients (mg kg^{-1}) Relevant to Most Soils of Pakistan

Micronutrient	Soil Test*	Low	Marginal	Adequate
Boron	Hot water	<0.5	0.5–1.0	>1.0
	HCl	<0.45	0.45–0.65	>0.65
Zinc	DTPA	<0.5	0.5–1.0	>1.0
	AB-DTPA	<1.0	1.0–1.5	>1.5
Copper	DTPA	<0.2	0.2–0.5	>0.5
	AB-DTPA	<0.2		>0.5
Iron	DTPA	<4.5	2.0–4.0	>4.5
	AB-DTPA	<2.0		>4.0
Manganese	DTPA	<1.0	1.0–2.0	>2.0
	AB-DTPA	<1.8		>1.8

Modified from Ryan et al., 2001.
* DTPA = Diethylenetriamine pentaacetic acid; AB-DTPA = Ammonium bicarbonate-Diethylenetriamine pentaacetic acid.

2002). Boron in the soil extracts is measured by a colorimetric procedure using Azomethine-H (Bingham, 1982). Therefore, since the mid-1990s, many laboratories in Pakistan have switched over to the dilute HCl method as the soil test for B. Soil testing for Zn and Cu is quite reliable for predicting their deficiencies. However, plant availability of Fe is difficult to assess accurately by soil testing (McFarlane, 1999). Therefore, often, recommendations for Fe fertilizer use are based on the sensitivity of a crop genotype and the soil type rather than the soil test Fe status. Soil testing is more practical than plant analysis and is a widely used technique for predicting micronutrient deficiencies in crops.

Generalized soil test interpretation criteria used in Pakistan for determining the likelihood of micronutrient deficiencies are presented in Table 2.4. The soils testing low, deficient, or below the critical levels will need fertilizer application to support optimum crop growth and yield.

2.5   PLANT ANALYSIS FOR MICRONUTRIENTS

Plant analysis is carried-out as a series of steps and encompasses sampling, sample preparation, laboratory analysis, and interpretation. Plant species, plant age, plant part, time sampled, and fertilizer applied are all variables that affect the interpretation of the laboratory results. Careful sampling is important by adopting the suggested guidelines. De-contamination to remove foreign substances may be necessary, particularly if the tissue is covered with dust or spray materials that contain elements included in the analysis. The washing, oven drying (~70°C), and sample size reduction (milling) prepare the sample for laboratory analysis. Then organic matter in the prepared sample is destroyed by either wet acid oxidation (for metal micronutrients, i.e., Zn, Cu, Fe, and Mn) or high-temperature dry ashing (for B). Appropriate analytical procedures are followed for determining the concentration of specific micronutrients in the aliquots. The atomic absorption spectrometry is used to analyze Zn, Cu, Fe, and Mn, and B is determined by a colorimetric method. All these micronutrients can be analyzed more efficiently by inductively coupled plasma mass spectrometry (ICP-MS).

Like many other parts of the world, information on the status of plant micronutrients in Pakistan is far less compared with the soil data. However, systematic nutrient indexing has provided comprehensive information about the micronutrient status of the farmer-grown cotton (Rafique et al., 2002), wheat, sorghum, rapeseed-mustard, peanut (Rashid and Qayyum, 1991; Rashid et al., 1997a, 1997b, 1997c, 1997d), citrus (Rashid et al., 1991; Rehman, 1989; Khattak, 1995; Siddique et al., 1994), and apple (Rehman, 1990). Pakistan's micronutrient plant analysis data are also available in some global nutrient indexing investigations (e.g., Sillanpää, 1982) as well as in numerous greenhouse and field experiments carried-out in the country (NFDC, 1998).

Generalized interpretation guidelines for micronutrient plant analysis data are presented in Table 2.5. In-country research has also established plant analysis diagnostic criteria for B and Zn for a number of locally-grown crop genotypes (Table 2.6). As expected, the critical levels/ranges vary between species, varieties within species, and plant parts. Most of the criteria are similar to values in the literature. The concentration range of B for cotton leaves (without petioles), i.e., 53–55 mg B kg^{-1} (Rashid and Rafique, 2002; Ahmed et al., 2013) is much higher than generally listed critical concentration in literature, i.e., 15–20 mg B kg^{-1} (Shorrocks, 1992; Reuter and Robinson, 1997).

For micronutrient plant analysis, Pakistani scientists predominantly rely on the plant tissue sampling guidelines and procedures suggested by Jones et al. (1991);

TABLE 2.5
Generalized Criteria for Interpreting Plant Analysis Micronutrient Data

Micronutrient	Deficient	Sufficient or Normal (mg kg^{-1})	Excessive or Toxic
B	5–30	10–200	50–200
Cl	<100	100–500	500–1000
Cu	2–5	5–30	20–100
Fe	<50	100–500	>500
Mn	15–25	20–300	300–500
Mo	0.03–0.15	0.1–2.0	>1000*
Zn	10–20	21–150	100–400
Ni**	0.01–0.14	0.22–10.0[a]	10–83

Data from Jones, 1991, with permission; *Havlin et al., 2017; and **for selected vegetables and crops–Brown, 2006.

[a] Toxic for sensitive plants like strawberry, fruit trees, pea, and onion.

TABLE 2.6

Interpretation Criteria Developed in Pakistan for Micronutrient Plant Analysis of Selected Crops

	Critical concentration/range (mg kg^{-1})				
	Zn			B	
Crop species	Whole Shoots[a]	Leaves[b]	Seeds	Whole Shoots	Leaves
Wheat	16–20	12–16	20–24	4–6	5–7
Rice	20	19	15		6–7.2
Cotton					53–55
Maize	18	24	18		
Sorghum	27–33	20–22	10–14	17–18	25–31
Chickpea					49
Soybean		22	43		
Cowpea		21	36		
Rapeseed	29	33	29	32	38
Mustard	35	41	33	41	49
Peanut				[d]33–42	[c]29
Sunflower					
4-week				46–63	
8-week				36	

Data from Rashid and Din, 1992; Rashid and Fox, 1992; Rashid et al., 1994a, 1994b; Rashid et al., 1997a, 1997b, 1997c, 1997d; Rashid et al., 2002a, 2002b, 2002c, 2002d; Rashid and Rafique, 2005; Rafique et al., 2006; Rashid, 2006a; Ahmed et al., 2013; Rafique et al., 2014; Rashid and Rafique, 2017.

[a] Whole shoots = Young whole shoots ($\leq$ 30 cm tall);

[b] Leaves = Youngest fully expanded leaf blades, at flowering/heading (unless stated otherwise);

[c] ~4 cm shoot terminals;

[d] Upper parts of peanut plants at early peg formation stage

Reuter and Robinson (1997); and Tandon (1995) **(see Annex 1)**. Despite appreciable accomplishments in developing plant analysis interpretation criteria for local crop genotypes, the unfinished agenda remains quite challenging.

2.5.1 Ferrous Analysis in Fresh Plant Tissue

In the case of Zn, Cu, Mn, and B, total concentrations of these micronutrients are determined and interpreted for diagnosing their nutritional status in crop plants. However, the concentration of total Fe in plant tissues is a marginal indicator of the Fe nutritional status of plants. Ferrous (Fe^{2+}) fraction of Fe in the plant tissues is a better indicator of Fe requirement of plants. Agricultural soils in Pakistan are predominantly calcareous in nature causing Fe chlorosis to be a widespread problem in susceptible field crops (peanut and chickpea) and fruit orchards (apple and citrus) (Rashid, 2005; Rashid and Din, 1992). The chlorosis of Fe in plants grown on calcareous soils is not induced by an absolute Fe deficiency; rather, it results from a physiological disorder that affects the mobility of Fe in the entire plant. Therefore, generally total Fe content in plant tissues is a misleading index of the Fe nutritional status of crops. In fact, in many instances, the total Fe concentration in chlorotic plant tissues will be higher than in the green plant tissues of the same plant species growing in the same field. Rashid and Din (1992) also observed that Fe concentration in chlorotic leaf tissues of two chickpea cultivars was more than in green leaf tissues. However, Fe^{2+} concentration in fresh leaves is related well to the severity of chlorosis. As plant analysis criteria for chlorosis-prone crops in Pakistan were lacking, efforts were made in this direction at NARC, Islamabad. The determined critical level of o-phananthroline extractable Fe^{2+} for deficiency diagnosis in fresh young chickpea and peanut leaves was 40 mg kg^{-1} for both species (Rashid and Din, 1992; Rashid et al., 1997d).

2.5.2 Grain Analysis for Micronutrients

Historically, analysis of mature crop grain was not considered a reliable index for diagnosing nutrient deficiencies. However, research in Pakistan has established that seed analyses is a good indicator for evaluating the Zn fertility status of soils (Rashid et al., 1994a, 1997c). The use of seeds has advantages in terms of ease of sampling, processing, and/or analyzing. Though seed analysis diagnoses past problems, it can help identify field sites where future crops will respond to micronutrient applications and regional trends in the micronutrient status of crops. The unpublished data of Munir Zia et al. pertaining to the micronutrient status of wheat grains collected from 75 field locations across Pakistan, including Azad Jammu and Kashmir (Pakistan-Administered Kashmir), reveals wide variations as the values were as follows: total Fe (Min 30, Max 233, Median 42 mg kg^{-1}); Cu (Min 3.0, Max 7.7, Median 4.3 mg kg^{-1}); Mn (Min 16.6, Max 60.3, Median 34.3 mg kg^{-1}); Zn (Min 15.1, Max 39.7, Median 24.5 mg kg^{-1}); Mo (Min 0.17, Max 1.1, Median 0.63 mg kg^{-1}); and Ni (Min 0.06, Max 7.5, Median 0.14 mg kg^{-1}). The data reflect wide variations in geochemistry of the wheat-grown soils across the country that might also get influenced because of the usage of phosphatic fertilizers by the farmers.

2.6 SOIL–PLANT MICRONUTRIENT ANALYSIS FACILITIES IN PAKISTAN

In the early 1970s, when micronutrient deficiencies in crop plants were first recognized in the country, professional expertise for analyzing micronutrients was very limited and the requisite analytical facilities were lacking. As stated earlier, the first-ever AAS equipment for agricultural research purposes was procured soon after the recognition of Zn deficiency disorder in rice crops, for Rice Research Institute, Kala Shah Kaku. However, this equipment never became functional.

TABLE 2.7

Analytical Facilities for Soil and Plant Micronutrient Analysis in Pakistan

Institution	Major Facilities	Soil Analyses	Plant Analyses	Remarks	Contact
[a]LRRI, NARC, Islamabad	AAS, ICP	B, Zn, Cu, Fe, Mn,	B, Zn, Cu, Fe, Mn	Fe^{2+} in green leaf tissue	Director, LRRI
Grain Quality Testing Laboratories, PARC (Islamabad and Karachi)	AAS, GC, HPLC-MS		B, Zn, Cu, Fe, Mn	Commercial analysis	Director, Food Science and Product Development Institute
AARI, Faisalabad	AAS, ICP	B, Zn, Cu, Fe, Mn	B, Zn, Cu, Fe, Mn	Commercial analysis	DG, AARI
NIAB, Faisalabad	AAS, ICP	B, Zn, Cu, Fe, Mn,	B, Zn, Cu, Fe, Mn	Commercial analysis	Director, NIAB
[b]ISES, UAF	AAS, ICP	B, Zn, Cu, Fe, Mn,	B, Zn, Cu, Fe, Mn		Director
Soil & Water Testing Lab, Soil Fertility Institute, Lahore	AAS, ICP	B, Zn, Cu, Fe, Mn,	B, Zn, Cu, Fe, Mn	Commercial analysis	Agri. Chemist
Dept. of Soil/Environ Sciences, The Univ. of Agriculture Peshawar	AAS	B, Zn, Cu, Fe, Mn,	B, Zn, Cu, Fe, Mn		Chairman
Soil & Plant Nutrition Institute, ARI, Tarnab, Peshawar	AAS, ICP	B, Zn, Cu, Fe, Mn,	B, Zn, Cu, Fe, Mn		DG, ARI, Tarnab
NIFA, Peshawar	AAS	B, Zn, Cu, Fe, Mn,	B, Zn, Cu, Fe, Mn		Director
Dept. of Soil Science, Sindh Agriculture University, Tandojam	AAS	B, Zn, Cu, Fe, Mn,	B, Zn, Cu, Fe, Mn		Chairman
Agricultural Research Institute, Tandojam	AAS	B, Zn, Cu, Fe, Mn,	B, Zn, Cu, Fe, Mn		Agri. Chemist
NIA, Tandojam	AAS	B, Zn, Cu, Fe, Mn	B, Zn, Cu, Fe, Mn		Director
Directorate of Agric. Research / Plant Protection, Sariab, Quetta	AAS	B, Zn, Cu, Fe, Mn	B, Zn, Cu, Fe, Mn	Graphite furnace facility exists	Director
[c]BARDC, Quetta	AAS -Out of order				DG
[d]PCSIR Laboratories	ICP, AAS	B, Cu, Fe, Mn, Mo, Zn	B, Cu, Fe, Mn, Mo, Zn,	Commercial analysis	Director, Center for Environ. Protection

TABLE 2.7 (CONTINUED)
Analytical Facilities for Soil and Plant Micronutrient Analysis in Pakistan

Institution	Major Facilities	Soil Analyses	Plant Analyses	Remarks	Contact
Fauji Fertilizer Company Ltd.,	AAS	Fe, Cu, Mn, Zn, B,		Free soil analysis	Manager, Agri. Services Deptt.
Engro Fertilizers	AAS	Zn, Fe, Mn, Cu, B		Analysis free for farmers	Zonal Manager

[a] LRRI = Land Resources Research Institute;
[b] UAF = University of Agriculture, Faisalabad;
[c] BARDC = Balochistan Agricultural Research and Development Centre of PARC;
[d] PCSIR = Pakistan Council for Scientific and Industrial Research, Lahore and Karachi

Thereafter, in 1973 an AAS became functional at NIAB, Faisalabad, which played a major role in the initiation of micronutrient research in the country. Subsequently, AAS equipment was obtained by many agricultural laboratories. During the late 1990s and early 2000s, some agricultural laboratories have even been equipped with ICP equipment. The current status of analytical facilities for micronutrient soils and plant micronutrient analysis in Pakistan is given in Table 2.7.

REFERENCES

Ahmed, N., M. Abid, A. Rashid, M. A. Ali, and M. Amanullah. 2013. Boron requirement of irrigated cotton in a Typic Haplocambid for optimum productivity and seed composition. *Communications in Soil Science and Plant Analysis* 44:1293–1309.

Bell, R. W. 1997. Diagnosis and prediction of boron deficiency for plant production. *Plant and Soil* 193:149–168.

Bell, R. W., and B. Dell. 2008. *Micronutrients for Sustainable Food, Feed, Fibre and Bioenergy Production.* International Fertilizer Association, Paris, France.

Berger, K. C., and E. Truog. 1940. Boron deficiency as revealed by plant and soil test. *American Society of Agronomy* 32:297–301.

Bhutto, M. A., Z. T. Maqsood, S. Arif, et al. 2013. Effect of zinc and boron fertilizer application on uptake of some micronutrients into grain of rice varieties. *American-Eurasian Journal of Agricultural and Environmental Sciences* 13:1034–1042.

Bingham, F.T. 1982. Boron. p. 431–448, In: Page, A.L. (Ed.), *Methods of Soil Analysis, Part 2: Chemical and Microbial Properties.* American Society of Agronomy, Madison, WI, USA.

Brown, P. H. 2006. Nickel. p. 395–410, In: Barker, A.V., and D. J. Pilbeam (Ed.), *Handbook of Plant Nutrition,* Boca Raton, FL: CRC Press, Taylor and Francis Group.

Chaudhry, F. M., S. M. Alam, A. Rashid, and A. Latif. 1977. Micronutrient availability to cereals from calcareous soils. IV. Mechanism of differential susceptibility of two rice verities to Zn deficiency. *Plant and Soil* 46:637–642.

Chaudry, E. H., V. Timmer, A. S. Javed, and M. T. Siddique. 2007. Wheat response to micronutrients in rain-fed areas of Punjab. *Soil and Environment* 26:97–101.

Havlin, J. L., S. L. Tisdale, W. L. Nelson, and J. D. Beaton. 2017. *Soil Fertility and Fertilizers: An Introduction to Nutrient Management*, 8th ed. Pearson, Chennai, India.

Hussain, S., M. A. Maqsood, and L. V. Miller. 2012b. Bioavailable zinc in grains of bread wheat varieties of Pakistan. *Cereal Research Communications* 40:62–73.

Hussain, S., M. A. Maqsood, Z. Rengel, and M. K. Khan. 2012a. Mineral bioavailability in grains of Pakistani bread wheat declines from old to current cultivars. *Euphytica* 18:1–11.

Hussain, S., M. A. Maqsood, Z. Rengel, T. Aziz, and M. Abid. 2013. Estimated Zn bio-availability in milling fractions of biofortified wheat grains and flours of different extraction rates. *International Journal of Agriculture and Biology* 15:921–926.

Imtiaz, M., B. J. Alloway, P. Khan, et al. 2006. Zinc deficiency in selected cultivars of wheat and barley as tested in solution culture. *Communications in Soil Science and Plant Analysis* 23:1703–1721.

Irshad, M., M. Gill, T. Aziz, and I. Ahmed. 2004. Growth response of cotton cultivars to zinc deficiency stress in chelator-buffered nutrient solution. *Pakistan Journal of Botany* 36:373–380.

Jones Jr, J. B. 1991. Plant tissue analysis in micronutrients. p. 477–521, In: Mortvedt, J.J. et al. (Ed.), *Micronutrients in Agriculture*, 2nd ed. SSSA, Madison, Wisconsin.

Jones Jr, J. B. 1994. *Plant Nutrition Manual*. Micro-Macro Publishing, Athens, Georgia, USA.

Jones Jr., J. B., B. Wolf, and H. A. Mills. 1991. *Plant Analysis Handbook: A Practical Sampling, Preparation, Analysis and Interpretation Guide*. Micro-Macro Publishing, Athens, Georgia, USA, 213 pp.

Kausar, M. A. 1998. Differential sensitivity of eleven rice varieties to low soil zinc and copper. *Pakistan Journal of Soil Science* 14:50–53.

Kausar, M. A., and M. Tahir. 1994. Susceptibility of eleven wheat cultivars to zinc and copper deficiency. p. 157–160, In: Efficient Use of Plant Nutrients. Proc. 4th National Congress of Soil Science, Islamabad, 24–26 May, 1992. Soil Science Society of Pakistan.

Kausar, M. A., M. Tahir, and A. Hamid. 1990. Comparison of three methods for the esti-mation of available boron for maize. *Pakistan Journal of Scientific and Industrial Research* 33:221–224.

Khattak, J. K. 1995. *Micronutrients in Pakistan Agriculture*. Pakistan Agricultural Research Council, Islamabad and NWFP Agricultural University, Peshawar. 135 pp.

Lindsay, W. L., and W. A. Norvell. 1978. Development of DPTA soil test for zinc, iron, manganese, and copper. *Soil Science Society of America Journal* 42: 422–428.

Liu, G., E. H. Simonne, and Y. Li. 2014. Nickel Nutrition in Plants. HS1191, one of a series of the Horticultural Sciences Document no. 1191, University of Florida, Institute of Food and Agricultural Sciences Extension, Available at http://edis.ifas.ufl.edu/hs1191 [Accessed Feb 03, 2017].

Maqsood, M. A., Rahmatullah, S. Kanwal, T. Aziz, and M. Ashraf. 2009a. Evaluation of Zn distribution among grain and straw of twelve indigenous wheat (*Tricticumaestivum* L.) genotypes. *Pakistan Journal of Botany* 41:225–231.

Maqsood, M. A., Rahmatullah, A. M. Ranjha, and M. Hussain. 2009b. Differential growth response and zinc utilization efficiency of wheat genotypes in chelator buffered nutrient solution. *Soil and Environment* 28:174–178.

McFarlane, J. D. 1999. Iron. p. 295–301, In: K. Peverill, I. Sparrow, and D. J. Reuter (Ed.), *Soil Analysis: An Interpretational Manual*. CSIRO Publishing, Melbourne.

NFDC. 1998. *Micronutrients in Agriculture: Pakistan perspective*. National Fertilizer Development Center, Islamabad, 51 pp.

NIAB. 1997. *Silver Jubilee of NIAB: 25 Year Report*. Nuclear Institute for Agriculture and Biology, Faisalabad.

Ponnamperuma, F. N., M. T. Caytan, and R. S. Lantin. 1981. Dilute hydrochloric acid as an extractant for available zinc, copper and boron in rice soils. *Plant and Soil* 61:297–310.

Rafique, E., M. Mahmood-ul-Hassan, M. Yousra, I. Ali, and F. Hussain. 2014. Boron nutrition of peanut grown in boron-deficient calcareous soils: Genotypic variation and proposed diagnostic criteria. *Journal of Plant Nutrition* 37:172–183.

Rafique, E., A. Rashid, A. U. Bhatti, G. Rasool, and N. Bughio. 2002. Boron deficiency in cotton grown in calcareous soils of Pakistan. I. Distribution of B availability and comparison of soil testing methods. p. 349–356, In: Goldbach, H. E. et al. (Ed.), *Boron in Plant and Animal Nutrition*. Kluwer Academic Publishers, New York.

Rafique, E., A. Rashid, J. Ryan, and A. U. Bhatti. 2006. Zinc deficiency in rainfed wheat in Pakistan: Magnitude, spatial variability, management, and plant analysis diagnostic norms. *Communications in Soil Science and Plant Analysis* 37:181–197.

Ram, H., A. Rashid, W. Zhang, et al. 2016. Biofortification of wheat, rice and common bean by applying foliar zinc fertilizer along with pesticides in seven countries. *Plant and Soil* 403:389–401.

Rashid, A. 1996a. Secondary and micronutrients. p. 341–385, Chapter 12, In: *Soil Science*, Rashid, A., and K. S. Memon (Managing Authors). National Book Foundation, Islamabad, Pakistan.

Rashid, A. 1996b. Nutrient Indexing of Cotton in Multan District and Boron and Zinc Nutrition of Cotton. Micronutrients Project, Annual Report, 1994–95. National Agricultural Research Center, Islamabad. 76 pp.

Rashid, A. 2005. Establishment and management of micronutrient deficiencies in Pakistan: a review. *Soil and Environment* 24:1–22.

Rashid, A. 2006a. Incidence, diagnosis and management of micronutrient deficiencies in crops: Success stories and limitations in Pakistan. In: *Optimizing Resource Use Efficiency for Sustainable Intensification of Agriculture*. IFA Agriculture Conference, Kunming, PR China. 27 February - 2 March, 2006. Available at:www.fertilizer.org/Public/Stewardship/Publication_Detail.aspx?SEQN= 3865&PUBKEY=6DCE3EF6-B7F5-45CF-90C8-A61F010D9462

Rashid, A. 2006b. Boron Deficiency in Soils and Crops of Pakistan: Diagnosis and Management. Pakistan Agricultural Research Council, Islamabad, Pakistan, viii+34 pp. ISBN: 969-409-184-5.

Rashid, A., N. Bughio, and N. Rafique. 1994a. Diagnosis of zinc deficiency in rapeseed and mustard by seed analysis. *Communications in Soil Science and Plant Analysis* 25:3405–3412.

Rashid, A., E. Rafique, and B. Bughio. 1994b. Diagnosing boron deficiency in rapeseed and mustard by plant analysis and soil testing. *Communications in Soil Science and Plant Analysis* 25:2883–2897.

Rashid, A., and J. Din. 1992. Differential susceptibility of chickpea cultivars to iron chlorosis grown on calcareous soils of Pakistan. *Journal of the Indian Society of Soil Science* 40:488–492.

Rashid, A., and R. L. Fox. 1992. Evaluating internal zinc requirements of grain crops by seed analysis. *Agronomy Journal* 84:469–474.

Rashid, A., F. Hussain, A. Rashid, and J. Din. 1991. Nutrient status of citrus orchards in Punjab. *Pakistan Journal of Soil Science* 6:25–28.

Rashid, A., M. M. Mahmud, and E. Rafique. 2007. Potato responses to boron and zinc application in a calcareous Udic Ustochrept. p. 103–116, In: F, Xu (Ed.) Advances in Plant and Animal Boron Nutrition. Proc. 3rd *Int. Symp. on All Aspects of Plant and Animal Boron Nutrition (Boron2005), Wuhan, China, 9–13 Sep 2005*. Springer, Dordrecht, The Netherlands. *ISBN*: 978-90-481-7356-3.

Rashid, A., S. Muhammad, and E. Rafique. 2002b. Genotypic variation in boron uptake and utilization by rice and wheat. p. 305–310, In: Goldbach, H. E. et al. (Ed.), *Boron in Plant and Animal Nutrition*. Kluwer Academic Publishers, New York.

Rashid, A., E. Rafique, and N. Bughio. 2002c. Boron deficiency in rainfed alkaline soils of Pakistan: Incidence and boron requirement of wheat. P. 371–379, In: Goldbach, H.E. et al. (Ed.), *Boron in Plant and Animal Nutrition*. Kluwer Academic Publishers, New York.

Rashid, A., and F. Qayyum. 1991. Micronutrient Status of Pakistan Soils and Their Role in Crop Production, Cooperative Program. Final Report, 1983–1990. National Agricultural Research Center, Islamabad. 84 pp.

Rashid, A., and E. Rafique. 2002. Boron deficiency in cotton grown in calcareous soils of Pakistan. II. Correction and criteria for foliar diagnosis. p. 357–362, In: Goldbach, H.E. et al. (Ed.), *Boron in Plant and Animal Nutrition*. Kluwer Academic Publishers, New York.

Rashid, A., and E. Rafique. 2005. Internal boron requirement of young sunflower plants: proposed diagnostic criteria. *Communications in Soil Science and Plant Analysis* 36:2113–2119.

Rashid, A., and E. Rafique. 2017. Boron deficiency diagnosis and management in field crops in calcareous soils of Pakistan: A mini review. *BORON* 2:142–152.

Rashid, A., E. Rafique, and N. Ali. 1997d. Micronutrient deficiencies in rainfed calcareous soils of Pakistan. II. Boron nutrition of the peanut plant. *Communications in Soil Science and Plant Analysis* 28:149–159.

Rashid, A., E. Rafique, and N. Bughio. 1997c. Micronutrient deficiencies in rainfed calcareous soils of Pakistan. III. Boron nutrition of sorghum. *Communications in Soil Science and Plant Analysis* 28:441–454.

Rashid, A., E. Rafique, N. Bughio, and M. Yasin. 1997b. Micronutrient deficiencies in rainfed calcareous soils of Pakistan. IV. Zinc nutrition of sorghum. *Communications in Soil Science and Plant Analysis* 28:455–467.

Rashid, A., E. Rafique, J. Din, S. N. Malik, and M. Y. Arain. 1997a. Micronutrient deficiencies in rainfed calcareous soils of Pakistan. I. Iron chlorosis in peanut. *Communications in Soil Science and Plant Analysis* 28:135–148.

Rashid, A., E. Rafique, S. Muhammed, and N. Bughio. 2002a. Boron deficiency in rainfed alkaline soils of Pakistan: Incidence and genotypic variation in rapeseed-mustard. p. 363–370, In: Goldbach, H.E. et al. (Ed.), *Boron in Plant and Animal Nutrition*. Kluwer Academic/Plenum Publishers, New York.

Rashid, A., E. Rafique, and J. Ryan. 2002d. Establishment and management of boron deficiency in field crops in Pakistan: A country report. p. 339–348, In: Goldbach, H.E. et al (Ed.), *Boron in Plant and Animal Nutrition*. Kluwer Academic Publishers, New York.

Rashid, A., and J. Ryan. 2004. Micronutrient constraints to crop production in soils with Mediterranean-type characteristics: A review. *Journal of Plant Nutrition* 27:959–975.

Rashid, A., M. Yasin, M. Ashraf, and R. A. Mann. 2004. Boron deficiency in calcareous soils reduces rice yield and impairs grain quality. *International Rice Research Notes* 29:58–60.

Rehman, H. 1989. Annual Report, Directorate of Soils and Plant Nutrition. 1988–89. Agricultural Research Institute, Tarnab, Peshawar, Pakistan.

Rehman, H. 1990. Annual Report, Directorate of Soils and Plant Nutrition. 1989–90. Agricultural Research Institute, Tarnab, Peshawar, Pakistan.

Rehman, H., T. Aziz, M. Farooq, A. Wakeel and Z. Rengel. 2012. Zinc nutrition in rice production systems: a review. *Plant and Soil* 361:203–226.

Rerkasem, B., and S. Jamjod. 2004. Boron deficiency in wheat: a review. *Field Crops Research* 89:173–186.

Reuter, D. J., and J. B. Robinson. 1997. *Plant Analysis – An Interpretation Manual*. CSIRO Publishing, Melbourne, Australia.

Ryan, J., G. Estefan, and A. Rashid. 2001. *Soil and Plant Analysis Laboratory Manual*, 2nd ed. International Center for Agricultural Research in the Dry Areas (ICARDA), Aleppo, Syria.

Shah, J. A., N. Rais, Z. Hassan, M. Abbas, and M. Y. Memon. 2016. Evaluating non-aromatic rice varieties for growth and yield under different rates of soil applied boron. *Pakistan Journal of Analytical and Environmental Chemistry* 17:1–7.

Shar, G. Q., L. A. Shar, P. M. Makhija, and S. B. Sahito. 2012. Evaluation of eleven macro and micro elements present in various hybrids of Millet (*Pennisetum glaucum*, or *P. americanum*). *Pakistan Journal of Analytical and Environmental Chemistry* 13:78–84.

Shorrocks, V. M. 1992. Boron – A global appraisal of the occurrence of boron deficiency. p. 39–53, In: Portch, S. (Ed.), *Proc. Int. Symp. Role of Sulphur, Magnesium, and Micronutrients in Balanced Plant Nutrition*. The Sulphur Institute, Washington, DC.

Shorrocks, V. M. 1997. The occurrence and correction of boron deficiency. *Plant and Soil* 193: 121–148.

Siddique, M. T., M. Rashid, and M. Saeed. 1994. Micronutrient status of soils/plants of citrus growing areas in Punjab. p. 355–360, In: Efficient Use of Plant Nutrients. Proc. 4th National Congress of Soil Science, Islamabad, May 24–26, 1992. Soil Science Society of Pakistan.

Sillanpää, M. 1982. Micronutrients and the Nutrient Status of Soils: A Global Study. FAO Soils Bulletin 48, FAO, Rome.

Soltanpour, P. N., and S. Workman. 1979. Modification of the NH_4HCO_3-DTPA soil test to omit carbon black. *Communications in Soil Science and Plant Analysis* 10:1411–1420.

Tandon, H. L. S. 1995. *Micronutrients in Soils, Crops and Fertilizers*. Fertilizer Development and Consultation Organization, New Delhi, India.

Tandon, H. L. S. 2003. *Micronutrient Handbook from Research to Application*. Fertilizer Development and Consultation Organization, New Delhi, India. 234 pp.

Zou, C. Q., Y. Q. Zhang, A. Rashid, et al. 2012. Biofortification of wheat with zinc through zincfertilization in seven countries. *Plant and Soil* 361:43–55.

3 Quality Issues with Micronutrient Research Data

3.1 INTRODUCTION: DATA QUALITY ISSUES

While writing this book, it was observed that a substantial segment of the Pakistani-origin micronutrient research data was of tainted quality. The quality issues were obvious both in the case of the reported magnitudes of crop yield increases as well as the laboratory analysis data. Such quality issues were observed in the case of certain published data, in peer-reviewed journals, as well as in unpublished data. For instance, some authors have reported crop yield increases with the application of micronutrients (i.e., Zn, B, and Mn) as high as >25%, without reporting the native status of micronutrients in soils and plant tissues. Such phenomenal yield increases with micronutrient fertilizer use, without valid correction of the results, cannot be endorsed (**see Annex 2**).

For instance, in field experiments on wheat over three years, phenomenal increases in grain yield with soil-applied Zn, i.e., 88–152%, reported by a provincial research institute are unheard of. Also, native DTPA soil Zn reported for these field trials was 1.39 mg kg^{-1}, which was way higher than the critical DTPA Zn level for wheat, i.e., 0.4–0.8 mg kg^{-1} (Zou et al., 2012; Prasad et al., 2013). The authors of this book also question the accuracy of Zn concentrations in wheat grains as low as 7.9, 9.8, 10.6, and 11.6 mg kg^{-1} reported by the provincial research institute; because grains of the prevalent wheat cultivars in Pakistan contain around 25 mg Zn kg^{-1} (Zou et al., 2012; Ram et al., 2016). Thus, wheat grain Zn concentrations below 15 mg kg^{-1} are hard to reconcile with even in the crop grown on Zn-deficient soils (Zia et al., unpublished data). This analytical error might be due to the low extraction efficiency of total Zn during the digestion of grains, especially in the absence of standard reference material. Also, appreciable wheat yield increases with Zn fertilization, when leaf Zn, without Zn fertilizer use, was 25 mg kg^{-1} (AARI, Annual Report 2011), cannot be endorsed, because the critical Zn concentration range in wheat leaves is 12–16 mg Zn kg^{-1} (Rafique et al., 2006). Moreover, soil Zn fertilization-driven unprecedented increases in wheat leaf Zn concentrations, i.e., from 15 to 30 mg kg^{-1} (Arshad, 2014) and from 25 to 37 mg kg^{-1} reported by a local agricultural research institute are highly questionable.

In the case of rice, astonishingly high increases in paddy yield, up to 106%, with soil-applied Zn to soils already containing 1.29 mg Zn kg^{-1} (DTPA extractable) in the year 2012, are equally unlikely (AARI, Annual Report 2012).

An equally astonishing aspect of the rice data is unbelievably high increases in rice grain Zn concentration after Zn fertilization, e.g., from 35 mg kg^{-1} to 97 mg kg^{-1} in the year 2012. Contrarily, based on results of a multi-year, multi-country field research under the HarvestZinc Project, Phattarakul et al. (2012) have stated that the maximum increase in rice grain Zn concentration with soil+foliar Zn fertilization was 19 to 35 mg kg^{-1}. The quality of some reported Zn data was equally questionable in the case of maize. For example, 52–86% increase in grain yield with Zn application in soils having DTPA extractable Zn more than critical level (i.e., 1.02 mg Zn kg^{-1}) (Malik, 2011; Annual Report of a provincial research institute) simply does not make sense. Strangely, it has also been reported that maize leaves grown on Zn-adequate soils contained extremely low Zn concentrations (i.e., 7.7 mg kg^{-1}). It was equally questionable that with Zn fertilization in 2013 maize leaf Zn concentration increased almost fivefold, i.e., from 7.7 to 35.5 mg kg^{-1} (Annual Reports of a Provincial Research Institute). The same provincial research institute reported a 156% increase in onion yield with soil Zn application. Also, another study reported a 73% increase in maize fodder yield as a result of Zn application @10 kg ha^{-1} along with the N that was applied at the rate of 150 kg ha^{-1} (Jamil et al., 2015).

In the case of B nutrition of crops as well, inconsistent data were observed in many cases. For example, Malik (2011) reported a 54% increase in maize yield with B fertilization of a soil containing 2.9 mg B kg^{-1}, which is almost six times higher than the critical level of soil B (0.5 mg B kg^{-1}; Ryan et al., 2001).

A recent open-access journal article (Ali et al., 2021) is another classic example of poor-quality micronutrient research data, where the authors compared the effect of i) unbranded zinc sulfate fertilizer (containing 33% Zn); ii) branded zinc sulfate fertilizer (Engro's product *Zingro*, containing 33% Zn); iii) bioactive zinc and zinc solubilizing bacteria-coated urea, named as "*Zabardast* Urea (ZU)"; and iv) a control treatment, on wheat productivity, wheat grain Zn density, and Zn use efficiency (ZUE) by the crop. It is reported that wheat grain yield increase, over control yield (4.364 tonnes ha^{-1}), was a mere 2.7% with unbranded zinc sulfate, 7.2% with *Zingro*, and 11.6% with ZU. Similarly, the reported increase in wheat grain Zn density, over grain Zn density with control treatment (21.0 mg kg^{-1}), was 56% with unbranded zinc sulfate (31.6 mg kg^{-1}), 82% with *Zingro* (38.2 mg kg^{-1}), and 171% with ZU (57.0 mg kg^{-1}). The reported agronomic Zn use efficiency (ZUE) by wheat was 8 kg grains kg^{-1} of unbranded zinc sulfate, 21 kg grains kg^{-1} of *Zingro*, and exponentially high 410 kg grains kg^{-1} of ZU.

These astonishing results raise serious logical questions. For instance, how come grain yield increase with *Zingro* was 2.7 times higher compared with unbranded zinc sulfate when both zinc sulfate products contained 33% elemental Zn? Similarly, what is the explanation for the 1.6 times more increase in grain Zn density with *Zingro* than with unbranded zinc sulfate, when both fertilizer products contained 33% inorganic Zn? The reported 262% higher agronomic ZUE with *Zingro* compared with unbranded zinc sulfate is equally astonishing. No logical reasoning has been offered for these unexplainable results in the article. As both zinc sulfate products have same composition and Zn concentration, the reported results are absolutely illogical.

Another astonishing aspect of the reported results is that increase in wheat grain Zn density, over control grain Zn density (21.0 mg kg^{-1}), was 10.6 mg kg^{-1} (50%) with unbranded zinc sulfate, 17.2 m kg^{-1} (82%) with *Zingro*, and exponentially high 36 mg kg^{-1} (171%) with ZU. No scientific reasoning for such unprecedented grain Zn density increases has been offered either.

The following aspects should have been considered by the journal's handling editor and the reviewers while reviewing such tall claims, which are so far unheard of:

i) The apparent conflict of interest, as all authors (except for one) have a vested interest in promoting the tested fertilizer products *Zingro* and ZU, as they are employees of the same two companies which developed and/ or are marketing *Zingro* and ZU.

ii) The handling editor should have got this manuscript reviewed by relevant research scientists having an evident track record of publications in this specific research area. In the instant case, all the mentioned reviewers lack relevant research experience and publication records.

iii) Based on multi-year, multi-country field research, Zou et al. (2012) and Joy et al. (2015) have reported an average increase of ~3 mg kg^{-1} in wheat grain Zn density (~5% over control) with soil-applied zinc sulfate fertilizer. Surprisingly, Ali et al. (2021) have reported an increase in grain Zn concentration of 10.2 mg kg^{-1} with unbranded zinc sulfate fertilizer and of 17.2 mg kg^{-1} with *Zingro* (both applied @ 5 kg Zn ha^{-1}), and 36 mg kg^{-1} with ZU, containing 12-folds less Zn compared with both zinc sulfate products.

iv) As the reported results are startling, the reviewers should have reviewed location-wise and replication-wise data as well, instead of the "treatment-wise average values for all locations" only. Detailed replication-wise data have not been published as supplementary material of the article either.

v) The reviewers should have asked for the scientific reasons for 262% higher agronomic ZUE of *Zingro* compared with unbranded zinc sulfate, both having the same Zn concentration.

As the reported results in this open access article are plagued with conflict of interest and poor peer-review, these cannot be endorsed.

3.2 SUGGESTED REMEDIAL MEASURES FOR QUALITY ASSURANCE OF MICRONUTRIENT DATA

The suggested measures to minimize the data quality issues are as follows:

3.2.1 REMEDIAL MEASURES FOR FIELD AND GREENHOUSE DATA

We expect that quality issues regarding field and greenhouse yield data might be related to the erroneous application of micronutrient fertilizer treatments, improper randomization, inadequate replications, and/or mixing-up of the harvested grain/

biomass samples. Some salient remedial measures to minimize such issues are as follows:

i) Based on prior soil testing, field (greenhouse) experiments must be conducted on the soils that are deficient. An alternate approach could be carrying out trials simultaneously at two sites with contrasting (deficient and sufficient) status of target micronutrient(s).

To be accurate, the experiments must be properly randomized in accordance with the appropriate experimental design. Also, the number of replications must be adequate (e.g., field experiments must have a minimum of four replications). The general good practices of field experiments must be followed, e.g., use of guard rows, adequate spacing between treatments to avoid cross-contamination, and observation and reporting of the crop condition at the early-, mid-, and late-development stage (e.g., losses due to pests and crop lodging).

ii) The micronutrient fertilizer products that are to be applied must be of good quality. This can be ensured by using micronutrient fertilizers marketed by major fertilizer companies and/or by laboratory analysis of the fertilizer products prior to using them in the experiments.

iii) Uniform field application of a small quantity of micronutrient fertilizer must be ensured by its prior mixing with about five times the volume of well-pulverized soil of the same field.

iv) Great care must be exercised in the harvesting, threshing, and labeling of biomass and grain samples of the individual experimental plots.

3.2.2 REMEDIAL MEASURES FOR LABORATORY/ANALYTICAL DATA

Some remedial measures suggested to minimize erroneous laboratory data are as follows:

i) Great care must be exercised in obtaining composite grain and biomass samples from the field experiments. The sampled plant tissues should be washed carefully with distilled water/EDTA and dried immediately with tissue paper and placed in a hot-air oven.

ii) Due to the possibility of micronutrient contamination during sampling, sample processing (washing, drying, grinding) must be avoided. For example, soil sampling tools should be stainless steel, and if soil samples are crushed in a soil grinder, its hammers/blades must be of stainless steel. In the case of plant samples, the grinding mills should have stainless steel cutting blades as well as stainless steel sieves.

iii) Ground soil, as well as plant samples, must be stored in good-quality plastic bottles/vials in a clean place devoid of excess moisture.

iv) The laboratory glassware must be of Pyrex quality. For B analysis, the use of glassware should be minimal, and the Pyrex glassware must be acid-treated.

v) The chemical reagents for soil extraction and plant digestion must be of good quality.

vi) Soil extractions must be performed at room temperature (25°C); therefore, air-conditioning of the laboratory must be ensured. While performing soil extraction, each batch of samples must have a blank sample as well as a reference soil sample with a known concentration of the target micronutrient(s).

vii) In performing plant digestion, each batch of the samples must have a blank sample as well as a standard reference plant material, e.g., of NIST, USA – preferably, these should be of the same plant part so that digestion is mimicked, and of a similar range of concentrations.

viii) In performing the measurement of micronutrients on AAS (or on Spectrophotometer in case of B), the instrument should be adequately calibrated and the standard solutions must be of reliable quality.

ix) Samples preparation laboratory for soil materials should not be used for plant materials samples processing.

REFERENCES

Ali, M. A., F. Naeem, N. Tariq, I. Ahmed, and A. Imran. 2021. Bioactive nutrient fortified fertilizer: A novel hybrid approach for the enrichment of wheat grains with zinc. *Frontiers in Plant Science* 12:743378. doi: 10.3389/fpls.2021.743378.

Arshad, M. 2014. Interactive effect phosphorus and zinc on wheat crop. M.Sc. (Hons) thesis, Department of Soil and Environmental Sciences, The University of Agriculture, Peshawar.

Jamil, M., A. Sajad, M. Ahmad, et al. 2015. Growth, yield and quality of maize (*Zea mays* L.) fodder as affected by nitrogen-zinc interaction in arid climate. *Pakistan Journal of Agriculture Sciences* 52:637–643.

Joy, E. J. M., A. J. Stein, S. D. Young, et al. 2015. Zinc-enriched fertilisers as a potential public health intervention in Africa. *Plant Soil* 389:1–24. doi:10.1007/s11104-015-2430-8.

Malik, Z. 2011. Effect of zinc, boron and sulfur on the yield and nutrient concentration of maize. MSc (Hons) thesis, Department of Soil and Environmental Sciences, The University of Agriculture, Peshawar.

Phattarakul, N., B. Rerkasem, L. J. Li, et al. 2012. Biofortification of rice grain with zinc through zinc fertilization in different countries. *Plant and Soil* 361:131–141.

Prasad, R., Y. S. Shivay, D. Kumar. 2013. Zinc fertilization of cereals for increased production and alleviation of zinc malnutrition in India. *Agricultural Research* 2:111–118.

Rafique, E., A. Rashid, J. Ryan, and A. U. Bhatti. 2006. Zinc deficiency in rainfed wheat in Pakistan: Magnitude, spatial variability, management, and plant analysis diagnostic norms. *Communications in Soil Science and Plant Analysis* 37:181–197.

Ram, H., A. Rashid, W. Zhang, et al. 2016. Biofortification of wheat, rice and common bean by applying foliar zinc fertilizer along with pesticides in seven countries. *Plant and Soil* 403:389–401.

Ryan, J., G. Estefan, and A. Rashid. 2001. *Soil and Plant Analysis Laboratory Manual*, 2nd ed. International Center for Agricultural Research in the Dry Areas (ICARDA), Aleppo, Syria.

Zou, C. Q., Y. Q. Zhang, A. Rashid, et al. 2012. Biofortification of wheat with zinc through zinc fertilization in seven countries. *Plant and Soil* 361:43–55.

4 Micronutrient R&D in Pakistan
Historical Perspective and Current Status

4.1 INTRODUCTION

The first-ever fertilizer use in Pakistan pertained to the application of N in field crops during the 1950s. Thereafter, P fertilizer use in the country was initiated in 1960 and K fertilizer use began in 1966.The first-ever micronutrient deficiency in Pakistan was identified by two scientists of the International Rice Research Institute (IRRI), Yoshida and Tanaka (1969). They diagnosed Zn deficiency as the cause of "*Hadda*" disease of rice crop in the Punjab province. Simultaneously, in India, this Zn deficiency caused rice growth disorder that was known as "*Khaira*" disease (Tandon, 2013).

4.2 HISTORICAL PERSPECTIVE OF MICRONUTRIENT RESEARCH IN PAKISTAN

Zinc was the first-ever micronutrient recommended for the use of field crops, starting with rice (Saleem et al., 1986; Rashid, 2005). Concurrently, a positive response of the cotton crop to B fertilizer application was observed in Sindh province by Chaudhary and Hisiani (1970). Micronutrient use gained importance in the early 1970s with the initiation of the Pakistan Agricultural Research Council (PARC)-sponsored micronutrient research projects at Nuclear Institute for Agriculture and Biology (NIAB), Faisalabad, Ayub Agricultural Research Institute (AARI), Faisalabad, and Agricultural Research Institute (ARI), Quetta. In the late 1970s, the scope of micronutrient research was enhanced with the launch of a Nationally Coordinated Research Program by PARC. During this era, KP and Sindh provinces were also made part of the project.

In the early 1970s, because of the lack of atomic absorption spectrometer (AAS) in any agricultural laboratory in the country, micronutrients (Zn, Cu, Fe, and Mn) were analyzed by colorimetric procedures. The colorimetric procedures were laborious, time-consuming, and prone to error. The access to AAS played a pivotal role in accelerating micronutrient research in Pakistan. The first-ever AAS instrument was procured for the Rice Research Institute (RRI), Kala Shah Kaku, but never got installed because the rats at the RRI laboratory ruined the instrument by damaging

DOI: 10.1201/9781003314226-4

its delicate wires/electrical circuit. Soon thereafter, in 1973, the first AAS instrument became functional at NIAB, Faisalabad, in the Micronutrients Laboratory of Dr. F.M. Chaudhry, who, on earning a Ph.D. from the University of Western Australia under the supervision of Professor Jack F. Loneragan (who inspired a generation of scientists around the world to work on micronutrients) had initiated a vigorous research program on micronutrient nutrition of rice and wheat. Over a short span of a couple of years, Dr. F.M. Chaudhry's research group produced about 20 peer-reviewed publications on micronutrient nutrition of crops, six in *Plant and Soil*, Vol. 45–47 (1976, 1977) (Kausar et al., 1976; Rahmatullah et al., 1976; Rashid et al., 1976; Chaudhry et al., 1977a, 1977b, 1977c). Thus, the credit for initiating a systematic research on micronutrients in Pakistan goes to Dr. F.M. Chaudhry. This pioneer research group also established Zn deficiency in rice, wheat, and maize.

In the late 1970s, AARI, Faisalabad also acquired an AAS. Later, many other institutions throughout the country equipped their laboratories with AAS. Micronutrient research at National Agricultural Research Center (NARC), Islamabad, was initiated in the early 1980s.With the passage of time, the role of micronutrients in crop production gained enhanced recognition and micronutrients research was taken up by almost all agricultural research and educational institutions in the country. Now, almost every agricultural institution in the country is involved in micronutrient R&D. Currently, all major agricultural research and educational institutions have this analytical facility. Also, many agricultural laboratories are equipped with far superior analytical facilities like inductively coupled plasma mass spectrometer (ICP-MS). An ICP performs simultaneous analysis for 30–50 elements and is more efficient compared with the AAS. Since the detection limit is much lower than the AAS, a very low concentration of micronutrients can be analyzed in agricultural materials without difficulty, while using ICP-MS. Conclusively, every institution has contributed toward enhancing understanding of the value of micronutrients.

4.2.1 ZINC DEFICIENCY

Most Pakistani soils under crop production are inherently low in availability of most micronutrients, including Zn. This is primarily because of the alkaline–calcareous nature of the soils arising from highly alkaline–calcareous parent material with low status of the micronutrients. The occurrence of Zn deficiency is common in alkaline–calcareous and problematic soils with no exception to the submerged soils (human-induced or naturally-occurring soils). At NIAB, Faisalabad, the early research focused on micronutrient problems and their management in cereal crops, i.e., rice, wheat, and maize. This research established widespread occurrence of Zn deficiency in flooded rice grown in the main rice-growing areas of Punjab (Chaudhry and Sharif, 1975; Kausar et al., 1976; Chaudhry et al., 1977a). Their extensive multi-year research established that rice was much more susceptible to Zn deficiency compared with wheat grown in the same fields (Kausar et al., 1976). They also postulated the mechanisms of higher susceptibility of rice than wheat to Zn deficiency (Kausar et al., 1976) and of coarse-grain rice (i.e., cv. IR-6) than fine-grain aromatic Basmati-370 rice (Chaudhry et al., 1977a). As a result of appreciable paddy yield increases and amelioration of Zn deficiency symptoms in rice with Zn fertilizer, rice was the first crop in Pakistan for which Zn use was recommended and adopted in the country during the early 1970s.

4.2.2 BORON DEFICIENCY

The first-ever report of B deficiency in Pakistan pertains to Sindh, where Chaudhary and Hisiani (1970) observed cotton yield increase with B fertilizer. Although B deficiency in crops was among the initial micronutrient disorders identified in the country, B research did not receive due emphasis until the mid-1980s, primarily because of the analytical constraints. The laboratory analysis of B in soils and plant tissues remained an unsolved problem for quite some time. The researchers remained afraid of B contamination from glassware, and kept waiting for B-free glassware, which was quite expensive and never became available in agricultural laboratories of Pakistan. The initial colorimetric method for determining B, which employed curcumin, had some problems. With the passage of time, the Pakistani researchers learned to carryout satisfactory B analysis with minimal use of concentrated acid-treated Pyrex glassware. Since the adoption of the azomethine-H method of color development (Gaines and Mitchell, 1979) in the 1980s, soil B analysis is now a routine soil test in many research laboratories. A pivotal role in this regard was played by Mr. M.A. Kausar of NIAB, after spending one year as an IAEA research fellow with an acclaimed B toxicity expert, Dr. K.G. Tiller, at CSIRO Land and Water in Adelaide, Australia.

Historically, hot-water extraction (HWE) is a widely used laboratory procedure for assessing B status in soils around the world. This method is quite effective and was adopted in the agricultural laboratories of Pakistan as well. Undoubtedly, the procedure is tedious and prone to error because of the difficulty in maintaining uniform temperature for the soil samples in a batch throughout the boiling period. Rashid et al. (1994) determined that the 0.05 M HCl extraction procedure was a better option for determining B in alkaline soils. The establishment of the dilute HCl method as an effective B soil test for alkaline soils helped in rapid adoption of B analysis for both soil samples and plant tissues across the country.

4.2.3 IRON CHLOROSIS

The Fe deficiency-induced problem in plants is the easiest to recognize by virtue of its peculiar symptoms, i.e., interveinal chlorosis on the young leaves. Nevertheless, its prognosis/diagnosis by soil testing and conventional plant analysis as well as its management by soil application of Fe fertilizer are quite tricky. Soil testing even by the most widely adapted micronutrient test for alkaline/calcareous soils, i.e., diethylene-triamine penta-acetic acid (DTPA) test, may or may not help in assessing the true Fe availability status of the soils. Similarly, plant tissue analysis for total Fe content is not effective in diagnosing the Fe nutritional status of crop plants. In fact, in many instances, the total Fe concentration in chlorotic leaf tissue is greater than that in green (healthy) leaves of the crop in the same field. The incidence of Fe chlorosis in crop plants is in patches within the field, and the entire crop plants are not affected equally. Thus, field sampling of soil as well plant tissues for Fe analysis is not simple and straightforward. Because of such complexities, for a long time, research on Fe chlorosis in Pakistan remained limited to the correction of the chlorosis by Fe fertilization through foliar feeding. It was only during the 1990s that the Micronutrient Research Group at NARC, Islamabad, developed plant analysis critical levels of Fe^{2+} by analyzing fresh leaves of legume crops (peanut and chickpea) (Rashid and Din, 1992; Rashid et al., 1997d). Laboratory analysis of soil and plant Fe has hardly changed so far in the country.

4.2.4　CONTRIBUTIONS OF VARIOUS PAKISTANI INSTITUTIONS IN MICRONUTRIENT R&D

The early micronutrient research at NARC (during 1980–1990) focused predominantly on the rainfed soils and crops of the Pothohar plateau (Rashid and Qayyum, 1991; Rashid et al., 1997a, 1997b, 1997c, 1997d; Rashid et al., 2002b, 2002c). However, the later research was expanded to the rice–wheat system (Rashid et al., 2002a, 2002d, 2007) and the cotton–wheat system (Rashid and Rafique, 2002; Rafique et al., 2002). Salient research contributions from NARC include the establishment and management of field-scale deficiencies of B and Zn in cotton (Rafique et al., 2002; Rashid and Rafique, 2002), establishment of deficiency of B in rice and its management (Rashid et al., 2004, 2007), establishment of Fe chlorosis in legume crops (Rashid and Din, 1992; Rashid et al., 1997d) and citrus (Rashid et al., 1991; Rashid and Rashid, 1994), calibration of more efficient and economical multi-element soil test for micronutrients, i.e., AB-DTPA (Rashid and Ahmad, 1994), determination of plant analysis diagnostic criteria for locally grown crop genotypes (Rashid and Rafique, 2002; Rashid et al., 1997a, 1997b, 1997c, 1997d, 2002a, 2002c), and determination of suitability of mature grains (seed) as index tissue for diagnosing micronutrient deficiency and establishment of soil test interpretation guidelines in the country (Rashid and Ahmad, 1994; Rashid, 2006a). The major contributions of AARI include the establishment and management of Zn deficiency in citrus (Siddique et al., 1994; Rashid and Rashid, 1994), and widespread micronutrient deficiencies in rice and wheat.

　　The agricultural institutions which have made major contributions micronutrient research include NARC, Islamabad; NIAB, Faisalabad; KP Agricultural University, Peshawar; AARI, Faisalabad. Agricultural research institutions of other provinces have also made valuable contributions in the specific areas of micronutrient nutrition of crop plants. The R&D by all concerned institutions over the past 50 years have generated a reasonable volume of information about the nature, extent, and severity of micronutrient deficiencies in major cropping systems throughout the country. However, most of the information still pertains to two provinces, i.e., Punjab and KP, and even in these provinces all the geographical areas and/or cropping systems have not received adequate attention.

Besides the assumption that rice is tolerant to B deficiency, one major constraint to R&D on B nutrition of rice was the non-availability of B-free glassware and requisite professional expertise for laboratory analysis of B in the country (Rashid et al., 2002d). Research at KP Agricultural University, led by Professor Jahangir Khan Khattak, revealed micronutrient deficiencies in a wide range of soils, crops, and fruit orchards in the NWFP (now KP) as well as in Balochistan and in Pakistan Administered Kashmir. Apart from making the micronutrient problems better known, Prof. Khattak played an effective leadership role by coordinating the National Research Program on Micronutrients, financed by PARC, as well as by organizing a national symposium on micronutrients in Peshawar in1987, and later publishing its proceedings (Khattak et al., 1988).

4.2.5 Salient Reviews Concerning Micronutrients in Pakistani Agriculture

A number of reviews regarding micronutrient deficiency problems in Pakistan have been published, since the initiation of micronutrient research in the country (Table 4.1). The first-ever review on micronutrients was published in 1975 by Chaudhry and Sharif (1975). Thereafter, more reviews were published. The most comprehensive status report on micronutrients in Pakistani agriculture was prepared by the lead author of this book, Dr. Abdul Rashid, and published by the National

TABLE 4.1

Some Published Reviews Regarding Micronutrient Deficiency Problems in Pakistan

Author(s)/Citation	Micronutrient(s)	Remarks
Chaudhry and Sharif (1975)	All micronutrients, with emphasis on Zn and Cu deficiency	Published in "*Isotope-aided Micronutrient Studies in Rice Production with Special Reference to Zinc Deficiency*", published by International Atomic Energy Agency (IAEA), Vienna
Tahir (1978)	All micronutrients	Published in a Technical Bulletin, commissioned by Pakistan Agricultural Research Council (PARC), Islamabad
NFDC (1979)	Zinc	Fertilizer situation report on crop requirements of zinc
Khattak and Perveen (1986)	All micronutrients	A review, published in Proceedings of "*XII International Forum on Soil Taxonomy and Agrotechnology Transfer*"
Khattak and Perveen (1988)	All micronutrients	A review on "Micronutrient status of NWFP soils" in Proceedings of "*National Seminar on Micronutrients in Soils and Crops in Pakistan*"
Rashid and Salim (1988)	Zinc	A review in Proceedings of "*National Seminar on Micronutrients in Soils and Crops in Pakistan*"
Khattak (1991)	All micronutrients	Micronutrients in KP agriculture

(*Continued*)

TABLE 4.1   (CONTINUED)
Some Published Reviews Regarding Micronutrient Deficiency Problems in Pakistan

Author(s)/Citation	Micronutrient(s)	Remarks
Khattak (1995)	All micronutrients	A review on *"Micronutrients in Pakistani Agriculture"*
Rashid (1996)	All micronutrients	A chapter on "Secondary and micronutrients" in *Soil Science* textbook, published by National Book Foundation (NBF)
NFDC (1998)	All micronutrients	National status report on *"Micronutrients in Agriculture: Pakistan Perspective"*, prepared by Dr. Abdul Rashid and published by National Fertilizer Development Center (NFDC), with support from the FAO of the United Nations (FAO-UN), Pakistan
Rashid et al. (2002d)	Boron	A country report on *Boron in Plant and Animal Nutrition*, published by Kluwer
Kausar and Rashid (2002)	All micronutrients	A review, published in *Technologies for Sustainable Agriculture*
Rashid and Ryan (2004)	All micronutrients	A review, published in *Journal of Plant Nutrition*
Rashid (2005)	All micronutrients	A review, published in *Soil and Environment*
Rashid (2006a)	All micronutrients	A review, published by International Fertilizer Association (IFA), Paris
Rashid (2006b)	Boron	A booklet, summarizing Research and Development information on B nutrition of crop plants in Pakistan
Rashid and Rafique (2017)	Boron	A review, published in *Boron* journal
Rehman et al. (2018)	Boron	A review on B nutrition of rice, published in *Agronomy for Sustainable Development* journal
Rehman et al. (2020)	Zinc	Review of research on agronomic enrichment of food grains with zinc in Pakistan

Fertilizer Development Center (NFDC, 1998). Later, some additional reviews were attempted to address one particular micronutrient or a certain specific aspect of micronutrients. For instance, Rashid et al. (2002d); Rashid (2006b); and Rashid and Rafique (2017) reviewed B deficiency in calcareous soils of Pakistan and its management aspects. However, the overall situation regarding micronutrient problems and their management was not documented.

4.3   SOIL CONDITIONS AND AGRONOMIC PRACTICES INDUCING MICRONUTRIENT DEFICIENCIES

Out of about 80 million hectares (Mha) of total geographical area of Pakistan, about 22 Mha are cultivated. About 75% of the total cultivated area is irrigated (mostly through canal/river water) and the rest is rainfed. Major cropping systems in the country are cotton–wheat (~3.0 Mha), rice–wheat (~2.2 Mha), mixed cropping (i.e., maize-based, sugarcane-based, etc.), rainfed cropping (wheat, sorghum, maize, chickpea, peanut, etc.; e.g., 1.82 Mha in rainfed Pothohar plateau in northern Punjab), oilseeds, pulses, and a wide array of vegetables and fruits. Unless moisture is a serious constraint, two crops a year is the norm.

Fertilizer use in Pakistan predominantly pertains to nitrogen (N) and phosphorus (P); potassium (K) use is confined to a few high-K requiring crops, like tobacco and potato. Compared with the actual crop requirements, micronutrient fertilizer use in the country is negligible (NFDC, 1998; Rashid, 2006a). Also, many recently introduced high-yield potential crop varieties are more susceptible to micronutrient deficiencies (Chaudhry et al., 1977a; Rashid and Din, 1992; Rashid et al., 2002a, 2002b, 2002c). Hence, soil conditions and agronomic practices in the country are conducive to the incidence of micronutrient deficiencies in crop plants. The soils across much of the cultivated areas were formed from calcareous alluvium and loess deposits and are low in OM, and many essential plant nutrients. The climate, except for some high mountains in the north, is mostly arid to semi-arid. Because of the calcareous nature of parent material and low rainfall, almost all the soils are young on the geological time scale. Hence the soils are alkaline and calcareous in nature due to the presence of alkaline earth cations (Ca^{2+} and Mg^{2+}). These cations are hardly leached down from the upper soil horizons or beyond the root zone. Some small pockets of moderately acid soils exist in the country, but these are predominantly confined to the high rainfall areas of Mansehra and Swat districts in KP province. The parent material of these soils is predominantly non-calcareous like granite (Rashid, 1993; Nizami et al., 1994). Even a vast majority of these acid soils suffer from deficiencies of Zn (Rashid, 1994).

A lot of R&D information on the identification/establishment of micronutrient deficiency problems and their management has been generated since the 1998 status report was published (NFDC, 1998). It was thought necessary to update the micronutrients scenario in the country, in the context of 4R nutrient stewardship.

4.4 ESTABLISHMENT OF FIELD-SCALE MICRONUTRIENT DEFICIENCIES IN PAKISTAN

The first-ever field-scale micronutrient problem established in Pakistan was that of Zn deficiency in flooded rice in Punjab during the 1970s (Kausar et al., 1976; Chaudhry et al., 1976, 1977a). During the early 1980s, micronutrient deficiencies were recognized in a wide range of soils, crops, and fruits throughout the country, i.e., in KP (North West Frontier Province), Punjab, Balochistan, Sindh, and Pakistan Administered Kashmir (Azad Jammu and Kashmir). Incidence of extensive B, Zn, and Fe deficiency in rainfed soils and crops of Pothohar plateau was also established during this period (Rashid and Qayyum, 1991). In addition to organizing a national seminar on micronutrients (Khattak et al., 1988), micronutrient reviews were also published (Chaudhry and Sharif, 1975; Khattak and Perveen, 1986). Salient research accomplishments during the 1990s included the identification and establishment of field-scale deficiencies of B and Zn in cotton, wheat, citrus, mango, and other crops (Rashid et al. 2002c; Rashid, 2005) and the establishment of Fe deficiency in apple and legume crops (like chickpea and peanut) (Rashid and Din, 1992; Rashid et al., 1997d), and citrus (Rashid et al., 1991; Rashid and Rashid, 1994).

A salient development during 2000–2005 was the identification and establishment of an alarming extent of the incidence of B deficiency in rice crop (fine-grain aromatic Basmati cultivars as well as coarse-grain IRRI-type rices) in the Punjab and Sindh provinces (Rashid et al., 2004; 2007). Consequently, on demonstration of substantial increases in paddy yield and grain quality improvements with B fertilization in both the provinces, formal B fertilizer use recommendations were made for rice crop by the Punjab Agriculture Department as well as by Sindh Agriculture Department. Later, in 2006, Pakistan Agricultural Research Council published a detailed overview of B deficiency R&D in the country entitled "*Boron Deficiency in Soils and Crops of Pakistan: Diagnosis and Management*".

An added dimension of micronutrient R&D during recent times is the realization that soil–plant micronutrient deficiencies underlie widespread micronutrient malnutrition (also called "*hidden hunger*") in the Pakistani population who are predominantly dependent on staple cereals to meet their caloric requirements. The consequent global effort, under the umbrella of the HarvestPlus program, has led to the development of wheat genotypes which are Zn-efficient (HarvestPlus, 2014). Whereas the predominant Zn concentration in Pakistani wheat grains is around 25.0 mg kg^{-1}, the six Zn-efficient wheat lines of HarvestPlus contain Zn of about 37.0 mg kg^{-1}. During the year 2016, one of these high-Zn wheat lines, namely NR-421, was approved as a variety, named "*Zincol-2016*", by the Punjab Seed Council. Later, two additional zinc biofortified wheat varieties"*Akbar-2019*" and "*Nawab-2021*" developed by AARI Faisalabad have also been approved by the Government of Punjab, Pakistan. The simultaneous development of Zn foliar fertilization technology for agronomic biofortification of Zn in wheat crop in Pakistan and elsewhere (Zou et al., 2012) is also a major development to address Zn malnutrition. Whereas the desirable Zn concentration in wheat grains to meet daily dietary requirement is 40–60 mg Zn kg^{-1}, the grains of high-Zn wheat developed by HarvestPlus in Pakistan contain around 37.0 mg Zn kg^{-1} (average of field trials at 12 sites). Agronomic biofortification of the high-Zn wheat lines can result in producing wheat grains with $\geq$ 50.0 mg Zn kg^{-1}. As foliar Zn can be applied effectively by mixing with some popular pesticide solutions, like *Confidor* insecticide (Ram et al., 2016), agronomic biofortification is an effective complementary approach to produce high-Zn density wheat grains in a cost-effective manner. The high-Zn wheat grains also result in better wheat productivity as a consequence of better seed germination and crop stand.

Fertilization with Zn in soil-deficient situations provides attractive economic returns due to yield improvements. In addition, the use of Zn fertilizers in Pakistan could almost halve the prevalence of inadequate dietary Zn intakes (Joy et al., 2017). Application of Zn fertilizers in Punjab and Sindh provinces could generate returns of US$ >800 M through increased yield at a benefit:cost ratio of ~15, and an additional return of US$ 405–1216 M due to health benefits in Punjab province alone. Thus, enhanced use of Zn fertilizer can improve crop productivity, farmer profitability, and human well-being.

REFERENCES

Chaudhary, T. M., and G. R. Hisiani. 1970. Effect of B on the yield of seed cotton. *The Pakistan Cottons* 1:13–15.

Chaudhry, F. M., F. Hussain, and A. Rashid. 1977a. Micronutrient availability to cereals from calcareous soils: Phosphorus-zinc interaction in rice. *Plant and Soil* 47:297–302.

Chaudhry, F. M., M. A. Kausar, A. Rashid, and Rahmatullah. 1977b. Mechanism of nitrogen effect on zinc nutrition of flooded rice. *Plant and Soil* 46:649–654.

Chaudhry, F. M., S. M. Alam, A. Rashid, and A. Latif. 1977c. Micronutrient availability to cereals from calcareous soils. IV. Mechanism of differential susceptibility of two rice verities to Zn deficiency. *Plant and Soil* 46:637–642.

Chaudhry, F. M., A. Latif, A. Rashid, and S. M. Alam. 1976. Response of the rice varieties to field application of micronutrient fertilizers. *Pakistan Journal of Scientific and Industrial Research* 19:134–139.

Chaudhry, F. M., and M. Sharif. 1975. Micronutrient problems of crops in Pakistan with special reference to zinc and copper deficiency. In: Isotope-aided Micronutrient Studies in Rice Production with Special Reference to Zinc Deficiency. Proc. Combined Panel, International Atomic Energy Agency (IAEA), Vienna, Sept 23–27, 1974. IAEA-TECDOC-172, IAEA, Vienna.

Gaines, T. P., and G. A. Mitchell. 1979. Boron determination in plant tissue by azomethine-H method. *Communications in Soil Science and Plant Analysis* 10:1099–1108.

HarvestPlus. 2014. Biofortification Progress Briefs. Accessible at www.HarvestPlus.org. (Accessed on Feb 23, 2019).

Joy, E. J. M., W. Ahmad, M. H. Zia, et al. 2017. Valuing increased zinc (Zn) fertilizer-use in Pakistan. *Plant and Soil* 411:139–150.

Kausar, M. A., F. M. Chaudhry, A. Rashid, A. Latif, and S. M. Alam. 1976. Micronutrient availability to cereals from calcareous soils. Comparative Zn and Cu deficiency and their mutual interaction in rice and wheat. *Plant and Soil* 45:397–410.

Kausar, M. A., and A. Rashid. 2002. Micronutrients in crop production. p. 161–166, In: Farooq-e-Azam et al. (Ed.), *Technologies for Sustainable Agriculture*. Nuclear Institute for Agriculture and Biology, Faisalabad, Pakistan.

Khattak, J. K. 1991. *Micronutrients in NWFP Agriculture*. Barani Research and Development Project, PARC, Islamabad.

Khattak, J. K. 1995. *Micronutrients in Pakistan Agriculture*. Pakistan Agricultural Research Council, Islamabad and NWFP Agricultural University, Peshawar. p. 135.

Khattak, J. K., and S. Perveen. 1986. Trace element status of Pakistan soils. p. 119–124, In: *Proceedings XII International Forum on Soil Taxonomy and Agro Technology Transfer*, Pakistan: 1st Volume (Technical Sessions), 9–23 October 1985. Soil Survey of Pakistan, Lahore.

Khattak, J. K., and S. Perveen. 1988. Micronutrient status of NWFP soils. p. 62–74, In: *Proceedings of National Seminar on Micronutrients in Soils and Crops in Pakistan*, 13–15 December 1987, NWFP Agricultural University, Peshawar.

Khattak, J. K., S. Perveen, and Farmanullah. 1988. Proceedings of National Seminar on Micronutrients in Soils and Crops in Pakistan, 13–15 December 1987, NWFP Agricultural University, Peshawar.

NFDC. 1979. Crop requirements of zinc and projected demand in Pakistan for period 1979/80 to 1985/86. National Fertilizer Development Center, Islamabad.

NFDC. 1998. *Micronutrients in Agriculture: Pakistan perspective*. National Fertilizer Development Center, Islamabad, 51 pp.

Nizami, M. M. I., M. S. Akhtar, A. Rashid, and J. Din. 1994. Soil diagnostic analysis for tea cultivation in Swat and Dir areas. p. 619–627, In: *Efficient Use of Plant Nutrients. Proc. Fourth National Congress of Soil Science*, Islamabad, 24–26 May, 1992. Soil Science Society of Pakistan.

Rafique, E., A. Rashid, A. U. Bhatti, G. Rasool, and N. Bughio. 2002. Boron deficiency in cotton grown in calcareous soils of Pakistan. I. Distribution of B availability and comparison of soil testing methods. p. 349–356, In: Goldbach, H.E. et al. (Ed.), *Boron in Plant and Animal Nutrition*. Kluwer Academic Publishers, New York.

Rahmatullah, F.M. Chaudhry, and A. Rashid. 1976. Micronutrient availability to cereals from calcareous soils. II. Effect of flooding on electrochemical properties of soils. *Plant and Soil* 45:411–420.

Ram, H., A. Rashid, W. Zhang, et al. 2016. Biofortification of wheat, rice and common bean by applying foliar zinc fertilizer along with pesticides in seven countries. *Plant and Soil* 403:389–401.

Rashid, A. 1993. Nutritional disorders of rapeseed-mustard and wheat grown in *Pothohar* area. Micronutrient Project, Annual Report, 1991–92. National Agricultural Research Center, Islamabad.

Rashid, A. 1994. Nutrient Indexing Surveys and Micronutrient Requirement of Crops. Micronutrient Project, Annual Report, 1992–93. National Agricultural Research Center, Islamabad.

Rashid, A. 1996. Secondary and micronutrients. p. 341–385, Chapter 12, In: *Soil Science*, Rashid, A., and K.S. Memon (Managing Authors). National Book Foundation, Islamabad, Pakistan.

Rashid, A. 2005. Establishment and management of micronutrient deficiencies in Pakistan: a review. *Soil and Environment* 24:1–22.

Rashid, A. 2006a. Incidence, diagnosis and management of micronutrient deficiencies in crops: Success stories and limitations in Pakistan. In: *Optimizing Resource Use Efficiency for Sustainable Intensification of Agriculture*. IFA Agriculture Conference, Kunming, PR China. 27 February - 2 March, 2006. Available at: www.fertilizer.org/Public/Stewardship/Publication_Detail.aspx?SEQN=3865& PUBKEY=6DCE3EF6-B7F5-45CF-90C8-A61F010D9462

Rashid, A. 2006b. Boron Deficiency in Soils and Crops of Pakistan: Diagnosis and Management. Pakistan Agricultural Research Council, Islamabad, Pakistan, viii+34 pp. ISBN: 969-409-184-5

Rashid, A., and N. Ahmad. 1994. Soil Testing in Pakistan: country report. p. 39–53, In: *Proc FADINAP Regional Workshop on Cooperation in Soil Testing for Asia and the Pacific*, 16–18 Aug 1993, Bangkok, Thailand. United Nations, New York.

Rashid, A., F. M. Chaudhry, and M. Sharif. 1976. Micronutrient availability to cereals from calcareous soils. III. Zinc absorption by rice and its inhibition by important ions of submerged soils. *Plant and Soil* 45:613–623.

Rashid, A., and J. Din. 1992. Differential susceptibility of chickpea cultivars to iron chlorosis grown on calcareous soils of Pakistan. *Journal of the Indian Society of Soil Science* 40: 488–492.

Rashid, A., F. Hussain, A. Rashid, and J. Din. 1991. Nutrient status of citrus orchards in Punjab. *Pakistan Journal of Soil Science* 6:25–28.

Rashid, A., and F. Qayyum. 1991. Micronutrient Status of Pakistan Soils and Their Role in Crop Production, Cooperative Program. Final Report, 1983–1990. NARC, Islamabad. p. 84.

Rashid, A., and E. Rafique. 2002. Boron deficiency in cotton grown in calcareous soils of Pakistan. II. Correction and criteria for foliar diagnosis: p 357–362, In: Goldbach, H.E. et al. (Ed.), *Boron in Plant and Animal Nutrition*. Kluwer Academic Publishers, New York.

Rashid, A., and E. Rafique. 2017. Boron deficiency diagnosis and management in field crops in calcareous soils of Pakistan: a mini review. *BORON* 2(3):142–152.

Rashid, A., E. Rafique, and N. Bughio. 1994. Diagnosing boron deficiency in rapeseed and mustard by plant analysis and soil testing. *Communications in Soil Science and Plant Analysis* 25:2883–2897.

Rashid, A., E. Rafique, J. Din, S. N. Malik, and M. Y. Arain. 1997a. Micronutrient deficiencies in rainfed calcareous soils of Pakistan. I. Iron chlorosis in peanut. *Communications in Soil Science and Plant Analysis* 28:135–148.

Rashid, A., E. Rafique, N. Bughio, and M. Yasin. 1997b. Micronutrient deficiencies in rainfed calcareous soils of Pakistan. IV. Zinc nutrition of sorghum. *Communications in Soil Science and Plant Analysis* 28:455–467.

Rashid, A., E. Rafique, and N. Bughio. 1997c. Micronutrient deficiencies in rainfed calcareous soils of Pakistan. III. Boron nutrition of sorghum. *Communications in Soil Science and Plant Analysis* 28:441–454.

Rashid, A., E. Rafique, and N. Ali. 1997d. Micronutrient deficiencies in rainfed calcareous soils of Pakistan. II. Boron nutrition of the peanut plant. *Communications in Soil Science and Plant Analysis* 28:149–159.

Rashid, A., E. Rafique, S. Muhammed, and N. Bughio. 2002a. Boron deficiency in rainfed alkaline soils of Pakistan: Incidence and genotypic variation in rapeseed-mustard. p. 363–370, In: Goldbach, H.E. et al. (Ed.), *Boron in Plant and Animal Nutrition*. Kluwer Academic/Plenum Publishers, New York.

Rashid, A., S. Muhammad, and E. Rafique. 2002b. Genotypic variation in boron uptake and utilization by rice and wheat. p. 305–310, In: Goldbach, H.E. et al. (Ed.), *Boron in Plant and Animal Nutrition*. Kluwer Academic Publishers, New York.

Rashid, A., E. Rafique, and N. Bughio. 2002c. Boron deficiency in rainfed alkaline soils of Pakistan: Incidence and boron requirement of wheat. p. 371–379, In: Goldbach, H.E. et al. (Ed.), *Boron in Plant and Animal Nutrition*. Kluwer Academic Publishers, New York.

Rashid, A., E. Rafique, and J. Ryan. 2002d. Establishment and management of boron deficiency in field crops in Pakistan: A country report. p. 339–348, In: Goldbach, H.E. et al (Ed), *Boron in Plant and Animal Nutrition*. Kluwer Academic Publishers, New York.

Rashid, A., and A. Rashid. 1994. Nutritional problems of citrus on calcareous soils of Pakistan. p. 190–197, In: *Proc. First International Seminar on Citriculture in Pakistan*. Faisalabad, 2–5 Dec 1992. University of Agriculture, Faisalabad.

Rashid, A., and J. Ryan. 2004. Micronutrient constraints to crop production in soils with Mediterranean-type characteristics: A review. *Journal of Plant Nutrition* 27:959–975.

Rashid, A., and M. Salim. 1988. Soil conditions and crop factors inducing zinc deficiency in plants. p. 94–117, In: *Proc. National Seminar on Micronutrients in Soils and Crops in Pakistan*. Peshawar, 13–15 Dec. 1988. NWFP Agricultural University, Peshawar.

Rashid, A., M. Yasin, M. A. Ali, Z. Ahmad, and R. Ullah. 2007. An alarming boron deficiency in calcareous rice soils of Pakistan: Boron use improves yield and cooking quality. p. 103–116, In: F Xu (Ed) Advances in Plant and Animal Boron Nutrition. Proc 3rd*Int Symp on All Aspects of Plant and Animal Boron Nutrition, Wuhan, China*, 9–13 Sep 2005. Springer, Dordrecht, The Netherlands.

Rashid, A., M. Yasin, M. Ashraf, and R. A. Mann. 2004. Boron deficiency in calcareous soils reduces rice yield and impairs grain quality. *International Rice Research Notes* 29:58–60.

Rehman, A., M. Farooq, A. Rashid, et al. 2018. Boron nutrition of rice in different production systems: a review. *Agronomy for Sustainable Development.* doi:10.1007/s13593-018-0504-8.

Rehman, A., M. Farooq, A. Ullah, F. Nadeem, and D. J. Lee. 2020. Agronomic biofortification of zinc in Pakistan: Status, benefits, and constraints. *Frontiers in Sustainable Food Systems* 4:591722. doi:10.3389/fsufs.2020.591722.

Saleem, M. T., N. Ahmad, and J. G. Divide. 1986. *Fertilizers and Their Use in Pakistan.* National Fertilizer Development Center, Islamabad.

Siddique, M. T., M. Rashid, and M. Saeed. 1994. Micronutrient status of soils/plants of citrus growing areas in Punjab. p. 355–360, In: *Efficient Use of Plant Nutrients.* Proc. 4th National Congress of Soil Science, Islamabad, May 24–26, 1992. Soil Science Society of Pakistan.

Tahir, M. 1978. Micronutrients for crops. p. 57–77. In: *Review on Potash, Micronutrients and Rhizobium. Technical Bulletin.* Pakistan Agricultural Research Council, Islamabad.

Tandon, H. L. S. 2013. *Micronutrient Handbook from Research to Application.* Fertilizer Development and Consultation Organization, New Delhi, India. 234 pp.

Yoshida, S., and A. Tanaka. 1969. Zinc deficiency of rice in calcareous soils. *Soil Science and Plant Nutrition*15:75–80.

Zou, C. Q., Y. Q. Zhang, A. Rashid, et al. 2012. Biofortification of wheat with zinc through zinc fertilization in seven countries. *Plant and Soil* 361:43–55.

5 Micronutrient Status of Soils and Crops in Pakistan

5.1 INTRODUCTION

This chapter pertains to total and plant available micronutrients in soils, as well as (theoretical) longevity of micronutrient supplying capacity of Pakistani soils. The local research concerning micronutrient status in crop plants and the identification of micronutrient deficiencies in crops is also summarized in this chapter. In addition to discussing possible relationships between soil types and micronutrient disorders in crops, spatial variability of micronutrients in soils and crop plants is also reported here.

5.2 TOTAL MICRONUTRIENTS AND SUPPLYING CAPACITY OF SOILS

Total micronutrient contents in soils are far greater than available contents. Only a very small fraction of the total amounts is represented by the available contents. Available micronutrients are only 0.5–5% of the total contents (Katyal and Deb, 1982). For example, Fe is the fourth most abundant element in the earth's crust, yet many crops growing in calcareous soils exhibit Fe chlorosis obviously because of the inadequate plant availability of Fe^{2+} in such soils. A soil can be very "rich" in the total content of a micronutrient but at the same time "poor" in terms of its availability to the plant. Generally, there is a poor relationship between total micronutrient content in soil and its available content. Therefore, laboratory analysis for total micronutrient contents is not a useful approach for assessing the micronutrient-supplying capacity of soils to crops. Nevertheless, it points out that any technology developed with the sole purpose of increasing the total concentration of soil micronutrients has only a limited value and cannot support sustainable crop production intensification. Realizing these aspects, very little data have been generated on total micronutrient contents in Pakistani soils. Wahab and Bhatti (1958) were perhaps the first ones to analyze total micronutrient contents in some Pakistani soils.

DOI: 10.1201/9781003314226-5

5.2.1 Longevity of Micronutrient Supply Capacity of Pakistani Soils

Micronutrients are taken up by plants from soil through roots unless supplied in foliar sprays. The concentration of micronutrients in various plant parts, total uptake, and removal depends on a large number of factors including soil properties, crop species and variety, crop management practices (like irrigation and its source, use of FYM and NPK fertilizers, etc.), crop yield, and the removal of crop residues and its recycling.

Micronutrient uptake by the crop is estimated by multiplying the micronutrient concentration in dry matter with the yield of dry matter produced. For example, maize crop removes 30.9 g Zn ha^{-1} in grains (when grain yield is 1.5 Mg ha^{-1}) and an additional 84.9 g ha^{-1} in straw (when straw yield is 3.0 Mg ha^{-1}). The total Zn uptake by maize crop at this yield level is 30.9 g ha^{-1} + 84.9 g ha^{-1} = 115.8 g ha^{-1}. If only grains are removed, Zn removal from the field is 30.9 g ha^{-1}only, but if both grains and straw are removed from the field, then Zn removal is almost equal to the total Zn uptake by the crop. In many crops, such as potato, sugarcane, and cotton, most of the leaf residue remains in the field and a certain amount of the micronutrient is returned to the soil. In fruit orchards and other perennial plantations, only a fraction of the total micronutrients taken up by the plants is removed from the field with the fruit harvest. Some nutrients are recycled back with leaf fall, and some amounts of the micronutrients remain tied up in the woody plant parts. Crop species differ in their micronutrient concentration in plant tissues as well as in partitioning between grain (or fruit) and straw (or stalks, etc.). Additionally, amounts of the micronutrient uptake and/or removal by the crop depend on the total biomass produced, and grain:straw ratio.

Local field experimental data on micronutrient uptake/removal by crops in dominant cropping systems are very limited. To our knowledge, such data of Pakistani origin pertains only to the rainfed wheat crop (Rafique et al., 2006; Rashid et al., 2011) and cotton–wheat cropping (Rafique et al., 2012). We have adjusted data from other published sources to the average crop yield levels in Pakistan. Estimates of micronutrient uptake/removal from the field for some selected crops indicate wide variations between uptake by various crop species and even by the same species grown at different locations (Table 5.1). Recently published five-year field experimental data pertaining to the cotton–wheat cropping in Pakistan revealed that total Zn uptake by cotton crop (in cotton seeds, burs, leaves, and stalks) grown in accordance with the farmers' fertilizer use practices (without Zn application) is very low~ 62 g Zn ha^{-1} only (Rafique et al., 2012), compared with the very high figure of 480 g ha^{-1} reported by Mortvedt (1983). Such a huge difference between total Zn uptake values by cotton crop reported by two different sources is not understandable. Regarding B uptake on a single crop basis, the maximum value listed in the literature is 120 g ha^{-1} for cotton and potato (Mortvedt, 1983; Tandon, 1995). A minimum B uptake of 5 g ha^{-1} by mustard is reported in literature (Tandon, 1995). These variations are also evident for other micronutrients (Table 5.1). Unlike Zn uptake by cotton

TABLE 5.1
Estimates of Micronutrient Uptake/Removal by Selected Crops

Crop	Yield Component	Yield[†] (t ha⁻¹)	Micronutrient uptake/Removal (g ha⁻¹)[‡] Zn	Cu	Fe	Mn	B
Wheat[4]	Grain	2.0	112	48	1248	140	96
Wheat[5]	Grain	2.0	66	34	232	216	-
Wheat[6]	Grain	2.0	209	50	1219	330	-
Wheat (irrigated)[8]	Grain	3.4	84	-	30		
Wheat (rainfed)[9]	Grain	2.9	78*	-	-	-	16*
Rice[4]	Paddy	2.9	116	52	444	1958	44
Rice[6]	Paddy	2.9	219	24	1182	529	-
Maize[4]	Grain	1.5	195	195	11800	480	14
Maize[5]	Grain	1.5	116	44	918	134	74
Maize[6]	Grain	1.5	75	29	701	194	-
Cotton[5]	Seed cotton	2.5	480	110	140	190	120
Cotton[8]	Seed cotton	2.2	62	-	96	-	-
Potato[4]	Tubers	14	126	168	2240	168	120
Tomato[7]	Fruit	48	60	608	535	95	28
Chickpea[4]	Grain	0.53	20	6	456	37	-
Peanut[4]	Pods	1.1	121	39	2517	102	-
Mustard[4]	Grain	0.23	22	4	251	21	5
Rapeseed[7]	Grain	3.0	50	17	150	90	50
Sunflower	Grain	1.25	58	48	1342	227	100
Alfalfa[4]	Green fodder	107	433	75	710	620	24
Citrus[7]	Fruit	48	60	120	600	140	120

[†] Average yields in Pakistan
[‡] Micronutrient uptake re-calculated for average yields in Pakistan
* Removal by grain + straw
[4] After Tandon (1995)
[5] After Mortvedt (1983)
[6] After Nambiar and Gosh (1984)
[7] After Shorrocks (1992a, 1992b)
[8] After Rafique et al. (2012), Rashid et al.'s unpublished results
[9] After Rafique et al. (2006) and Rashid et al. (2011)

crop, B uptake by cotton in Pakistan's cotton–wheat system is reported to be 96 g ha⁻¹ which is in agreement with the values reported in international literature (Rashid et al. unpublished results).

The magnitude of variation in uptake of various micronutrients by the same crop species is also great. For example, for the 2 tonnes grain ha⁻¹ yield level, wheat removes 66–209 g Zn ha⁻¹ and 140–330 g Mn ha⁻¹. Similar variations are also evident for some other crops (Table 5.1). The reasons could be many, including micronutrient fertility level of the soil, crop's micronutrient acquisition capacity, and the quantum of biomass produced per unit field area. In the absence of adequate field and agronomic information, variable data of this kind are confusing for making good estimates of micronutrient uptake or apparent soil balance sheets. The current micronutrient fertilizer use in Pakistan is inadequate and restricted to certain selected crops. We have considered only crop removals for the purpose of this book. Studies for determining the longevity of the soil productivity with emerging micronutrient deficiencies are limited to the cotton–wheat cropping system.

It is believed that the benchmark soils could be totally exhausted after a period in terms of the availability of micronutrients if the micronutrient fertilizers are not made part of the management practices, and similar land uses are prolonged. According to Takkar (1996), some benchmark soils of India grown to various cropping systems would be exhausted for total Zn in a time span of 165–320 years, and Fe in the duration of 10,000 to 24,000 years without the application of micronutrient fertilizers. In a long-term study by the authors, maize yield started declining after 10 annual cropping cycles (wheat-maize–cowpea) because of the depletion of DTPA-extractable Zn from adequate to the deficient level (1.1 mg kg^{-1} to 0.68 mg kg^1). The DTPA-extractable Zn was depleted from 2.54 mg kg^{-1} to 0.97 mg kg^{-1} in a rice–wheat–cowpea concentrated land use on a Hapludoll of Patanager (India) after 12 annual cycles of cropping (Takkar, 1996). A similar attempt was made by Dr. Sajida Perveen of The University of Agriculture, Peshawar (per. com.), for determining the longevity of micronutrient supplying capacity of the Peshawar Valley soils under the rice–wheat and maize–wheat rotations. According to her estimates, Mn would clearly be the most limiting micronutrient as it would be totally exhausted in 490–1340 years followed by B in 570 years, Cu in 582–1420 years, Zn in 1330–1800 years, and Fe in 19,000–30,000 years. However, some recent publications by the Micronutrient Research Group at NARC have revealed that apparent soil balances of Zn and B under the irrigated cotton–wheat cropping system are positive even with the prevalent fertilizer use practice (Rafique et al., 2012; Rashid et al. unpublished results). Therefore, the authors of this book believe that the above-stated theoretical exercises about the longevity of micronutrient supplying capacity of soils may be of little or no importance for the reported variations in the irrigated cotton–wheat cropping system.

5.3   AVAILABLE MICRONUTRIENT STATUS OF SOILS

The diagnostic techniques used for assessing micronutrient fertility of soils include:

a)   Deficiency symptoms in plants
b)   Soil testing for micronutrients
c)   Plant analysis for micronutrients
d)   Greenhouse and field experimentation

It is virtually impossible to estimate the actual amount of soil micronutrients that will be available to crop plants. Each diagnostic technique has its advantages as well as limitations. The diagnostic techniques, like soil testing, at the best, can provide an index value telling the adequacy or otherwise of a micronutrient the field soil can provide to the crop plants. Deficiency symptoms are a useful indicator of micronutrient status, particularly in disorders like Fe chlorosis. Greenhouse and field experimentation is quite a valuable technique but impossible to practice for each field location because of temporal and economic considerations. Therefore, only soil testing and, to a lesser extent, plant analysis are the practical approaches for diagnosing micronutrient disorders on a routine basis and at a wide scale. All these diagnostic techniques have been elaborated in detail in Chapter 2 of this book.

5.3.1 MICRONUTRIENT STATUS OF CROP PLANTS

Although soil testing is an effective and useful technique for the prognosis and diagnosis of micronutrient disorders, plant composition is a better indicator of micronutrient fertility of soils. In general, soil testing is quite reliable for micronutrients like B, Zn, and Cu. However, because of the overwhelming influence of soil/rhizosphere pH on the oxidation status of Fe and Mn, soil testing alone is not very effective to assess soil fertility status for these micronutrients. Moreover, plant analysis is considered more useful for perennial species like fruit orchards.

Like many other countries of the world, plant analysis information in Pakistan is far less compared with soil analysis data. However, systematic nutrient indexing studies carried out by NARC and some other agricultural research institutions provided comprehensive information about the micronutrient status of farmer-grown citrus (Rashid et al., 1991; Rehman, 1989; Khattak, 1991, 1995; Siddique et al., 1994; Rashid and Rashid, 1994), apple (Rehman, 1990), rapeseed-mustard (Rashid et al., 1994), wheat (Rashid et al., 2011), rice (Zia, 1993), sorghum (Rashid and Qayyum, 1991; Rashid et al., 1997b, 1997c), peanut (Rashid et al., 1997a, 1997d), and cotton (Rashid and Rafique, 1997; Rafique et al., 2002). Plant micronutrient analysis data are also available in many other nutrient indexing investigations (e.g., Sillanpää, 1982) and many field research investigations (e.g., Ahmed et al., 2013; Hussain, 2006).

5.3.2 NATURE AND EXTENT OF ZINC, BORON, AND IRON DEFICIENCIES

To know where micronutrient fertilizer use is needed, it is necessary to know the areas in which soils and/or crop plants are deficient in specific micronutrients. A lot of research information has been generated in Pakistan on the micronutrient status of soils and crops by way of systematic nutrient-indexing of farmer-grown crops (by sampling and analyzing diagnostic plant tissues and associated soils) and conducting field experiments on many crops at farmers' fields and at government research farms. Although most of the data pertain to two provinces of the country, i.e., Punjab and KP, some research reports also help in assessing the nature and extent of micronutrient deficiencies in Sindh, Balochistan, and Pakistan Administered Kashmir. Most of the micronutrient status information is restricted to soil analysis data; however, nutrient-indexing studies of irrigated cotton and rice crops and rainfed crops in the Pothohar plateau (sorghum, rapeseed-mustard, wheat, peanut), and many types of fruit orchards (citrus, apple, mango) have helped in assessing the nature and extent of micronutrient deficiencies in these crop plants. The results of numerous field experiments have been used to augment this crucial segment of the book on micronutrients. Initially, in 1986, a bulletin containing maps of the micronutrient status of Pakistan soils was published by Pakistan Agricultural Research Council (PARC). These maps delineated geographical areas affected by deficiencies of Zn, Cu, Fe, Mn, and B. This exercise revealed the province-wise predominant deficiencies in Punjab and KP (Zn and B); Sindh (Zn); and Balochistan (Zn and Fe). The later-compiled comprehensive national status report on micronutrients, published by NFDC in 1998, revealed that field-scale micronutrients deficiencies of economic significance only pertained to Zn, B, and Fe (NFDC, 1998).

Detailed, province-wise, information regarding the deficiency of Zn in various field crops, vegetables, and fruit orchards is presented in **Annex 3**. Similar information about B, Fe, Cu, and Mn deficiencies is given in **Annex 4** to **Annex 7**. Summarized information about the extent of deficiencies of Zn, B, and Fe is presented in Table 5.2. An elaboration of the nature and extent of micronutrient deficiencies is given in the following paragraphs.

5.4 EXTENT OF ZINC, BORON, AND IRON DEFICIENCIES

The existing information reveals that extents of Zn and B deficiencies in the country are about the same, i.e., 51–53% sites are deficient (Table 5.2). For instance, in the Punjab province, 52% of agricultural fields are Zn-deficient and 53% are B-deficient. In KP province, 41% of fields are Zn-deficient and 42% are B-deficient. In Sindh province, the extent of deficiency of both the micronutrients is about the same, i.e., 62–63%. In Balochistan province, Zn deficiency has been diagnosed in 75% of fields and B deficiency in 61% of fields. In Azad Jammu and Kashmir (AJK), Zn deficiency prevails in 28% of fields and B deficiency in 45% of fields. Zinc deficiency has been well established almost in all field crops (i.e., wheat, rice, cotton, maize, sugarcane, sorghum, rapeseed-mustard, and peanut) and all fruit plants (i.e., citrus, mango, apple, pear, peach, plum, persimmon, and banana). However, in the case of vegetables, such information is limited to only a few crops, i.e., potato, tomato, etc.

5.4.1 ZINC DEFICIENCY

Zinc deficiency is a widespread nutritional disorder in almost all field crops and fruit orchards. According to the research information summarized in Table 5.2, 56% field area of the most important staple grain crop, wheat, grown on around 9 Mha is affected by Zn deficiency. The geometric mean for total Zn in wheat

TABLE 5.2

Nature and Extent of Micronutrient Deficiencies Based on Crop Responses to Micronutrient Fertilization

Crop/Soil	Zinc Sites	Zinc DeficientSites (%)	Boron Sites	Boron DeficientSites (%)	Iron Sites	Iron DeficientSites (%)
Pakistan	7317	51	5436	53	5305	8
Punjab	4163	52	2336	53	2331	10
Wheat	407	56	357	46	61	0
Cotton	420	41	433	51	420	0
Rice	401	76	212	35	110	0
Maize	65	60	65	50	--	--
Sugarcane	234	57	234	56	111	17
Sorghum (rainfed)	255	67	255	67	255	3
Rapeseed-mustard (rainfed)	120	80	120	68	120	20
Peanut (rainfed)	100	60	100	50	100	50
Potato	80	40	89	55	--	--
Mango	1074	29	100	45	174	0
Citrus	252	74	100	45	152	71
Soils	755	55	271	61	828	4
KP	1814	41	1399	42	1928	6
Apple, Peach, Persimmon, and Plum	340	50	90	46	140	0
Apple	102	50	102	30	302	38
Citrus	241	41	147	51	241	0
Soils (including acid soils)	1131	37	1060	42	1245	8
Sindh	1036	63	990	62	807	0
Cotton	100	48	100	56	--	--
Sugarcane	123	60	123	45	123	0
Banana	65	57	65	60	--	--
Soils	748	66	702	66	684	3
Balochistan	259	75	666	61	239	30
Apple	70	80	520	61	70	80
Soils	189	73	146	60	169	9
Pakistan Administered Kashmir	45	28	45	45	--	--
Soils	45	28	45	45	--	--

-- Data not available

fields (86 soils) across Pakistan was reported to be 66.7 mg kg⁻¹ (min 34, max 110 mg kg⁻¹) (Zia et al. unpublished data). For the wheat grains grown over these sites across Pakistan (n=75), average Zn concentrations were recorded at 24.4 mg kg⁻¹. The bioavailability of Zn consumed via grains is very low due to phytic acid contents. Therefore, the majority of the population of this country is deficient of this essential mineral.

The other crops widely affected by Zn deficiency are rice, 76%; rapeseed-mustard, 80%; maize and peanut, 60%; sugarcane, 57%; and cotton, 41% in Punjab and 48% in Sindh province. Maximum extent of Zn deficiency (i.e., 75% sites) has been observed in Balochistan, followed by 63% sites in Sindh. The obvious reason is unfavorable soil chemistry in Balochistan– because of the arid climate, the soils in this province are highly calcareous. Most fruit plants, including citrus, mango, deciduous fruits (apple, peach, plum, pear, and grape), and persimmon also suffer from Zn deficiency (Table 5.2).

In a nutrient indexing study on sugarcane, Ahmad and Rafique (2008) sampled diagnostic leaf tissues and associated soils from the districts Jhang (63 field sites) and Sargodha (48 field sites). Using 10 mg kg⁻¹ as the critical level of Zn in sugarcane leaf sheaths (Humbert, 1968; Clements, 1980; Reuter and Robinson, 1997), 51% of sugarcane fields of district Jhang and 56% of district Sargodha were categorized as Zn-deficient. However, using 0.9 mg kg⁻¹ of AB-DTPA-extractable Zn as the critical level (Soltanpour, 1985), a much greater percentage of sugarcane soils appeared to be Zn-deficient, i.e., 92% of fields in Jhang and 100% of fields in Sargodha. As sugarcane is known to be much less susceptible to Zn deficiency compared with many other field crops (like maize, rice, and wheat), the discrepancy between the plant and soil analysis-based interpretations is understandable.

Recently, Arain et al. (2017) have also reported that Zn deficiency disorder prevails in 24%of sugarcane fields of Thatta district in Sindh province. Sugarcane is less prone to Zn deficiency disorder than many other crops like maize, sorghum, millet, soybean, and cowpea. Despite its tolerance, the prevalence of Zn deficiency of this magnitude calls for remedial measures for optimizing the productivity of this important cash crop in the country.

Another piece of practically important information on micronutrient deficiency problems pertains to mango orchards. In a comprehensive nutrient indexing study comprising >800 leaf samples collected from mango orchards in Khanewal, Multan, Jhang, Faisalabad, and Sargodha districts of Punjab province, Asif et al. (1998) observed that 21% of samples were deficient in Zn (using a critical level of 20 mg Zn kg⁻¹), suggested by Jones et al. (1991) and Bharghava and Raghupathi (1993). The overall range of foliar Zn concentration in this study was 17–31 mg kg⁻¹. However, in a similar nutrient indexing study comprising 174 mango leaf samples collected from three southern districts of Punjab, Siddique et al. (1998) observed a much higher extent of Zn deficiency in mango orchards as 47% of leaf samples were Zn-deficient using 20 mg kg⁻¹ as the critical level of Zn. In a nutritional survey of fruit orchards conducted in 2007 in KP, Khattak and Hussain (2007) reported that 80% of samples of the leaves collected from three sites were deficient in Zn. In another investigation, Shah and Shahzad (2008) noticed that 37% of apple orchards in Swat Valley had a lower level of Zn in the apple leaves. The lowest Zn concentration in the leaves was found to be 13.30 μg g^{-1}. Ahmad et al. (2010) tested 13 apple orchards in Murree and showed that all the orchards were deficient in Zn (Zn was less than 25 mg kg^{-1} dry matter). Similarly, in a nutrient indexing for apple orchards in northern Punjab (Murree), 80% of the collected samples of apple leaves were found deficient in Zn (Ahmad et al., 2014).

As 52% of the randomly collected soils (1478 soils out of 2868) across Pakistan were diagnosed deficient in Zn, large-scale areas under many other corps must be affected by Zn deficiency. In particular, the crops known to be susceptible to Zn deficiency (**Table 2.1**) must be suffering from this micronutrient disorder. Zinc deficiency in many field crops, including wheat, is well established for decades in Pakistan. However, a recent significant development in this regard is much higher increases in wheat yield with a relatively higher dose of soil-applied Zn fertilizer (i.e., 10 kg Zn ha^{-1}) and enrichment of wheat grains with Zn by foliar sprays of Zn fertilizer. As Zn fertilizer use in wheat is highly cost-effective, this research information will have far-reaching positive consequences by increasing wheat productivity and improving Zn density in wheat grains on millions of acres with the enhancement of stakeholder awareness.

5.4.2 BORON DEFICIENCY

Boron deficiency is also a widespread plant nutritional problem affecting Pakistani agriculture. Likely 53% or even more of fields under many crops (including cotton, wheat, sugarcane, maize, sorghum, rapeseed/mustard, and peanut) are affected by B deficiency. This could also be confirmed from unpublished data of FFC Farm Advisory Laboratories, where more than 14,000 soils samples have been analyzed since 2004. The results of FFC labs suggest that B deficiency in soils is about 60%, which is close to the above referred average figure of 53%. Similarly, 30–61% of field area under fruit orchards (apple, banana, citrus, mango, plums, and apricot) is affected by B deficiency (Table 5.2; **Annex 4**). Though the percentage of B-deficient field area under rice crop is relatively smaller (i.e., 35%); the consequences of this micronutrient disorder in this crop are very serious as both crop yield and quality are badly hampered. As 42–62% of the randomly collected soils in different provinces have been diagnosed to be deficient in B, large-scale areas under many other corps must also be affected by B deficiency. In particular, the crops known to be susceptible to B deficiency (Table 2.1) must be suffering from this micronutrient disorder.

In the 1998 status report on micronutrients in Pakistan's agriculture (NFDC, 1998), the extent of Zn deficiency was stated to be relatively more than B deficiency. However, research studies undertaken during the last two decades have revealed that this is not true. The deficiency of Zn, being the first-ever micronutrient disorder diagnosed in Pakistan (Yoshida and Tanaka, 1969), received much early and better research attention compared with the B deficiency partly because of the limited laboratory capacity for B analysis in soils and plant tissues. It was much later, in the mid-1980s, that some research laboratories in the country (specifically NIAB, Faisalabad, and NARC, Islamabad) initiated analyzing soils and plant tissues for B. Later on, many other agricultural laboratories also started B analysis. The fact remains that even now Zn analysis is performed by many more laboratories in the country compared with B analysis. Among the deficient micronutrients of immense practical significance, research information about B is minimal. For example, in two nutrient indexing studies of mango in the Punjab province, Asif et al. (1998) and Siddique et al. (1998) did not study B. However, based on a systematic nutrient indexing investigation, Arain et al. (2017) reported that B deficiency disorder prevails in 21% of sugarcane fields in Thatta district in Sindh province. Therefore, to the understanding of these authors, the extent of B deficiency in Pakistan's agricultural fields is not lesser than that of Zn deficiency. Rather, B deficiency appears to be either at par or a more widespread micronutrient problem compared with Zn deficiency. The only limitation appears to be much less research information about B compared with Zn.

5.4.3 Extent of Iron Deficiency

Research information about the Fe status of soils and plants in Pakistan is limited. This is because soil analysis is hardly effective for the prognosis of the Fe deficiency problem. Also, unlike Zn and B, plant analysis for total Fe is not useful for diagnosing the Fe nutritional status of plants. Only the analysis for physiologically active Fe fraction (i.e., Fe^{2+}) in fresh leaf tissues can help effective diagnosis of Fe deficiency. This analytical procedure for determining active Fe fraction is too difficult to adopt as a routine laboratory procedure. Thus, Fe deficiency diagnosis by soil and plant analysis is almost an impractical proposition. Out of the total soils and vegetation samples (n=5,305), about 8% were found deficient in Fe. Its deficiency was found to be of more concern in apple orchards in Balochistan and KP (80%; and 38% deficient, respectively); citrus orchards in Punjab (71% deficient); and peanut crop in rainfed areas of Punjab (50% deficiency). In a nutrient indexing study comprising 174 mango leaf samples collected from three southern districts of Punjab, Siddique et al. (1998) observed that all leaf samples collected from five districts of Punjab contained sufficient Fe (i.e., >50 mg Fe kg^{-1}). In a similar but more comprehensive nutrient indexing study comprising >800 leaf samples collected from mango orchards of Punjab province, Asif et al. (1998) did not analyze Fe concentration in mango leaves and associated soils. In a study (Zia et al. unpublished data), 86 soils were collected from wheat-growing regions across Pakistan. The geometric mean for total Fe in soils was reported to be 32,945 mg kg^{-1} (min 18,600, max 53,211 mg kg^{-1}). For the wheat grains grown over these sites across Pakistan (n=75), average Fe concentrations were recorded at 45 mg kg^{-1}.

As identification of Fe deficiency in plants is mostly limited to Fe chlorosis, symptoms and crop response to fertilization information regarding this micronutrient is very sketchy.

TABLE 5.3

Multiple Deficiencies of Micronutrients for Different Crops, Vegetable, and Fruits

Wheat, Rice, Maize, Sorghum	Deficiencies of Zn and B are well established. Some reports of localized Cu deficiency suggest further research before using Cu fertilizer because indiscriminate Cu fertilization can induce Zn deficiency in soils marginal in Zn supplies.
Cotton	Deficiency of B is most important; Zn deficiency is relatively less severe. Localized Fe chlorosis requires research attention – by way of screening the cotton germplasm for tolerance to Fe chlorosis.
Rapeseed-mustard	Boron and Zn deficiencies are prevalent.
Peanut	Iron chlorosis and B deficiency are well established. Zinc deficiency is also a problem.
Potato, tomato	Zinc and B deficiencies are very obvious.
Chili	Affected with Zn and B deficiencies. Also, many varieties are susceptible to Fe chlorosis – especially when grown in light-textured soils.
Citrus	Zinc and Fe deficiencies are very obvious.
Deciduous fruits	Iron chlorosis is very obvious. Zinc deficiency is also a serious problem. Boron deficiency is highly suspected.

5.5 MULTIPLE MICRONUTRIENT DEFICIENCIES

The available research information has revealed that many field crops, vegetables, and fruit plants in Pakistan are inflicted with multiple micronutrient deficiencies. Such information regarding crops and micronutrient deficiencies is given in the following text (Table 5.3). Therefore, the use of single micronutrient fertilizer is not the right approach. It is not only more laborious (by way of field broadcast, foliar solution application, etc.) but also more expensive. Rather, application of compound fertilizers (i.e., micronutrient-fortified NPK fertilizer products) or fertilizer blends containing the required micronutrients for specific crops/vegetables/fruit orchards are needed. However, in the case of a single micronutrient deficiency, the use of multi-micronutrient fertilizers must be avoided as their use would amount to unnecessary extra cost and may also lead to an induced deficiency of other micronutrients because of nutrient antagonism.

5.6 DEFICIENCIES OF OTHER MICRONUTRIENTS

Whereas widespread deficiencies of Zn, B, and Fe in a number of field crops, vegetables, and fruit crops are well established in Pakistan, this is not the case with any other micronutrient.

5.6.1 COPPER

Research information about the status of Cu in soils and crop plants is summarized in **Annex 6**. The FAO global study on micronutrients found that all the 242 soil samples collected from throughout Pakistan were adequate in Cu fertility (Sillanpää, 1982). Subsequent research by local scientists also revealed that the incidence of soil Cu deficiency is of very limited magnitude: e.g., 5% in agricultural areas in Punjab province, 7% in KP, and 1% each in Sindh and Balochistan provinces **(Annex 6)**. Zia et al. (2006) analyzed orchards' soils from across Punjab (158 samples), Sindh (94 samples), and Balochistan (72 samples) and observed Cu deficiency (DTPA-extractable Cu) in 2% soils collected from apple and citrus orchards and in 7% soils from guava orchards. Similarly, nutrient indexing investigations based on plant tissue and soil analyses also revealed that the incidence of Cu deficiency was of very small magnitude: 1% in the cotton belt of Punjab and 1–6% in rainfed wheat, rapeseed-mustard, sorghum, and peanut grown in Pothohar plateau of Punjab. Earlier, extensive field experiments on micronutrient nutrition of wheat, using the missing element technique, also revealed only very localized incidence of Cu deficiency. For example, in 50 wheat experiments carried out at farmers' fields in districts Jhang, Muzaffar Garh, and Vehari, there was no indication of Cu deficiency. However, in 40 similar field experiments on rice at farmers' fields in the rice belt of Punjab, paddy yield increases were observed at 7% of the field sites. In a study (Zia et al. unpublished data), 86 soils were collected from wheat-growing regions across Pakistan. The geometric mean for total Cu in soils was reported to be 22.6 mg kg^{-1} (min 10, max 55 mg kg^{-1}).

A detailed nutrient indexing investigation in Mansehra and Swat districts of KP province involving 232 surface soils (0–15 cm) of Mansehra (139 acid soils and 84 alkaline soils) and 166 surface soils of Swat (75 acid soils and 88 alkaline soils) revealed that all the soil samples contained adequate levels of Cu (Rashid, 1994). In a nutrient indexing study comprising of >800 leaf samples collected from mango orchards in five districts of Punjab province, Asif et al. (1998) observed that 10% of leaf samples were Cu-deficient (contained <7 mg Cu kg^{-1}). Overall range of foliar Cu concentration in this study was very wide, i.e., 5–90 mg kg^{-1}. In a similar nutrient indexing study comprising 174 mango leaf samples collected from three southern districts of Punjab, Siddique et al. (1998) observed that 8% of leaf samples were Cu-deficient.

The only report of wide-scale Cu deficiency pertains to citrus orchards in Sargodha district (Siddique et al., 1994) which contradicts the results of a similar nutrient indexing study indicating that only 5% of the surveyed citrus orchards in districts Sargodha, Sahiwal, and Faisalabad were affected by Cu deficiency (Rashid et al., 1991). In the absence of field experimental data on the response of citrus to Cu fertilization, it is hard to affirm the wide-scale prevalence of Cu deficiency in citrus. Thus, the incidence of Cu deficiency in Pakistani soils is reported to be very limited and is of localized nature. For the wheat grains grown over 75 sites across Pakistan, average Cu concentrations were recorded at 4.4 mg kg^{-1} (Zia et al. unpublished data). Based on per capita wheat flour consumption of 330 g a day, this would provide a daily intake of 1.5 mg Cu for a 70 kg individual against recommended daily intake of 0.9 mg. Hence, there seems to be no problem with Cu malnutrition in humans.

5.6.2 Manganese

The soil and plant analysis data summarized in **Annex 7** indicate that the magnitude of Mn deficiency is much smaller than those of Zn, B, and Fe except for Punjab orchards, especially citrus and mango where up to 29% and 75% of vegetation samples were found deficient in Mn. For instance, the extent of Mn deficiency is only 2% each in the cotton belt of Punjab and5% in rainfed fields of peanut (Pothohar plateau) in Punjab. In the 1998 report about micronutrients status in Pakistan, there was no indication of Mn deficiency at all in KP, Sindh, and Balochistan provinces. However, the updated status that included post-1998 literature reveals that 26, 11, 1.6, and 3% of sites (soils/vegetation) are deficient in Punjab, KP, Sindh, and Balochistan (soils only) provinces, respectively. In a study (Zia et al. unpublished data), 86 soils were collected from wheat-growing regions across Pakistan. The geometric mean for total Mn in soils was reported to be 648 mg kg^{-1} (min 367, max 1018 mg kg^{-1}). For the wheat grains grown over 75 sites, average Mn concentrations were recorded at 35 mg kg^{-1}. Based on per capita wheat flour consumption of 330 g a day, this would provide a daily intake of 12.2 mg Mn for a 70 kg individual against recommended daily intake of 2.3 mg.

The only exceptions are a few reports of wide-scale Mn deficiency, i.e., 27% in citrus orchards of Sargodha (Siddique et al., 1994), 75% in mango leaves (Asif et al., 1998), 96% in citrus orchards of Swat and Malakand (Shah et al., 2012), 58% in apple orchards' leaves of Swat Valley (Shah and Shahzad, 2008), and 60% in KP (Khattak, 1994). However, another publication based on nutrient indexing of citrus plants and associated soils tells that only 4% of citrus orchards in Sargodha, Sahiwal, and Faisalabad suffered from Mn deficiency (Rashid et al., 1991). As securing reliable fruit yield data is a well-known issue in our country, the validity of yield-based diagnosed Mn deficiency in citrus orchards of KP is in doubt. This is particularly because no subsequent study has affirmed such a widespread incidence of Mn deficiency in any crop in the country so far. Therefore, field experimentation on Mn fertilization of the suspected Mn-deficient citrus orchards is warranted to resolve this ambiguity.

Asif et al. (1998) have reported that 75% of the >800 mango leaf samples collected from Khanewal, Multan, Jhang, Faisalabad, and Sargodha districts of Punjab province contained low Mn levels (i.e., <50 mg Mn kg^{-1}, the critical Mn level for mango suggested by Jones et al., (1991), and Bharghava and Raghupathi (1993). In this extensive nutrient indexing study, the overall range of Mn concentration in mango leaves was 25–127 mg kg^{-1}. They observed that all leaf samples of the Sindhri variety contained adequate Mn. However, all other varieties (*Chaunsa, Langra, Dasehri, Anwar Rataul, Fajri, Malda, Kala Chaunsa, Chaunsa Late, Sobey-de-Ting*, and *Mehmood Shah Wala*) contained deficient as well as adequate Mn concentrations. In a similar nutrient indexing study comprising 174 leaf samples from 29 orchards in the districts Vehari, Multan, Muzaffar Garh, Rahim Yar Khan, and Bahawalpur, Siddique et al. (1998) observed that only 13% of the 174 leaf samples contained Mn concentration <25 mg kg^{-1}. The authors attributed this very low concentration of Mn as the critical Mn level, and perhaps misinterpreted the work of Jones et al. (1991). Jones et al. (1991) defined 50–250 mg Mn kg^{-1} as sufficient and 25–49 mg Mn kg^{-1} as a low concentration in most recently matured leaves of mango, at the post-flowering growth stage (please see page 162 of Jones et al., 1991). Thus, the use of 25 mg Mn kg^{-1} as the critical level by Siddique et al. (1998) is an error. It is worth emphasizing that, despite using 50 mg kg^{-1} as the critical level, Asif et al. (1998) diagnosed Mn deficiency in 75% of leaf samples; while using 25 mg kg^{-1} as the critical level, Siddique et al. (1998) observed Mn deficiency only in 13% mango leaf samples. It is true that we still lack appropriate plant analysis interpretation criteria for fruit plants (as well as for many field crops and vegetables), but the use of 25 mg kg^{-1} as the critical Mn level in mango leaves without any valid reference is highly questionable. Also, the high magnitude of Mn deficiency reported by Asif et al. (1998) (i.e., 75%) is unheard of in soils and field crops of Pakistan. Therefore, the widespread Mn deficiency reported cannot be endorsed unless verified through field experimentation by way of fruit yield increases with Mn application in Mn-deficient orchards.

5.6.3 Molybdenum, Chlorine, and Nickel

Unlike B and the micronutrient cations (Zn, Cu, Fe, and Mn), deficiencies of Mo, HCl, and Ni are not expected in Pakistani soils because of the alkaline nature of the soils. In fact, unlike micronutrient cations, the solubility of Mo is high in alkaline soil environments. This was affirmed by the FAO global study on micronutrients as there was no indication of low Mo in the 242 soil samples collected from all provinces of Pakistan (Sillanpää, 1982). Moreover, Mo concentrations in pot-grown wheat plants using these Pakistani soils were the highest among the 29 countries included in this global study (Sillanpää, 1982). In a study, 86 soils were collected from wheat-growing regions across Pakistan (Zia et al. unpublished data). The geometric mean for total Mo in soils was reported to be 0.44 mg kg^{-1} (min 0.21, max 0.81 mg kg^{-1}). The geometric mean for total Ni in wheat fields during the same study was recorded at 39.7 mg kg^{-1} (min 19, max 98 mg kg^{-1}). For the wheat grains grown over these sites, average Ni and Mo concentrations were recorded at 0.16, and 0.6 mg kg^{-1}, respectively (n = 75). This depicts that the population of Pakistan that meets more than 50% of its caloric needs from wheat flour is not deficient in Mo. There is no possibility of low Cl in alkaline–calcareous soil environments in the country as majority of our agricultural soils are salt-affected and inherently high in Cl. Similarly, Ni is not accepted to be deficient in our soils because Ni is a component of clay silicates, and plants' requirement of Ni is very low compared with all other micronutrients.

Due to obvious reasons, therefore, no extensive research efforts have been made to ascertain the status of Mo, Cl, and Ni in soils and plants of Pakistan. Molybdenosis (Mo toxicity) can occur in forages, grown on soils with relatively prolonged wetness, high pH, and under saline-sodic conditions. The toxicity of Mo operates mainly by inducing Cu deficiency, and the affected animals respond to Cu supplementation. Therefore, this aspect of Mo merits research attention in the country.

5.6.4 IODINE

Most of the soil-iodine is derived from the volatilization of methylated forms from seawater which then enters the soil–plant system via rainfall and dry deposition. It could be predicted that the iodate form of iodine might be prevalent in Pakistan due to high pH and carbonate contents in the soils. The concentrations of iodine in alluvial soils, like most agricultural soils of Pakistan, have been reported in the range of 3.65–9.82 mg kg^{-1} (Watts and Mitchell, 2009). The origins of widespread iodine deficiency in the population of Pakistan are dietary. Cereal grains are poor sources of many micronutrients including iodine. Wheat provides the staple diet for the country's poorest people; they are most vulnerable to micronutrient malnutrition and the consequent deficiency diseases.

Zia et al. (2015) determined the iodine status of 86 field sites from wheat-growing regions across Pakistan. Tetra-methyl ammonium hydroxide (TMAH)-extractable iodine concentrations generally revealed low levels of iodine in the soils, with the exception of some areas where high soil organic carbon prevails, like all field locations in the Pakistan Administered Kashmir highlands and KP and Balochistan provinces where temperate climate prevails. The TMAH-extractable soil-iodine ranged from 0.19 to 9.59 mg kg^{-1}, with a geometric mean concentration of 0.66 mg kg^{-1}. Wheat grains in Pakistan are low in TMAH-extractable iodine, i.e., 0.01 to 0.03 µg I g^{-1} wheat flour with a mean and median concentration of 0.013 µg I g^{-1} and 0.01 µg I g^{-1}, respectively, compared to a worldwide mean of 0.56 µg I g^{-1}. Iodine concentration in flag leaves of wheat ranged from 0.12 to 0.47 µg g^{-1}, with a geometric mean of 0.22 µg I g^{-1}. The maximum concentration of iodine in wheat grains (0.03 µg g^{-1}) was observed on a soil with the highest TMAH-extractable iodine at 9.59 µg g^{-1} in the Kochlaak district of Balochistan province where a temperate climate prevails. In most of the wheat grain samples (i.e., 62 of 84 field locations), TMAH-extractable iodine was below the analytical detection limit of ICP-MS.

Under HarvestZinc Project, Cakmak et al. (2017) investigated the possibility of using iodine-containing fertilizers for agronomic biofortification of cereal (wheat, rice, and maize) grains with iodine. This study was conducted in Turkey, Pakistan, Brazil, and Thailand, and revealed that soil application of iodine products did not increase iodine in wheat grains. In contrast to the soil application, foliar sprays of KIO$_3$ enhanced grain iodine concentrations up to five- to ten-fold without affecting grain yield. Adding KNO$_3$ and a surfactant to the iodine-containing foliar spray further increased grain iodine concentration. Field experiments conducted in different countries confirmed that foliar application with increasing rates of iodine significantly increased grain iodine concentrations in wheat, brown rice, and maize. This increase was also found in the iodine concentration of the endosperm part of wheat grains and in polished rice. Thus, it was concluded that foliar application of iodine-containing fertilizers is highly effective in increasing grain iodine concentrations in wheat, rice, and maize. Spraying KIO_3 up to the rate of 0.05% w/v is suggested as the optimal form and rate to be used in agronomic biofortification with iodine. The substantial increase in grain iodine concentrations could contribute to the prevention of iodine deficiency in human populations with low dietary iodine intake.

Per capita iodine intake from staple foods, based on average wheat consumption @ 350 g per day with a mean iodine concentration of 0.01 µg g^{-1}, would be 3.5 µg per day. As on average wheat (flour) contributes 65% of daily caloric needs in Pakistan, iodine intake from this staple cereal is extremely limited and is far below the minimum recommended iodine intake. An iodine intake of 25.4 µg a day has been estimated based on median iodine concentrations observed in wheat grains and groundwater against the World Health Organization (WHO) recommendations of a daily intake of 150 µg iodine (WHO, 2007). National Nutrition Survey (2018) estimated that on average, 80% of Pakistani households consume iodized salt with the lowest usage in KP, i.e., 31.6%. However, the number of the households that consume iodized salt have changed in a survey that was conducted in 2018 (NNS, 2019). With less than 100 µg I day^{-1}, a series of thyroid functional and developmental abnormalities occur in which symptoms could be as goiter and/or reduced mental and physical development of children. An iodine-deficient population might suffer from an IQ reduction of 10–15% on a national scale (Stewart et al., 2003). This situation demands diversification of dietary intake by including iodine-rich sources like fish, milk, fruits, and iodized salt. For example, by consuming 5 g of iodized salt (having 15 µg iodine per g of salt), an individual's additional intake of iodine might be about 75 µg a day, but one has also to take into account iodine losses during the process of cooking.

5.7 MAPPING OF SOIL–PLANT MICRONUTRIENT-DEFICIENT AREAS

The micronutrient research carried out by various institutions of the country helped in determining the nature and extent of micronutrient deficiencies problems in agricultural soils and in a variety of crop plants. The spatial variability maps of certain geographic regions were prepared and supported for delineating the areas affected by the deficiencies of specific micronutrients. Based on systematic extensive soil–plant nutrient indexing data, the Micronutrient Research Group at NARC prepared spatial distribution maps for the deficient micronutrients of the cotton belt in Punjab, a geographical district in Sindh province, and for many rainfed crops grown in Pothohar plateau in northern Punjab (Table 5.4).

In general, soil and plant micronutrient maps are in good agreement. For example, localized B-deficient cotton areas within Lodhran district in Punjab province are evident in Figure 5.1. Similarly, Zn-deficient rainfed wheat areas within Attock district of Pothohar plateau are delineated in Figure 5.2.

TABLE 5.4
Spatial Variability Maps Prepared to Delineate Micronutrient-Deficient Areas in Pakistan

Crop/Micronutrient	Cropping system/Region/District	Reference
All crops (Pakistan)/ Zn	Irrigated land across Pakistan	Zia et al. (2018)
Potato/ Cu, Fe, Zn, and Mn	Gilgit, Skardu, and Ghanche in Gilgit-Baltistan	Hussain et al. (2016)
Wheat (rainfed)/ B and Zn	Pothohar plateau: Districts Chakwal and Attock	Rashid et al. (2002); Rafique et al. (2006); Rashid et al. (2011)
Wheat (irrigated)/Zn	Southern Punjab	Maqsood et al. (2015)
Citrus/ Zn, Cu, Fe, Mn, and B	Malakand Agency, KPK	Noor et al. (2013)
Cotton (irrigated)/B and Zn	Cotton belt in Punjab: Districts Khanewal, Multan, Lodhran, Vehari, Bahawalpur, and Rahim Yar Khan	Rashid and Rafique (1997, 1999), Rafique (2010)
Boron (rainfed)/ Zn, Cu, Fe, Mn, B,	Kohat district	Wasiullah et al. (2010)
Cotton (irrigated)/B and Zn	Cotton–Wheat System in Sindh: Nawab Shah	Rashid and Rafique (1999)
Sorghum (rainfed)/B, Zn, and Fe	Pothohar plateau: Districts Jhelum and Chakwal	Rashid and Qayyum (1991); Rashid et al. (1997b, 1997c)
Rapeseed-mustard (rainfed)/B and Zn	Pothohar plateau: Districts Chakwal and Attock	Rashid et al. (1994); Rashid (1994)
Peanut (rainfed)/B	Pothohar plateau: Districts Chakwal and Attock	Rashid et al. (1997a)

In the absence of an effective soil advisory service for micronutrients, such maps can help identify the areas needing micronutrient fertilizer use. Further, these maps can direct future R&D efforts for devising adaptable technologies and strategies for the deficient micronutrients in respective geographical areas.

Another approach in geo-statistical prediction of nutrient concentration at un-sampled locations is kriging. Geo-statistical prediction of a spatial variable requires an estimate of its variogram function, which describes its spatial dependence. When a parametric model of the variogram of a spatial variable is estimated, it could be used to obtain predictions of that variable at un-sampled sites, by a linear combination of the data which minimizes the expected squared error of the predictions. This method is known as kriging (Webster and Oliver, 2007).

Recently, R&D Department of Fauji Fertilizer Company (FFC) Limited developed a country-scale geo-spatial map for Zn (Figure 5.3) while using its

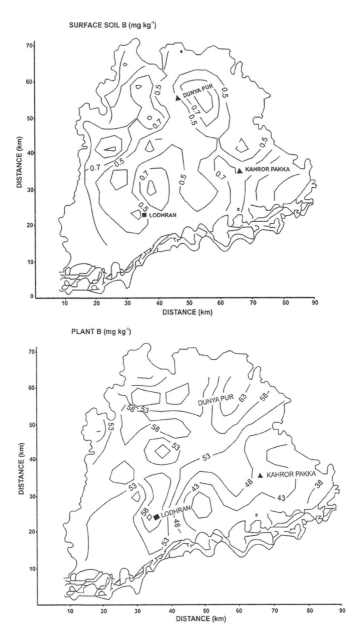

FIGURE 5.1   Spatial variability of B in soils and associated cotton leaves within Lodhran district in Punjab province. (Source: Rashid et al., 2002, reproduced with permission.)

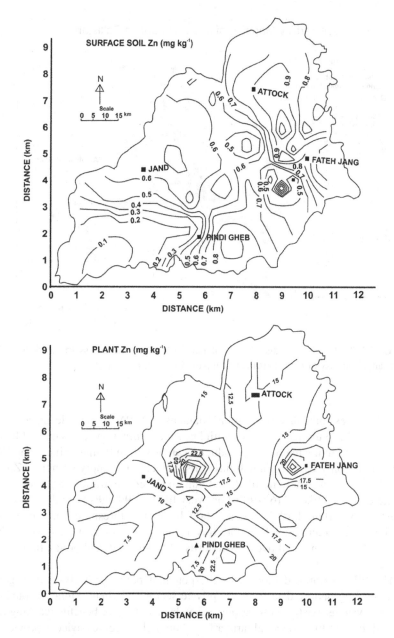

FIGURE 5.2 Spatial variability of Zn in soils and associated wheat plants within Attock district in the Punjab province. (Rafique et al., 2006, reproduced with permission.)

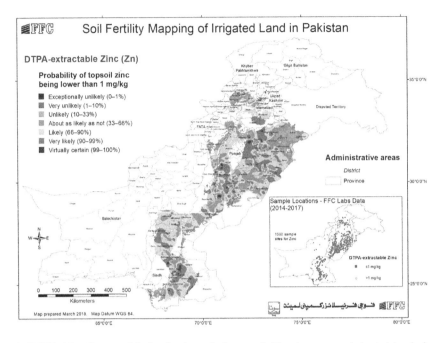

FIGURE 5.3   Geo-spatial distribution of plant-available zinc across irrigated lands in Pakistan. (Courtesy of Fauji Fertilizer Company Limited, Pakistan.)

Agri. Services soil analysis program's database (DTPA-extractable Zn concentrations of 1206 unique locations with known coordinates). The sampled locations fell well in the main grain production areas within Punjab, Sindh, and KP. In the map, only eastern Punjab has soils expected to supply sufficient Zn to cereal crops – elsewhere in Sindh and Punjab the concentrations are lower and predicted to be "likely" (67–90%) or "very likely" (90–99%) to fall below the threshold of 1.0 mg kg^{-1}. Where soil falls below this threshold, Zn-containing fertilizers are recommended, and this research shows that this would be expected to be beneficial over the majority of the main cereal production areas studied.

The FFC research demonstrates the potential power of well-curated, geo-referenced agronomy data from the private sector (or where public, or public-private, systems exist). These resources can have collective benefits reaching far beyond those to the individual farmer for whom field-specific advice is provided. The value of collecting location details and maintaining a consistent database of results and sampling information should be developed more widely, to allow such spatial assessments to be implemented more frequently. Regional or national predictive modeling can deliver strategic information to support food security in terms of both yield and micronutrient concentrations, including in small-scale agriculture situations.

5.8 SOIL CLASSIFICATION AND MICRONUTRIENT DEFICIENCIES

Out of 12 Soil Orders of the *US Soil Taxonomy System*, only six Soil Orders have been identified in Pakistan. These Soil Orders are Aridisol, Entisol, Inceptisol, Alfisol, Vertisol, and Mollisol (Rafiq, 1996). As micronutrient attributes are altered with the soil management practices, like other soil fertility parameters, these are not among the diagnostic criteria used in soil classification. However, attempts have been made from time to time in various parts of the world to find relationships between micronutrient status and soil types. For example, Shorrocks (1997) reported the Soil Orders of the *US Soil Taxonomy System* in the regions with prevalent B deficiency. According to this exercise, B deficiency disorder was observed in Pakistan-prevalent three Soil Orders, i.e., Inceptisol, Entisol, and Alfisol. Later, Fageria et al. (2002) proposed an association between major soil groups of the *US Soil Taxonomy System* and potential micronutrient deficiencies, compiled from various sources (Bell and Dell, 2008). According to this compilation, a wide range of Soil Orders express micronutrient deficiencies, but among the Soil Orders prevalent in Pakistan only Alfisol, Entisol, and Mollisol seem to represent the greatest risk of multiple micronutrient deficiencies.

In Pakistan, only the Micronutrient Research Group at NARC has attempted to document such possible relationships. For instance, in a nutrient indexing investigation of rainfed peanut crop grown in Chakwal and Attock districts of the Pothohar region, Rashid et al. (1997a) observed that the extent of B deficiency had no relationship with the parent materials (i.e., alluvium, loess, or residuum from sandstone) of the sampled soils in 100 random field locations, but was somewhat related to the soil types. They reported that, among the field soils belonging to Ustochrepts, B deficiency was observed in 50% of fields of Typic Ustochrepts, 50–75% of fields of Aridic Ustochrepts, 33–100% of fields of Calcic Ustochrepts, and 83–100% of fields of Fluventic Ustochrepts. Similarly, B deficiency existed in 50–100% of fields of Typic Haplustalfs, 33% of Typic Ustochrepts, and 100% of Lithic Torripsamments. In a similar nutrient indexing study on Zn nutrition of rainfed wheat in Pothohar plateau, Rafique et al. (2006) observed that 82% of the sampled fields were Zn-deficient. They observed that various soil series under wheat crop varied in their soil test Zn status. Out of the total 21 soil series sampled, 16 soil series exhibited a mean Zn level less than the critical level (i.e., 0.9 mg Zn kg^{-1} soil) with a minimum of 0.19 mg Zn kg^{-1} in Murat soil series, followed by 0.32 mg Zn kg^{-1} both in Dhulian series and in Qazian series. As much as 92% of the fields of Missa, Qutbal (W), and Talagang series were deficient in Zn. Overall, the soils belonging to Soil Order Alfisol contained far less AB-DTPA-extractable Zn than those soils representing Entisol and Inceptisol.

REFERENCES

Ahmad, S., and E. Rafique. 2008. Nutrient indexing and integrated nutrient management for sustaining sugarcane yields. Agricultural Linkages Program (ALP) Project, Final Technical Report. Sugarcane Crops Research Program, Crop Sciences Institute, National Agricultural Research Center, Islamabad. 104 pp.

Ahmad, H., M. T. Siddique, S. Ali, et al. 2014. Micronutrient indexing in the apple orchards of Northern Punjab, Pakistan using geostatistics and GIS as diagnostic tools. *Soil and Environment* 33:7–16.

Ahmad, H., M. T. Siddique, I. A. Hafiz, and Ehsan-ul-Haq. 2010. Zinc status of apple orchards and its relationship with selected physico-chemical properties in Murree tehsil. *Soil and Environment* 29:142–147.

Ahmed, N., M. Abid, A. Rashid, M. A. Ali, and M. Amanullah. 2013. Boron requirement of irrigated cotton in a TypicHaplocambid for optimum productivity and seed composition.*Communications in Soil Science and Plant Analysis* 44:1293–1309.

Arain, M.Y., K. S. Memon, M. S. Akhtar, and M. Memon. 2017. Soil and plant nutrient status and spatial variability for sugarcane in lower Sindh. *Pakistan Journal of Botany* 49:531–540.

Asif, M., K. Daud, M. Ashraf, et al. 1998. Nutrient status of mango orchards in Punjab. *Pakistan Journal of Soil Science* 15:1–6.

Bell, R.W., and B. Dell. 2008. *Micronutrients for Sustainable Food, Feed, Fibre and Bioenergy Production*. International Fertilizer Industry Association, Paris, France.

Bharghava, B. S., and H. B. Raghupathi. 1993. Analysis of plant materials for macro and micronutrients. p. 49–82, In: Tandon, H.L.S. (Ed.) *Methods of Analysis of Soil, Plants, Waters and Fertilizers*. Fertilizer Development and Consultation Organization, New Delhi, India.

Cakmak, I., C. Prom-u-thai, L. R. G. Guilherme, et al. 2017. Iodine biofortification of wheat, rice and maize through fertilizer strategy. *Plant and Soil* 418:319–335.

Clements, H. F. 1980. *Sugarcane Crop Logging and Crop Control: Principles and Practices*. The University Press of Hawaii, Honolulu, HI.

Fageria, N. K., V. C. Baligar, and R. B. Clark. 2002. Micronutrients in crop production. *Advances in Agronomy* 77:185–268.

Humbert, R. P. 1968. *The Growing of Sugarcane*. Elsevier Publishing, 219 pp.

Hussain, F. 2006. Soil Fertility Monitoring and Management in Rice-Wheat System. Agricultural Linkages Program (ALP) Project, Final Report, 2002–2006. Land Resources Research Program, National Agricultural Research Center, Islamabad, 83 pp.

Hussain, A., M. S. Awan, S. Ali, et al. 2016. Spatial variability of soil micronutrients (Cu, Fe, Zn and Mn) and population dynamic of mycoflora in potato fields of CKNP Region Gilgit-Baltistan Pakistan. *Pakistan Journal of Agriculture Sciences* 53:541–550.

Jones Jr, J. B., B. Wolf, and H. A. Mills. 1991. *Plant Analysis Handbook: A Practical Sampling, Preparation*, Analysis and Interpretation Guide. Micro-Macro Publishing Inc., Athens, Georgia, USA, 213 pp.

Katyal, J. C., and D. L. Deb.1982. Nutrient transformations in soils – Micronutrients. p. 146–159. In: *Review of Soils Research in India. Part-I*. Indian Society of Soil Science, New Delhi.

Khattak, J. K. 1991. *Micronutrients in NWFP Agriculture*. Barani Research and Development Project, PARC, Islamabad.

Khattak, J. K. 1994. Effect of foliar application of micronutrients in combination with urea on the yield and fruit quality of sweet oranges. Final Technical Report, 1991–1994. NWFP Agricultural University, Peshawar.

Khattak, J. K. 1995. Micronutrients in Pakistan Agriculture. Pakistan Agricultural Research Council, Islamabad and NWFP Agricultural University, Peshawar, 135 pp.

Khattak, R. A., and Z. Hussain. 2007. Evaluation of soil fertility status and nutrition of orchards. *Soil and Environment* 26:22–32.

Maqsood, M. A., S. Hussain, M. A. Naeem, et al. 2015. Zinc indexing in wheat grains and associated soils of southern Punjab. *Pakistan Journal of Agricultural Sciences* 51:1–8.

Mortvedt, J. J. 1983. Impact of acid deposition on micronutrient cycling in ago-ecosystems. *Environmental and Experimental Botany* 23:243–249.

Nambiar, K. K. M., and A. B. Gosh. 1984. Highlight of research of a long-term fertilizer experiment in India (1971–1982). All India Coordinated Research Project on Long Term Fertilizer Experiments; LTFE Res. Bull No. 1. Indian Agricultural Research Institute, New Delhi.

NFDC. 1998. *Micronutrients in Agriculture: Pakistan perspective.* National Fertilizer Development Center, Islamabad, 51 pp.

NNS. 2019. National Nutrition Survey, 2018. Key Findings Report. Ministry of National Health Services, Regulation and Coordination, Government of Pakistan, Islamabad, 52 pp.

Noor, Y., Subhanullah, and Z. Shah.2013. Spatial variability of micronutrients in citrus orchard of north western Pakistan. *Sarhad Journal of Agriculture* 29: 387–394.

Rafiq, M. 1996. Soil Resources of Pakistan. p. 439–469. In: *Soil Science.* Rashid, A., and K. S. Memon (Managing Authors). National Book Foundation, Islamabad.

Rafique, E. 2010. Integrated Nutrient Management for Sustaining Irrigated Cotton-Wheat Productivity in Aridisols of Pakistan. PhD dissertation, Quad-i-Azam University, Islamabad.

Rafique, E., A. Rashid, A. U. Bhatti, G. Rasool, and N. Bughio. 2002. Boron deficiency in cotton grown in calcareous soils of Pakistan. I. Distribution of B availability and comparison of soil testing methods. p. 349–356, In: Goldbach, H.E. et al. (Ed.), *Boron in Plant and Animal Nutrition.* Kluwer Academic Publishers, New York.

Rafique, E., A. Rashid, and M. Mahmood-ul-Hassan. 2012. Value of soil zinc balances in predicting fertilizer zinc requirement for cotton-wheat cropping system in irrigated Aridisols. *Plant and Soil* 361:43–55.

Rafique, E., A. Rashid, J. Ryan, and A. U. Bhatti. 2006. Zinc deficiency in rainfed wheat in Pakistan: Magnitude, spatial variability, management, and plant analysis diagnostic norms. *Communications in Soil Science and Plant Analysis* 37:181–197.

Rashid, A. 1994. Nutrient Indexing Surveys and Micronutrient Requirement of Crops. Micronutrient Project, Annual Report, 1992–93. National Agricultural Research Center, Islamabad.

Rashid, A., F. Hussain, A. Rashid, and J. Din. 1991. Nutrient status of citrus orchards in Punjab. *Pakistan Journal of Soil Sciences* 6:25–28.

Rashid, A., and F. Qayyum. 1991. Micronutrient Status of Pakistan Soils and Their Role in Crop Production, Cooperative Program. Final Report, 1983–1990. NARC, Islamabad, 84 pp.

Rashid, A., and E. Rafique. 1997. Nutrient Indexing and Micronutrient Nutrition of Cotton and Genetic Variability in Rapeseed-Mustard and Rice to Boron Deficiency. Micronutrient/Nutrient Management in cotton in Relation to CLCV. Annual Project Report 1995–96. NARC, Islamabad, 111 pp.

Rashid, A., and E. Rafique. 1999. Nutrient indexing and soil fertility mapping in cotton. Annual Report 1998–99, Pak Kazakh Joint Research Fund "Soil Fertility and Fertilizer Use in Cotton and Wheat Production". NARC, Islamabad, 56 pp.

Rashid, A., E. Rafique, A. U. Bhatti et al. 2011. Boron deficiency in rainfed wheat in Pakistan: Incidence, spatial variability, and management strategies. *Journal of Plant Nutrition* 34:600–613.

Rashid, A., E. Rafique, and N. Bughio. 1994. Diagnosing boron deficiency in rapeseed and mustard by plant analysis and soil testing. *Communications in Soil Science and Plant Analysis* 25:2883–2897.

Rashid, A., E. Rafique, J. Din, S. N. Malik, and M. Y. Arain. 1997a. Micronutrient deficiencies in rainfed calcareous soils of Pakistan. I. Iron chlorosis in peanut. *Communications in Soil Science and Plant Analysis* 28:135–148.

Rashid, A., E. Rafique, N. Bughio, and M. Yasin. 1997b. Micronutrient deficiencies in rainfed calcareous soils of Pakistan. IV. Zinc nutrition of sorghum. *Communications in Soil Science and Plant Analysis* 28:455–467.

Rashid, A., E. Rafique, and N. Bughio. 1997c. Micronutrient deficiencies in rainfed calcareous soils of Pakistan. III. Boron nutrition of sorghum. *Communications in Soil Science and Plant Analysis* 28:441–454.

Rashid, A., E. Rafique, and N. Ali. 1997d. Micronutrient deficiencies in rainfed calcareous soils of Pakistan. II. Boron nutrition of the peanut plant. *Communications in Soil Science and Plant Analysis* 28:149–159.

Rashid, A., E. Rafique, and J. Ryan. 2002. Establishment and management of boron deficiency in field crops in Pakistan: A country report. p. 339–348, In: Goldbach, H. E. et al. (Ed.), *Boron in Plant and Animal Nutrition.* Kluwer Academic Publishers, New York.

Rashid, A., and A. Rashid. 1994. Nutritional problems of citrus on calcareous soils of Pakistan. p. 190–197, In: Proc. First International Seminar on Citriculture in Pakistan. Faisalabad, 2–5 Dec 1992. University of Agriculture, Faisalabad.

Rehman, H. 1989. Annual Report, Directorate of Soils and Plant Nutrition. 1988–89. Agricultural Research Institute, Tarnab, Peshawar, Pakistan.

Rehman, H. 1990. Annual Report, Directorate of Soils and Plant Nutrition. 1989–90. Agricultural Research Institute, Tarnab, Peshawar, Pakistan.

Reuter, D. J., and J. B. Robinson (Ed.) 1997. *Plant Analysis – An Interpretation Manual.* CSIRO Publishing, Melbourne, Australia.

Shah, Z., M. Z. Shah, M. Tariq, et al. 2012. Survey of citrus orchards for micronutrient deficiency in Swat valley of north western Pakistan. *Pakistan Journal of Botany* 44:705–710.

Shah, Z., and K. Shahzad. 2008. Micronutrients status of apple orchards in Swat valley of North West Frontier Province of Pakistan. *Soil and Environment* 27:123–130.

Shorrocks, V. M. 1992a. Boron – A global appraisal of the occurrence of boron deficiency. p. 39–53, In: Portch, S. (Ed.), *Proc. Int. Symp. Role of Sulphur, Magnesium, and Micronutrients in Balanced Plant Nutrition.* The Sulphur Institute, Washington, DC.

Shorrocks, V. M. 1992b. Micronutrients – Requirements, use and recent developments. p. 391–412, In: Portch, S. (Ed.), *Proc. Int. Symp. Role of Sulphur, Magnesium, and Micronutrients in Balanced Plant Nutrition.* The Sulphur Institute, Washington, DC.

Shorrocks, V. M. 1997. The occurrence and correction of boron deficiency. Plant and Soil 193: 121–148.

Siddique, M. T., M. Rashid, and M. Saeed. 1994. Micronutrient status of soils/plants of citrus growing areas in Punjab. p. 355–360, In: *Efficient Use of Plant Nutrients. Proc. 4th National Congress of Soil Science*, Islamabad, May 24–26, 1992. Soil Science Society of Pakistan.

Siddique, T., M. Rashid, and M. Saeed. 1998. Nutrient status of mango orchards in southern Punjab. *Pakistan Journal of Soil Sciences* 14:78–84.

Sillanpää, M. 1982. Micronutrients and the Nutrient Status of Soils: A Global Study. FAO Soils Bulletin 48, FAO, Rome.

Soltanpour, P. N. 1985. Use of ammonium bicarbonate-DTPA soil test to evaluate elemental availability and toxicity. *Communications in Soil Science and Plant Analysis* 16:323–338.

Stewart, A. G., J. Carter, A. Parker, and B. J. Alloway. 2003. The illusion of environmental iodine deficiency. *Environmental Geochemistry and Health* 25:165–170.

Takkar, P. N. 1996. Micronutrient research and sustainable productivity in India. *Journal of the Indian Society of Soil Science* 44:562–581.

Tandon, H. L. S. 1995. *Micronutrients in Soils, Crops and Fertilizers.* Fertilizer Development and Consultation Organization, New Delhi, India.

Wahab, A., and H. M. Bhatti. 1958. Trace element content of some West Pakistan soils. *Soil Science* 86: 319–332.

Wasiullah, A. U. Bhatti, F. Khan, and M. Akmal. 2010. Spatial variability and geo-statistics application for mapping of soil properties and nutrients in semi-arid district Kohat of Khyber Pakhtunkhwa (Pakistan). *Soil and Environment* 29:159–166.

Watts, M. J., and C. J. Mitchell. 2009. A pilot study on iodine in soils of Greater Kabul and Nangarhar provinces of Afghanistan. *Environmental Geochemistry and Health* 31:503–509.

Webster, R. and M. A. Oliver. 2007. *Geostatistics for Environmental Scientists*. John Wiley and Sons.

WHO (World Health Organization). 2007. *Assessment of Iodine Deficiency Disorders and Monitoring Their Elimination: A Guide for Program Managers*. 3rd ed., Geneva. http://whqlibdoc

Yoshida, S., and A. Tanaka. 1969. Zinc deficiency of rice in calcareous soils. *Soil Science and Plant Nutrition* 15:75–80.

Zia, M. S. 1993. Fertilizer Use Efficiency Project and Soil Fertility ARP-II. Annual Report, 1992–93. National Agricultural Research Center, Islamabad.

Zia, M. H., R. Ahmad, I. Khaliq, A. Ahmad, and M. Irshad. 2006. Micronutrients status and management in orchards soils: applied aspects. *Soil and Environment* 25:6–16.

Zia, M. H., M. Lark, L. Ander, M. Watts, and M. Broadley. 2018. Linking geospatial zinc in soils to predict zinc deficient regions and cereals across Pakistan. Presented at 3rd Annual Agriculture, Nutrition and Health (ANH) Academy Week, 25–29 June, 2018, Accra, Ghana.

Zia, M. H., M. J. Watts, A. Gardner, and S. R. Chenery. 2015. Iodine supply potential of soils, grain crops, and irrigation waters in Pakistan. *Environmental Earth Sciences* 73:7995–8008.

6 Crop Yield Increases with Micronutrient Fertilization and Its Economics

6.1 INTRODUCTION

The foremost benefit the growers seek from investing in micronutrients fertilizers is an appreciable increase in income due to the increased yield. In severely deficient soils, the application of micronutrient fertilizer makes an adequate difference in crop yield. Like any other profession in life, agriculture is also a business for the farmers to earn living for their families. Unless an investment at the farm generates improvements in yield or produce quality (or both) with attractive returns, the farmers are not expected to invest money and time in micronutrient fertilization. Unlike NPK fertilizers, micronutrient fertilizers are still at the nascent stage of adoption by farmers in Pakistan. The economic returns on micronutrient fertilizer products (and the cost of their application) must be attractive enough to convince the farmers to adopt this technology. Numerous field trials conducted at government farms and at farmers' fields provide very convincing information about the beneficial impact of using micronutrient fertilizers by way of increases in crop yields. A comprehensive summary of crop yield increases with Zn, B, and Fe fertilizers is included in the status report on micronutrients published by NFDC in 1998. Later, Rashid (2005, 2006) and Rashid and Rafique (2017) also summarized reported cases of crop yield increases with Zn, B, and Fe fertilizer use in soils of Pakistan.

One of the most spectacular, relatively recent cases of micronutrient fertilizer use having a major impact on crop production has been the adoption of Zn fertilizer in Turkey starting from nil in 1994 to 350,000 tonnes in 2006 (Cakmak, 2008a). Following the first field trial evidence in 1992 (Cakmak, 2008b) that Zn fertilizer resulted in substantial wheat yield increases in Central Anatolia; there has been rapid adoption of Zn fertilizer in Turkey. The application of Zn is now treated as an additive in NKP or NP fertilizers during the granulation process at the concentration of 1% Zn. The economic benefit of the use of Zn fertilizer is estimated to be US$ 100 M (Cakmak, 2004). In addition, the substantial increases in grain Zn density resulting from the widespread use of Zn fertilizers may have led to human health benefits by boosting Zn availability in the diet (Bell and Dell, 2008; Joy et al., 2017).

6.2   CROP YIELD INCREASES AND PROFITABILITY WITH MICRONUTRIENT FERTILIZATION

Initial identification and diagnosis of micronutrient deficiencies in crop plants were generally made in pot culture/greenhouse studies and/or by nutrient indexing of agricultural soils and farmers-grown crops. On identifying a micronutrient deficiency problem in the greenhouse, field experiments are conducted at research farms as well as at farmers' fields for verifying the problem and developing its management strategies. A salient example is the initial identification of Zn deficiency in Pakistani rice crop by two IRRI scientists who conducted a pot culture experiment by using the soil taken from a *Hadda*-affected rice field in the Kala Shah Kaku area (Yoshida and Tanaka, 1969). Similarly, the incidence of severe B deficiency in rice in the country was also initially identified in a greenhouse experiment conducted at NARC (Rashid et al., 2002a, 2005).

The most convincing diagnosis of deficiency of any nutrient (and for that matter any micronutrient) is attained by observing yield increases with the applied micronutrient fertilizer treatment in field situations. In fact, this was a mandatory "confirmatory test" for convincing the agricultural extension staff and growers for adopting the use of any fertilizer which has not been used previously, like B use in cotton or B use in rice. For the purpose, normally, simple field trials with only a few micronutrient treatments, i.e., control (no micronutrient applied) and an adequate dose of the micronutrient (like 1.0 kg B ha^{-1}) are conducted at multiple farmers' fields in the major crop-growing area. Adequate rates of all other deficient nutrients are applied as basal fertilizers. If appreciable and cost-effective yield increases occur in response to the applied micronutrient, the grower and extension staff are convinced to use the micronutrient fertilizer. Unless farmer participatory large-scale field demonstrations of cost-effective yield increases are organized in the major crop-growing areas, adoption of any micronutrient fertilizer use cannot be expected.

Therefore, once the deficiency of any micronutrient has been identified, a large number of field trials at farmers' fields and at research stations are conducted to establish the prevalence of the micronutrient deficiency. Most of the research information pertains to field crops. Though field trials on fruit orchards were also conducted (e.g., on citrus by AARI, Faisalabad in Sargodha area and on apple by ARI, Quetta at Sariab); research data on fruit yields as affected by micronutrient application is almost non-existent and attributed to the practical constraints in obtaining the fruit yield data. Most of the yield response information is on a single-crop basis, with very little information about cropping systems and/or long-term residual/cumulative effect of micronutrient fertilization.

The most convincing diagnosis of a deficiency is attained by alleviation of the constraint as a result of micronutrient application. In fact, this is a mandatory requirement for convincing the stakeholders (i.e., agricultural extension staff and growers) for adopting fertilizer micronutrient use in a crop for which it is not used previously. The approach is rather costly and time-consuming, and impractical to perform at every site. The research for establishing field-scale micronutrient deficiencies in the country involved greenhouse as well as field experimentation. On identifying a micronutrient deficiency, on the basis of soil test and/or plant analysis, initially, the deficiency is verified in the greenhouse situation. Subsequently, crop responses to micronutrient application are tested in field situations, in the major crop-growing regions of the country. Only on obtaining consistent positive crop responses to the micronutrient use, in farmers' fields, the agricultural extension service and the fertilizer industry are involved in extensive demonstrations of the technology. Two salient examples are the establishment of B and Zn deficiency in cotton and B deficiency in rice. Information about crop yield increases with Zn and B fertilizer use, and the economics of micronutrient fertilizer use in field situations is summarized in Table 6.1. Detailed information on crop responses to micronutrient fertilization (i.e., crops, location of field experiments, soil test micronutrient levels, yield without and with micronutrient use, and value:cost ratios of micronutrient fertilizer use is given in **Annex 8 to Annex 11**.

6.2.1 Yield Increases and Profitability with Zinc Fertilization

Appreciable crop yield improvements have been observed with Zn fertilization, in a number of field crops at research stations as well as in farmers' fields (Table 6.1). Average yield increases with Zn use are wheat, 13.2%; rice (Basmati-type cvs.), 11.5% and (IRRI-type cvs.), 13.3%; cotton, 6.7%; maize, 16%; sugarcane, 7.4%; rapeseed (Canola), 15.1%; sunflower, 19%; and potato, 20%. The value:cost ratios listed in Table 6.1 and **Annex 8** indicate that in Zn-deficient soil situations, the use of Zn fertilizer is highly profitable for all crops. For instance, the minimum value:cost ratio (VCR) of 4:1 has been worked out for soil-applied Zn in rapeseed (Canola) and the maximum VCR of 25:1 for foliar-applied Zn in wheat. Though field experiments have also been conducted on some fruit crops, yield data have not been reported. It has been reported that the number of fruits per tree, citrus fruit size, fruit weight, and vitamin-C content are improved with Zn use. In general, 5 kg Zn ha^{-1} is adequate for 3–4 crop seasons. Foliar application of Zn in grain crops is also of high worth compared to the soil application strategy as it enhances grains' Zn concentration significantly. However, at the same time, soil application of Zn is also warranted to maintain soil Zn equilibrium. Another emerging strategy of farmers' interest is seed priming with zinc and other useful minerals solutions. In maize hybrids (Pioneer 30-Y-87 and DK-919), seed priming in 2.0% Zn solution coupled with foliar spray of 2.0% Zn solution resulted in improved crop growth, grain productivity (20% increase in grain yield), and enhancement of grain Zn density from 19.8 mg kg^{-1} to 28.5 mg kg^{-1} (Mohsin et al., 2014).

6.2.2 Yield Increases and Profitability with Boron Fertilization

Average yield increases with B fertilizer use are wheat, 10.8%; rice (Basmati types), 17.4%, and (IRRI types), 14.9%; cotton, 15.7%; potato, 18.2; peanut, 19.6%; and maize and sugarcane, each 23% (Table 6.1 **and Annex 9**). Research

TABLE 6.1

Summary of Information Regarding Crop Yield Increases with Zn, and B Fertilizer Use and Profitability of Micronutrient Fertilization in Field Situations

Crop/Variety	Province	Zn/B Applied (kg ha⁻¹)	Control Yield (t ha⁻¹)	Yield Increase (%)	Value:Cost Ratio
		Zinc			
Wheat					
Various cvs.	Punjab, KP, Sindh	2–22.5*	3.5	13.2	4.2:1
Various cvs.	Punjab	Foliar sprays (0.30–0.88)	3.92	19.0	25.1:1
Rice					
Basmati-types	Punjab	Soil and/or foliar (2.5–10)	3.49	11.5	7:1
IRRI-types	Punjab, Sindh, KP	7.5–10	5.5	13.3	5:1
Cotton	Punjab, Sindh	5	2.21	6.7	6:1
Maize	KP, Punjab	5–6	3.35	16.1	6:1
Sugarcane	KP	10–15*	85.5	7.4	5.5:1
Rapeseed(Canola)	Punjab	8	1.52	15.1	4:1
Sunflower	Punjab, Sindh, Islamabad	3–15*	2.04	19	4.4:1
Potato	KP, Punjab	5–10	16.50	20.3	13:1
		Boron			
Wheat	Punjab, Sindh, KP	1–2	3.87	10.8	9.7:1
Rice					
Basmati-types	Punjab	1.0–2.0	3.63	17.4	23.6:1
IRRI-types	Sindh, Punjab, KP	0.5–2.0	4.90	14.9	17:1
Cotton	Punjab, Sindh, KP	0.5–2.5	2.3	15.7	19.6:1
	Punjab	Foliar sprays	2.29	11.4	33:1
Maize	Punjab, KP	1–1.5	2.8	23.1	14.3:1
Mungbean	Punjab	1.0	0.71	25.3	37:1
Sugarcane	Punjab, KP	Soil and/or foliar (0.25–1.0)	69	23.4	33:1
Potato	KP	1.5–2.0	13.44	18.2	20:1
Peanut	Punjab	1.0	1.73	19.6	32:1

For calculating value:cost ratios of micronutrient fertilizer use in various crops, the prices of various micronutrient fertilizers products are given in **Annex 12**, and prices of agricultural commodities, prevalent in the country during December 2016, are listed in **Annex 13**. More information about micronutrient fertilizer brands available in central Punjab, and their sale prices are given in **Annex 14**.
* Zinc dose of >10 kg ha⁻¹ is definitely excessive for the crop requirement.

on rice has revealed that improved B nutrition of this crop can improve its productivity as well as grain quality tremendously (Rashid et al., 2004, 2007; Rehman et al., 2014, 2015). In case B fertilizer cannot be applied at the time of sowing of a crop, application of B by foliar spray or through seed priming is also quite effective and, a more cost-effective option (Rehman and Farooq, 2013; Rehman et al., 2018a). Initially, widespread substantial paddy yield increases with B fertilization were observed during the years 2003 and 2004 by Rashid et al. (2004, 2007). In very simple initial field trials at farmers' fields in the Punjab rice belt, using 0 and 1.0 kg B ha^{-1} rates, paddy yield increases with B fertilization were 17–20% in cv. Super Basmati, 20–23% in cv. Basmati-385, and 14% in cv. KS-282 (P=0.05; **Annex 9**). Yield increases were consistent in rice cultivars at all field sites during both experimental years. In a more detailed field experiment carried out during the years 2003 and 2004 at 14 field locations in the Punjab and Sindh provinces, employing graded doses of B fertilizer (0, 0.5, 1.0, 1.5, and 2.0 kg B ha^{-1}), paddy yield increases with B application were substantial in both the rice cultivars (Super Basmati and IR-6) at all field sites (P=0.05; **Annex 9**). In all cases, maximum paddy yield was mostly obtained with 1.0 kg B ha^{-1}, and no further yield increase was obtained with higher B rates. Paddy yield increases with B fertilization were 11–31% during 2003 and 7–30% during 2004 in cv. Super Basmati and 18–34% in cv. IR-6. Mean paddy yield increases with B fertilizer use were 26% in cv. Super Basmati and 27% in cv. IR-6. However, near-maximum (i.e., 95% of maximum) paddy yield was associated with 0.75 kg B ha^{-1} for cv. Super Basmati and 0.85 kg B ha^{-1}.

In rice, initially observed paddy yield increases with B application in Pakistan were 10–35% (mean increase of 14%) in cvs. Basmati-370 and IR-6, in the Punjab province (Chaudhry et al., 1976). Later on, paddy yield increases of 14–30% (mean 20%) were observed in cvs. Super Basmati, Basmati-385, KS-282, and IR-6 grown in the Punjab and Sindh provinces (Table 6.2). The yield increases with B accrue because of the increased number of productive tillers per hill, substantial increase in rice ear length (Figure 6.1), drastic reduction in panicle

TABLE 6.2

Effect of Boron Nutrition on Rice Panicle Sterility and Productive Tillers

Rice Cultivar	Paddy Yield (t ha^{-1})		Panicle Sterility (%)		Productive Tillers per Hill	
	Control	+B	Control	+B	Control	+B
Basmati-385	3.77	4.72	28	16	14.3	16.1
Super Basmati	3.23	3.89	23	14	18.4	20.1
KS-282	4.82	5.48	15	12	12.0	15.0
IR-6	4.34	5.64	41	27	11.8	16.1
Mean	**4.04**	**4.93**	**27**	**17**	**14.1**	**16.8**

Modified from Rashid et al. (2007).

FIGURE 6.1   Substantial improvement in paddy (cv. Super Basmati) yield with B use in a calcareous soil of Pakistan. (Rashid et al., 2007, reprinted with permission.)

FIGURE 6.2   Alleviation of rice panicle sterility and post-harvest shedding by using B fertilizer in a calcareous soil of Pakistan. (Rashid et al., 2007, reprinted with permission.)

sterility on the lower portion of the ears (Figure 6.2), and reduced post-harvest shedding of kernels (Rashid et al., 2007; Tables 6.2 and 6.3). Also, uneven crop maturity, caused by B deficiency, is cured with B application. In subsequent field experiments, Rehman et al. (2014) also observed substantial improvement in kernel yield and kernel quality with B application to rice crop grown under flooded conditions as well as under alternate wetting and drying soil conditions. Boron application through soil application, foliar spray, or seed priming was equally effective. The researchers observed that rice yield increase occurred because of a reduction in panicle sterility coupled with an increase in kernel size. Obviously, foliar spray and seed priming are economical options to improve crop growth and attain improved kernel quality and yield.

Despite being considered tolerant to B deficiency (Lucas and Knezek, 1972; Savithri et al., 1999), rice suffers from this nutritional disorder in Pakistan and

TABLE 6.3

Rice Quality Improvement with Boron Use in Calcareous Soil Environment

	Basmati-385		Super Basmati	
Grain Characteristic	Control	+ B	Control	+ B
Total milled rice (%)	71.1	73.1	70.4	72.0
Head rice (%)	54.3	57.6	52.9	56.5
Kernel thickness (mm)	1.52	1.53	1.53	1.54
Kernel length:breadth ratio	4.13	4.15	4.56	4.53
Quality Index (length/breadth x thickness)	2.70	2.69	2.97	2.94
Elongation ratio upon cooking	1.94	1.98	1.97	2.00
Bursting upon cooking (%)	11	8	10	7
Alkali spreading (Score 1–7)[a]	4.5	4.8	4.7	5.0

Modified from Rashid et al., 2004, 2007.

[a] Alkali spreading value: 4–5 score = Intermediate G.T.-type rice.

elsewhere (Chaudhry et al., 1977; Shorrocks, 1997; Rashid et al., 2004). Though appreciable rice yield increases with B application in Pakistan were initially observed about three decades ago (Chaudhry et al., 1977), it is only recently that this micronutrient disorder has received attention. Besides the assumption that rice is tolerant to B deficiency, major constraints to R&D on B nutrition of rice were the non-availability of B-free glassware and requisite professional expertise for laboratory analysis of B in the country (Rashid et al., 2002bb).

In an experiment at a farmer's field in Multan, Hussain et al. (2012) studied the impact of soil-applied and foliar-fed B on rice (cv. Super Basmati) at different crop growth stages. Soil-applied 1.5 kg B ha^{-1} at the transplanting stage increased grain yield by 13% over control (3.61 t ha^{-1} to 4.08 t ha^{-1}). A maximum grain yield increase of 20% was observed where soil B was applied at the flowering stage. Foliar-applied B (at tillering, flowering, or grain formation stage) was less effective, as increases in grain yield with foliar-fed B were8–10% over control. Both soil application and foliar feeding of Bat all growth stages were highly cost-effective. Therefore, Hussain et al. (2012) suggested that for improving rice productivity and maximizing net economic returns, B fertilizer may be soil-applied at the flowering stage.

Rehman and Farooq (2013) observed that seed priming of rice cvs. Super Basmati and Shaheen Basmati (@ 2.0 g B kg^{-1} seed) resulted insignificant increases in paddy yield of both the cultivars, primarily by a reduction in panicle sterility. Leaf and kernel B contents were increased with an increase in B concentration in seed coating (**Annex 9**). In another two-year field study on rice grown in a calcareous soil by adopting three different cropping systems (i.e., aerobic rice, alternate wetting and drying (AWD) system, and flooded rice), Rehman et al. (2014) observed that in addition to improvement in kernel yield (from 3.0 to 3.7 t ha^{-1}), B application also improved kernel quality (i.e., enhanced head rice recovery and kernel length as well as increased protein content and kernel B concentration). Foliar application of B under the AWD system gave maximum net returns than under the aerobic and flooded conditions. Seed priming with B and foliar feeding of B also improved rice productivity (P=0.05; **Annex 9**). This study suggested that B application through foliar spray or seed priming may be an economically viable option to reduce panicle sterility, and improve kernel quality, rice growth, and paddy yield. Improvement in rice yield by B application was attributed to an increase in kernel size and a decrease in panicle sterility. In a follow-up field study with fine-grain rice, Rehman et al. (2015) observed that soil application of B to rice improved its tillering, leaf elongation, panicle fertility, paddy yield, and grain B concentration.

In a two-year field study on B nutrition of short-duration Basmati rice (cv. Shaheen Basmati) in water-saving production systems, Rehman et al. (2015) observed that adequate B nutrition significantly improved productivity, quality, and profitability of flooded rice and aerobic rice. In this study, B application by either method (i.e., seed priming (0.1 mM B), foliar spray (200 mM B), or soil application (1.0 kg B ha^{-1})) improved rice growth, water relations, morphology, yield-related traits, panicle fertility, grain yield, grain quality, and grain B contents; B application as seed priming proved superior and cost-effective with maximum marginal rate of return. Boron moves from roots along the transpiration stream and is generally accumulated in leaves and shoot apex. Through this mechanism, B regulates the meristematic activities in the immediate vicinity and improves plant growth (Lovatt,1985; Ahmed et al., 2009; Herrera-Rodriguez et al., 2010). Reproductive structures are more sensitive to B fertilization (Brown et al., 2002), and any B deficiency during reproductive and grain-filling phases may cause pollen abortion (Dordas, 2006), resulting in the development of sterile panicles. Adequate B supply ensures better germination of pollen grains and tube growth (Mozafar, 1993; Subedi et al.,1998), which causes better fertilization, seed-set, and grain development (Noppakoonwong et al., 1997; Rerkasem and Jamjod,1997; Wang et al., 2007), and contributes for increase in seed/kernel weight and yield.

In a recent field study on wheat in Pakistan, B application by seed priming, soil application, and foliar sprays enhanced grain yield as well as profitability of the crop planted in plow tilled and zero tilled soils. Seed priming (in 0.01 *M* B solution) and soil application (@ 1 kg ha^{-1}) were the most cost-effective methods of B application to wheat grown in zero tillage and plow tillage, respectively (Nadeem et al., 2019). In cotton, the average increase in seed cotton yield was 14% (Rashid and Rafique, 2002), caused by reduced boll shedding and increased boll weight upon B fertilization. Soil application and foliar feeding of B are equally effective.

6.2.3 Yield Increases and Profitability with Iron Fertilization

The observed average yield increases with Fe use are peanut, 21%; rapeseed, 19%; potato, 59%; chickpea, 10.5%; rice, 31%; wheat,11%; and maize, 11.7%. Fruit size and quality are improved with Fe use. Inorganic Fe salts and even chelates of

Fe, except for *Sequestrene*, are generally ineffective in soil application. However, repeated foliar sprays of $FeSO_4$ or chelated Fe are helpful to cure chlorosis.

In summary, soil application and foliar feeding of Zn and B are equally effective, the latter being substantially more cost-effective (Table 6.1). However, foliar feeding is a practical option only for fruits, vegetables, and other high-value crops or in field crops like cotton requiring multiple sprays of pesticide(s). In Pakistan, the crop response information mostly pertains to single crop seasons, with very little information on the impact of micronutrient use in various cropping systems and/or the residual and cumulative effect of micronutrient fertilization. Also, because of practical constraints, hardly any fruit yield data have been recorded as affected by the application of micronutrient fertilizers.

6.3 PRODUCE QUALITY IMPROVEMENTS WITH MICRONUTRIENT FERTILIZATION

Deficiencies of micronutrients in plants may also deteriorate the crop produce quality apart from causing yield reductions, and, hence, lowering the price of the grain/fruit produced inmicronutrient-deficient soil situations. Some salient examples of produce quality deterioration with micronutrient deficiencies in crop plants are provided below.

6.3.1 BORON NUTRITION OF CROP PLANTS AND PRODUCE QUALITY

Boron deficiency in plants is known to result in the low quality of harvested products in a range of species, as follows:

- Internal fruit necrosis in mango (Ram et al., 1989)
- Hollow heart in peanut
- Brownish discoloration of cauliflower curds (Kotur, 1991)
- Premature staining of the testa in avocado (Harkness, 1959)
- Internal and external corking of fruit tissue in apple (Shorrocks and Nicholson, 1980)
- Lesions in storage organs of sugarbeet (Vlamis and Ulrich, 1971)
- Rice grain quality impairment with B deficiency (Rashid et al., 2004, 2007)

Micronutrient research in Pakistan has also revealed that the use of micronutrient fertilizers not only enhanced crop yield but, in many instances, the quality of crop production was also improved. A salient local experience is an appreciable improvement in the cooking quality of rice grains with fertilizer B use. NARC carried-out extensive field experiments for rice (cvs. Super Basmati, Basmati-385, KS-282, and IR-6), in the rice belt of Punjab (collaborated with Agricultural Extension and Adaptive Research), and Sindh (collaborated with Engro Fertilizers). The soil application of B fertilizer enhanced paddy yields, substantially, increased kernel milling recovery and head rice recovery, as well as improved cooking quality traits (increased elongation, reduced bursting, and reduced stickiness upon cooking). The improvement in rice quality with better B nutrition of plants is attributed to better grain filling and uniform crop maturity. Additional benefits of investment in B fertilization in rice include enhancement in milling return as well as in head rice recovery (Tables 6.2 and 6.3; Rashid et al., 2004; 2007). Recently, a similar beneficial impact of B fertilizer on rice grain quality was observed by Nadeem et al. (2021). Contrary to rice, no significant improvement in quality parameters of four advanced lines of brassica seed (oil content, protein, GSL µmol g^{-1}, moisture, oleic acid, enolenic acid, and erucic acid) could be observed in three-year trials at Nuclear Institute for Food and Agricultue (NIFA), Peshawar (Ali et al., 2016). However, in a B-deficient field study, B application (either soil-applied or foliar-fed) increased canola (cv. Hyola, Punjab Sarson, and Bulbul) yield under saline field conditions in the range of 10.7–14.9% over control (Abid et al., 2014). In the case of only a foliar B spray (0.1% B), VCR was noted in the range of 1.9–6.3 depending on the tested canola cultivar.

In Swat area of KP province, during the late 1980s low plant B status in apple was observed to cause lesser fruit numbers, smaller size, loosen fruit firmness, and deteriorated other fruit quality parameters. Foliar feeding of B improved fruit quality appreciably (Rehman, 1990).

6.3.2 ZINC DEFICIENCY IN WHEAT IMPAIRS GRAIN QUALITY

Cereal grains are inherently low in Zn concentration (Cakmak, 2008a). Whereas the optimum level of Zn in staple cereals, for catering adequate Zn for human nutrition, should be 40–60 mg kg^{-1}, the grains of prevalent cultivars of wheat in Pakistan contain around 25 mg Zn kg^{-1} (Zou et al., 2012). This low Zn in wheat grains is believed to be a major cause of Zn malnutrition in about one-half of the Pakistani population, which rely on wheat bread as a source of their daily caloric requirement. Zinc malnutrition leads to serious health hazards. As resource poors in a country like ours can't afford to diversify their diet (by consuming milk, fish, eggs, fruits, etc.) nor can they afford to take mineral pills, the only viable remedy appears to be an enrichment of the staple cereal(s) with Zn. Research in Pakistan and elsewhere have established that Zn density in wheat grains can be increased by genetic biofortification as well as by agronomic biofortification (Zou et al., 2012; HarvestPlus, 2014). Wheat genotypes grown in Pakistan are genetically less diverse with regard to grain Zn density, owing to the continuous focus of the breeding programs on the development of high-yielding varieties only (Rehman et al., 2018b). However, biofortification research in the country indicates that enrichment of wheat grains with Zn to the desired level, i.e., around 40–50 mg Zn kg^{-1}, is achievable by Zn fertilizer application to soil coupled with Zn foliar sprays, to Zn-efficient wheat genotypes developed under the HarvestPlus project in Pakistan. However, because of the site-specific factors that affect the ability of the biofortified wheat (like *Zincol-16*) to accumulate zinc in grains, wheat growers in Pakistan would need enabling environments to adopt these technologies (Zia et al., 2020).

6.4 ECONOMIC BENEFIT OF MICRONUTRIENT FERTILIZER USE

A farmer will not use a micronutrient fertilizer simply because it will increase crop yield, but the additional yield should be sufficient to pay the fertilizer (and application) cost and bring in a substantial net profit. A good indicator of the economics of fertilizer use is obtained by calculating VCR, i.e., the gross return per unit expenditure on fertilizer. In fertilizer use, one Pakistani Rupee (PKR) spent should produce an extra crop worth at least PKR 2 to 3. Wherever there is a good crop response to a micronutrient, its economics ought to be attractive for the farmer simply because micronutrient fertilizers are applied in small doses and leave beneficial residual effects on subsequent crops in the same field (Rafique et al., 2012; Abid et al. 2013; Nadeem et al., 2020, 2021).

Field trials conducted throughout the country, by almost all agricultural research and educational institutions, have adequately established that micronutrient use in a variety of crops is highly cost-effective, more so as foliar feeding. The VCRs reported in **Annex 8 to Annex 11** for Zn, B, Fe, and Cu are increases over the crop yields obtained with NP or NPK application, referred to as control yields.

6.4.1 ZINC FERTILIZER USE

The VCR of Zn fertilizer use in several crops (**Annex 8**), in general, presents a very rosy picture. For example, in 61 wheat experiments, VCR varied from 3:1 to 3.7:1 at Faisalabad and D.I. Khan to 10:1 in Sukkur and 12:1 in Pothohar. The three exceptions of low economic return were Thal Desert in Layyah (VCR, 2:1); Chakwal, Attock, and Jhelum (VCR, 2.2:1); and Rawalpindi rainfed (VCR, 2.5:1).

The VCR of Zn fertilizer in rice varied from 3.1:1 for cv. IR-6 to 21:1 for Basmati-515. In coarse varieties of rice, tested at 119 locations (IR-6, KS-282, JP-5), VCR ranged from 3.1:1 in Punjab to 10.5:1 in Malakand and Swat. In general, the VCR was better in fine varieties of rice tested at 81 locations, especially cv. Basmati-385, and cv. Basmati-515 as it varied from 5.7:1 to 21:1 (**Annex 8**). The average VCR was 5:1 in coarse rice and 7:1 in fine-grain rice.

The economics of Zn use in maize was still better as the VCR ranged from 3.5:1 in rainfed maize in Rawalpindi to 7.8:1 in Peshawar. The average VCR in 17experiments was 6:1. The economic benefit of Zn use in cotton was also high; VCR ranged from 4.8:1 to8.5:1 in Punjab, with an average of 6:1 in 19 field experiments. Thus, Zn application in cotton proved highly beneficial in the fields where soil Zn was 0.6–1.0 mg kg^{-1} soil, i.e., ≤ 1.0 mg kg^{-1} soil (**Annex 8**).

Zinc fertilizer use in sugarcane was also very profitable. The VCR ranged from 4.7:1 for Zn applied in Faisalabad to 6.2:1 for Zn applied in Peshawar. The highest VCR of Zn use pertains to that of potato in Nowshehra (VCR 23.2:1), with an average of 13:1 when tested across six locations. This was followed by sunflower where the highest VCR of 19:1 was noted in Faisalabad followed by 16:1 in Islamabad. Application of Zn to mungbean also appears highly beneficial while depicting a VCR of 28:1, enhancing grain Zn from 32.5 mg kg^{-1} (control grain) to 38.0 mg kg^{-1} in zinc-applied crop (**Annex 8**).

Although very limited local information is available regarding the residual effect of soil-applied Zn (Rafique et al., 2012; Nadeem et al., 2020), the long-term Indian studies and elsewhere on alkaline soil have established that one good soil application of Zn lasts for 3–4 crop seasons. Long-term research in the US even revealed that 10–20 kg Zn ha^{-1} was effective for about eight crop seasons. Therefore, actual soil-applied Zn fertilizer cost could be only one-third to one-fourth of the ones used in **Annex 8**, resulting in substantial improvement of VCR (about three times) for the entire crop.

6.4.2 BORON FERTILIZER USE

The VCR values in **Annex 9** indicate that by spending 1.0 PKR on B fertilizer, the value of increased crop yield was worth PKR 4.0–17.0 (mean PKR 9.7:1 for 48 sites) in wheat; PKR 10.0–41.0 in Basmati-type rice and PKR 5.0–29.0 in IRRI-type rice; PKR 10.0–28.0 with soil-applied B of cotton and PKR22.0–37.0with foliar feeding of B in cotton; PKR 6.60–19.7 in maize; PKR 5.0–64.0 in sugarcane; PKR 8.1–34.0 in potato; and PKR 17.0–43.0 in peanut (**Annex 9**). In medium- to heavy-textured soil, the residual effect of B applied @ 1.0 kg ha^{-1} is expected to benefit the succeeding crop for two years (Nadeem et al., 2021). Therefore, the actual VCRs are speculated to be two- to four fold the ones worked out in **Annex 9**. Thus, at a national level, correction of B deficiency in Pakistan's soil-crop system should be highly profitable to the growers and beneficial to the national economy.

6.4.3 IRON FERTILIZER USE

As inorganic sources of Fe are not very effective in curing Fe chlorosis in calcareous soils, one cost-effective approach for field crops has been to exploit genetic variability for tolerance to Fe deficiency. In fact, this approach has been used quite successfully for some crops in many countries including Pakistan. For example, the AARI scientists had developed chickpea varieties after screening the germplasm for tolerance to Fe chlorosis (Ali et al., 1988).Although yield data in fruit crops are not available, the extent and severity of Fe chlorosis in citrus and deciduous fruits are good evidence of the possible yield-loss because of Fe deficiency stress. In addition, fruit quality is badly affected by Fe deficiency. Therefore, foliar sprays of Fe in fruit trees are expected to be economically beneficial.

An estimated economic benefit of micronutrient use in selected crops indicates that their judicious use in deficient soils will result in enhanced farmers' income and improved national economy. However, the economics of micronutrient use will change with a change in fertilizer prices, product prices, application rates, and response ratios. It will also change if a new more-efficient crop variety becomes available or a cost-saving application method is employed.

The economics of using Fe fertilizer in agronomic crops might not always be highly attractive everywhere (**Annex 10**). The average VCR is11:1 for wheat. In maize, VCR is 11.7:1 and in rice, the mean VCR is 31:1 (for soil-applied FeSO$_4$), and 7:1 to 35:1 in peanut. The highest-reported VCR of $FeSO_4$ use was in potato in Nowshehra. The lowest economic return of Fe fertilizer appears to be in chickpea, i.e., average VCR 10.5:1. Thus, the use of micronutrient fertilizer in field crops must be suggested in consideration of past economic benefits in similar soil-crop situations. In addition to bringing immediate economic benefit by crop yield improvement, residual beneficial effects of micronutrient fertilizer bring additional economic benefits (Abid et al., 2013; Nadeem et al., 2020, 2021). For example, in our experience, soil-applied B fertilizer to rice crop enhanced the paddy yield of the current crop and also increased the grain/paddy yield of the succeeding two wheat and two rice crops. Therefore, actual VCRs for soil-applied B were, in fact, at least two to three times those given in Tables 6.1 and 6.2. This must be true in the case of soil-applied other micronutrients, like Zn, as well. Additionally, micronutrient use optimizes the use efficiency of other farm inputs, including N, P, and K fertilizers, and, thus, improves farm economy indirectly as well.

6.4.4 VALUING CROP YIELD INCREASE AND HEALTH BENEFITS OF ZINC FERTILIZER USE IN PAKISTAN

It is well established that, in soil Zn-deficient situations, the use of Zn fertilizer to staple cereal crops in Pakistan, like wheat, is cost-effective by increasing crop yields (Rashid, 2005) as well as by enhancing grain Zn density (Zou et al., 2012). However, despite the widespread occurrence of Zn deficiency in soils and crops, Zn fertilizers are underutilized in countries like Pakistan. In an effort to determine the value of Zn fertilizer use in wheat crop in Pakistan, Joy et al. (2017) simulated increased Zn fertilizer use scenarios for wheat crop in the Punjab and Sindh provinces of Pakistan. According to this study, application of Zn fertilizer to the area currently under wheat crop in Punjab and Sindh provinces, at current soil:foliar usage ratios, could increase dietary Zn supply from ~12.6 to 14.6 mg capita^{-1} d^{-1}, and almost halve the prevalence of Zn malnutrition in humans, assuming no other changes to food consumption. With enhanced Zn fertilizer use, gross wheat production could increase by 2.0 Mt grain year^{-1} in Punjab and 0.6 Mt grain year^{-1} in Sindh, representing an additional return of International Dollars (I$) >800 M and an annual increased grain supply of 19 kg capita^{-1}. According to these researchers, benefit:cost ratios (BCRs) for wheat yield alone are 13.3 in Punjab and 17.5 in Sindh. If each disability-adjusted life year (DALY) is monetized at one- to threefold Gross National Income per capita based on purchasing power parity (GNI$_{PPP}$), full adoption of Zn fertilizer for wheat crop in Punjab province alone provides an additional annual return of I$ 405–1216 M, at a cost per DALY saved of I$ 461–619. Thus, these authors concluded that there are huge potential market- and subsidy-based incentives to increase Zn fertilizer use in Pakistan (Joy et al., 2017).

To our knowledge, this is the first-ever study that has combined both staple crop yield and health-based benefits to quantify the overall economic benefits of an increased (projected) supply of a micronutrient fertilizer, i.e., Zn, which is a cause of human malnutrition because of wide-spread deficiency in soils and crop plants. The results of the study (Joy et al., 2017) have implications for food and nutritional security, especially micronutrients in general and Zn in specific. Wheat being the predominant staple cereal is a source of micronutrient nourishment, especially in resource-poor segments of the society in Pakistan.

6.5 PRACTICAL IMPLICATIONS OF MICRONUTRIENT RESEARCH INFORMATION

In Pakistan, deficiencies of N and P are known to occur in almost all agricultural soils/fields and crops. Incidence of K deficiency is also recognized, at least in high K-requiring crops and in specific soil types. By now a lot is known about the nature, extent, and severity of micronutrient disorders and crop responses to micronutrient fertilizer use in selected crops and agricultural regions of the country. However, many crops susceptible to specific micronutrient deficiencies and some important agricultural areas of the country have not received adequate R&D attention so far. For example, cauliflower, turnip, radish, sugarbeet, alfalfa, pomegranate, rose, and ornamentals are known to be highly susceptible to B deficiency, but these crops did not receive due R&D emphasis in the past. Similarly, mango fruit necrosis in India has been attributed to B deficiency. Though similar symptoms are common in Pakistani mango fruit as well (especially in a very popular variety *Chaunsa*), this important fruit has received very inadequate research attention regarding its micronutrient nutrition. Consequently, research information is hardly available on B nutrition of these crops. The same is true in the case of Zn nutrition of some sensitive fruit crops like apple, grape, peach, and pear. Field-level R&D information is also inadequate regarding Fe nutrition of highly susceptible fruits like deciduous fruits (apple, pear, plum, etc.), citrus, strawberry, and walnut. In general, fruit crops grown in alkaline soils are susceptible to deficiencies of certain micronutrients like Zn, Fe, and B. For example, internal necrosis and short shelf life in mango is a common observation, particularly in some varieties like *Chaunsa*.

In field crops, most of the R&D information pertains to single-crop experiments rather than long-term cropping-system studies. Long-term field studies in major cropping systems are warranted for studying the residual/cumulative effect of micronutrient fertilizer use and its economics. Along with generating basic information regarding the identification and establishment of micronutrient problems in all important field crops, vegetables, and fruits, much greater emphasis is needed on management and development aspects of micronutrient problems.

REFERENCES

Abid, M., N. Ahmed, M. F. Qayyum, M. Shabaan, and A. Rashid. 2013. Residual and cumulative effect of fertilizer zinc on cotton-wheat productivity in an irrigated Aridisol. *Plant, Soil and Environment* 59:505–510.

Abid, M., M. M. H. Khan, M. Kanwal, and M. Sarfraz. 2014. Boron application mitigates salinity effects in canola (*Brassica napus*) under calcareous soil conditions. *International Journal of Agriculture and Biology* 16:1165–1170.

Ahmed, N., F. Ahmad, N. Abid, and M. A. Ullah. 2009. Impact of zinc fertilization on gas exchange characteristics and water use efficiency of cotton crop under arid environment. *Pakistan Journal of Botany* 41:2189–2197.

Ali, M., W. Muhammad, and I. Ali. 2016. Yield of oil seed Brassica (napus and juncea) advanced lines as influenced by boron. *Soil and Environment* 35:30–34.

Ali, A., C. M. Yousaf, and M. Tufail. 1988. Screening of desi and Kabuli chickpea types for iron-deficiency chlorosis. *International Chickpea Newsletter* 18:5–6.

Bell, R. W., and B. Dell. 2008. *Micronutrients for Sustainable Food, Feed, Fibre and Bioenergy Production.* International Fertilizer Industry Association, Paris, France.

Brown, P. H., N. Bellaloui, M. A. Wimmer, et al. 2002. Boron in plant biology. *Plant Biology* 4:205–223.

Cakmak, I. 2004. Identification and correction of wide-spread zinc deficiency in Turkey – a success story (a NATO-Science for Stability Project). *International Fertilizer Society* 552:1–26.

Cakmak, I. 2008a. Zinc deficiency in wheat in Turkey. In: B. J. Alloway (Ed.), *Micronutrient Deficiencies in Global Crop Production*, p. 181–200. Springer, Dordrecht.

Cakmak, I. 2008b. Marschner Review. Enrichment of cereal grains with zinc: Agronomic or genetic biofortification? *Plant and Soil* 302:1–17.

Chaudhry, F. M., S. M. Alam, A. Rashid, and A. Latif. 1977. Micronutrient availability to cereals from calcareous soils. IV. Mechanism of differential susceptibility of two rice verities to Zn deficiency. *Plant and Soil* 46:637–642.

Chaudhry, F. M., A. Latif, A. Rashid, and S. M. Alam. 1976. Response of the rice varieties to field application of micronutrient fertilizers. *Pakistan Journal of Scientific and Industrial Research* 19:134–139.

Dordas, C. 2006. Foliar boron application improves seed set, seed yield, and seed quality of alfalfa. *Agronomy Journal* 98:907–913.

Harkness, R. W. 1959. Boron deficiency and alternate bearing in Avocados. *Proceedings of the Florida State Horticultural Society* 72:311–317.

HarvestPlus. 2014. Biofortification Progress Briefs. Available at www.HarvestPlus.org. (Accessed on Feb 23, 2019).

Herrera-Rodriguez, M. B., A. Gonzalez-Fontes, J. Rexach, et al. 2010. Role of boron in vascular plants and response mechanisms to boron stresses. *Plant Stress* 4:115–122.

Hussain, M., M. A. Khan, M. B. Khan, M. Farooq, and S. Farooq. 2012. Boron application improves growth, yield and net economic return of rice. *Rice Science* 19:259–262.

Joy, E. J. M., W. Ahmad, M. H. Zia, et al. 2017. Valuing increased zinc (Zn) fertilizer-use in Pakistan. *Plant and Soil* 411:139–150.

Kotur, S. C. 1991. Effect of boron, lime and their residue on yield of cauliflower, leaf composition and soil properties. p. 349–354. In: R. J. Wright (Ed.), *Plant-Soil Interactions at Low pH*. Kluwer Academic Publishers, Dordrecht.

Lovatt, C. J. 1985. Evolution of xylem resulted in a requirement for boron in the apical meristems of vascular plants. *New Phytologist* 99:509–522.

Lucas, R. E., and B. D. Knezek. 1972. Climatic and soil conditions promoting micronutrient deficiencies in plants. In: Mortvedt, J. J., et al. (Ed.) *Micronutrients in Agriculture*. Soil Science Society of America, Madison, p. 265–288.

Mohsin, A. U., A. U. H. Ahmad, M. Farooq, and S. Ullah. 2014. Influence of zinc application through seed treatment and foliar spray on growth, productivity and grain quality of hybrid maize. *Journal of Animal and Plant Sciences* 24:1494–1503.

Mozafar, A. 1993. Role of boron in seed production. p. 185–206. In: Gupta, U.C. (Ed.), *Boron and its Role in Crop Production*. CRC Press, Boca Raton, FL, USA.

Nadeem, F., M. Farooq, B. Mustafa and A. Nawaz. 2021. Influence of soil residual boron on the rice performance and soil properties under conventional and conservation rice-wheat cropping systems. *Crop and Pasture Science* 72:335–347.

Nadeem, F., M. Farooq, B. Mustafa, A. Rehman and A. Nawaz. 2020. Residual zinc improves soil health, productivity and grain quality of rice in conventional and conservation wheat-based systems. *Crop and Pasture Science* 71:322–333.

Nadeem, F., M. Farooq, A. Nawaz and R. Ahmad. 2019. Boron improves productivity and profitability of bread wheat under zero and plough tillage on alkaline calcareous soil. *Field Crops Research* 239:1–9

NFDC. 1998. *Micronutrients in Agriculture: Pakistan Perspective.* National Fertilizer Development Center, Islamabad, 51 pp.

Noppakoonwong, R. N., B. Rerkasem, R. W. Bell, B. Dell, and J. F. Loneragan. 1997. Diagnosis and prognosis of boron deficiency in black gram (*Vigna mungo* L. Hepper) in the field by using plant analysis. In: *Boron in Soils and Plants*, pp. 89–93. Bell, R.W., and Rerkasem, B. (Ed.). Kluwer Academic Publishers, Netherlands.

Rafique, E., A. Rashid, and M. Mahmood-ul-Hassan. 2012. Value of soil zinc balances in predicting fertilizer zinc requirement for cotton-wheat cropping system in irrigated Aridisols. *Plant and Soil* 361:43–55.

Ram, S., L. D. Bist, and S. C. Sirohi. 1989. Internal fruit necrosis of mango and its control. *Acta Horticulturae* 231:805–813.

Rashid, A. 2005. Establishment and management of micronutrient deficiencies in Pakistan: a review. *Soil and Environment* 24:1–22.

Rashid, A. 2006. *Boron Deficiency in Soils and Crops of Pakistan: Diagnosis and Management.* Pakistan Agricultural Research Council, Islamabad, Pakistan, viii+34 pp. ISBN: 969-409-184-5.

Rashid, A., S. Muhammad, and E. Rafique. 2002a. Genotypic variation in boron uptake and utilization by rice and wheat. p. 305–310, In: Goldbach, H.E. et al. (Ed.), *Boron in Plant and Animal Nutrition.* Kluwer Academic Publishers, New York.

Rashid, A., E. Rafique, and J. Ryan. 2002b. Establishment and management of boron deficiency in field crops in Pakistan: A country report. p. 339–348, In: Goldbach, H.E., et al. (Ed.), *Boron in Plant and Animal Nutrition.* Kluwer Academic Publishers, New York.

Rashid, A., S. Muhammad, and E. Rafique. 2005. Rice and wheat genotypic variation in boron use efficiency. *Soil and Environment* 24:98–102.

Rashid, A., and E. Rafique. 2002. Boron deficiency in cotton grown in calcareous soils of Pakistan. II. Correction and criteria for foliar diagnosis: p. 357–362, In: Goldbach, H.E., et al. (Ed.), *Boron in Plant and Animal Nutrition.* Kluwer Academic Publishers, New York.

Rashid, A., and E. Rafique. 2017. Boron deficiency diagnosis and management in field crops in calcareous soils of Pakistan: A mini review. *BORON* 2:142–152.

Rashid, A., M. Yasin, M. A. Ali, Z. Ahmad, and R. Ullah. 2007. An alarming boron deficiency in calcareous rice soils of Pakistan: Boron use improves yield and cooking quality. p. 103–116, In: Xu, F., et al. (Ed.), *Advances in Plant and Animal Boron Nutrition.* Proc 3rd *Int Symp on All Aspects of Plant and Animal Boron Nutrition, Wuhan, China, 9–13 Sep 2005.* Springer, Dordrecht.

Rashid, A., M. Yasin, M. Ashraf, and R. A. Mann. 2004. Boron deficiency in calcareous soils reduces rice yield and impairs grain quality. *International Rice Research Notes* 29:58–60.

Rehman, H. 1990. *Annual Report, Directorate of Soils and Plant Nutrition. 198990.* Agricultural Research Institute, Tarnab, Peshawar, Pakistan.

Rehman, A., and M. Farooq. 2013. Boron application through seed coating improves water relations, panicle fertility, kernel yield, and biofortification of fine grain aromatic rice. *Acta Physiologiae Plantarum* 35:411–418.

Rehman, A., M. Farooq, A. Nawaz, and R. Ahmad. 2014. Influence of boron nutrition on rice productivity, kernel quality and biofortification in different production systems. *Field Crops Research* 169:123–131.

Rehman, A., M. Farooq, A. Nawaz, and R. Ahmad. 2015. Improving the performance of short-duration basmati rice in water-saving production systems by boron nutrition. *Annals of Applied Biology* 168:19–28.

Rehman, A., M. Farooq, A. Rashid, et al. 2018a. Boron nutrition of rice in different production systems. A review. *Agronomy for Sustainable Development* 38:25. doi:10.1007/s13593-018-0504-8.

Rehman, A., M. Farooq, A. Nawaz, et al. 2018b. Characterizing bread wheat genotypes of Pakistani origin for grain zinc biofortification potential. *Journal of the Science of Food and Agriculture* 98:4824–4836.

Rerkasem, B., and S. Jamjod. 1997. Boron deficiency induced male sterility in wheat (*Triticum aestivum* L.) and implications for plant breeding. *Euphytica* 96:257–262.

Savithri, P., R. Perumal, R. Nagarajan. 1999. Soil and crop management technologies for enhancing rice production under micronutrient constraints. In: Balasubramanian, V., et al. (Ed.), *Resource Management in Rice Systems: Nutrients*. Kluwer Academic, Dordrecht, p. 121–135.

Shorrocks, V. M. 1997. The occurrence and correction of boron deficiency. *Plant and Soil* 193:121–148.

Shorrocks, V. M., and D. D. Nicholson. 1980. The influence of boron deficiency on fruit quality. p. 103–108. In: Atkinson, D., et al. (Ed.), *Mineral Nutrition of Fruit Trees*. Butterworth, London.

Subedi, K. D., P. J. Gregory, R. J. Summerfield, and M. J. Gooding. 1998. Cold temperature and boron deficiency caused grain set failure in spring wheat (*Triticum aestivum* L.). *Field Crops Research* 57:277–288.

Vlamis, J., and A. Ulrich. 1971. Boron nutrition in the growth and sugar content of sugar beet. *Journal of the American Society of Sugar Beet Technologists* 16:428–439.

Wang, Y., L. Shi, X. Cao, and F. Xu. 2007. Plant boron nutrition and boron fertilization in China. p. 93–101. Xu, F. et al. (Ed.), In: *Advances in Plant and Animal Boron Nutrition*. Springer Publishers, Dordrecht.

Yoshida, S., and A. Tanaka. 1969. Zinc deficiency of rice in calcareous soils. *Soil Science and Plant Nutrition* 15:75–80.

Zia, M. H., I. Ahmed, E. H. Bailey, et al. 2020. Site-specific factors influence the field performance of a Zn-biofortified wheat variety. *Frontiers in Sustainable Food Systems Special Issue*. doi:10.3389/fsufs.2020.00135.

Zou, C. Q., Y. Q. Zhang, A. Rashid, et al. 2012. Biofortification of wheat with zinc through zinc fertilization in seven countries. *Plant and Soil* 361:43–55.

7 Micronutrient Fertilizer Sources, Availability, and Quality in Pakistan

7.1 INTRODUCTION

Amelioration of micronutrient deficiency disorders in crop plants is predominantly accomplished by applying their fertilizers. This, in turn, depends on the adequate availability of good-quality fertilizer products in the affected crop areas at required times. However, in many developing countries, like Pakistan, micronutrient fertilizer availability in time and space as well as the quality of such fertilizer products remain problematic. Therefore, the availability and quality aspects of micronutrient fertilizers in the country are discussed in this chapter. An attempt has also been made to assess the market volume of predominant micronutrient fertilizer products.

7.2 MICRONUTRIENT FERTILIZER SOURCES

Micronutrient fertilizers can be grouped broadly into two categories: i) inorganic salts and ii) chelates. Within each category, there are single as well as multi-micronutrient fertilizers. The commonly used "single-nutrient" micronutrient fertilizers are listed in Table 7.1, and some important conversion factors for micronutrient fertilizers are given in **Annex 18**.

7.2.1 INORGANIC MICRONUTRIENT SALTS

This is the most widely used type of micronutrient fertilizer. The most common ones are sulfate salts (i.e., zinc sulfate, copper sulfate, ferrous sulfate, etc.), which contain sulfate along with the specific micronutrient. Fertilizers such as borax, boric acid, *Solubor*, and *Granubor* are also inorganic salts. *Solubor* and *Granubor* are readily soluble; *Solubor* is used for foliar sprays and *Granubor* for soil application. Boric acid can also be solubilized, with some difficulty, and is used for foliar sprays; however, borax is sparingly soluble and is suitable for soil application. Another category of inorganic salts, which are less soluble, are materials such as zinc oxide.

For foliar sprays, the soluble inorganic salts of micronutrients are generally as effective as synthetic chelates and, thus, are a preferred choice because of lower costs. Only a few studies in other countries, however, have reported the relative superiority of chelated forms in the case of foliar sprays, e.g., for Fe (Basiouny et al., 1970) and Zn (Hsu, 1986; Shazly, 1986; Correia et al., 2008; Karak et al., 2006). Contrarily, in multi-year field experiments on Zn nutrition of wheat in

DOI: 10.1201/9781003314226-7

TABLE 7.1

Common Single Nutrient/"Straight" Micronutrient Fertilizer Products

Source	Formula	Micronutrient Concentration
Zinc		
Zinc sulfate	$ZnSO_4.H_2O$	35% Zn
	$ZnSO_4.7H_2O$	22% Zn
Zinc chelates	Na_2Zn-EDTA	12–14% Zn
Zinc oxide	ZnO	78% Zn
Boron		
Borax (Fertilizer borate)	$Na_2B_4O_7.10H_2O$	11% B
Boric acid	H_3BO_3	17% B
Solubor	$Na_2B_8O_{13}.4H_2O$	20.9% B
Granubor Natur	$Na_2B_4O_7.5H_2O$	15% B
Iron		
Ferrous sulfate	$FeSO_4.7H_2O$	20% Fe
	$FeSO_4.H_2O$	33% Fe
Sequestrene	$NaFe$-EDDHA	6% Fe
Iron chelate	Fe-EDTA	12% Fe
Copper		
Copper sulfate	$CuSO_4.5H_2O$	25% Cu
	$CuSO_4.H_2O$	35% Cu
Copper chelate	Na_2Cu EDTA	13% Cu
	$NaCu$ EDTA	9% Cu
Manganese		
Manganese sulfate	$MnSO_4.xH_2O$	24–30% Mn
Manganese chelate	Mn-EDTA	5–14% Mn

Modified from Rashid, 1996a.

Pakistan and many other countries, inorganic $ZnSO_4$ fertilizer was as effective in curing Zn deficiency as chelated Zn products (I. Cakmak, per. com).

7.2.2 Chelated Micronutrient Fertilizers

True chelates are compounds containing ligands that can combine with a single metal ion (e.g., Zn^{2+}) to form a well-defined, relatively stable cyclic structure called a chelation complex. These fertilizer properties are particularly important and useful in agricultural regions with basic (i.e., high pH) and/or calcareous soils. Well-known chelating compounds include ethylenediaminetetraacetic acid (EDTA), diethylenetriaminepentaacetic acid (DTPA), (N-(hydroxyethyl)-ethylenediaminetriacetic acid (HEDTA), ethylenediaminedi(o-hydroxy-p-methylphenylacetic) acid (EDDHMA), ethylenediaminedi(2-hydroxy-5-sulfophenylacetic) acid (EDDHSA), ethylenediaminedi(5-carboxy-2-hydroxyphenylacetic) acid (EDDCHA), ethylenediaminedisuccinic acid (S,S-EDDS), (N,N'-bis(2-hydroxyphenyl) ethylenediamine-N,N'-diacetic acid) (HBED), imidodisuccinic acid (IDHA), ethylenediamine-N,N'-bis(2-hydroxyphenylacetic) acid (o,o-EDDHA), and polyhydroxyphenylcarboxylate (PHP). In the chelated form, metal(s) ions are less likely to react with and be immobilized by the soil and are more likely to be "delivered" to plant roots. For example, newly identified Fe-HBED chelates (Bolikel® XP and Dissolvine®X60) being marketed by Nouryon, the Netherlands, are 10,000 times more stable in alkaline–calcareous soils than that of the already-known most stable Fe-chelate in such soil environments, i.e., Fe-EDDHA. Fe-HBED chelates Bolikel® XP and Dissolvine®X60 contain 6% water-soluble iron (Fe) and have an equally high ortho-ortho level of 6% Fe[o,o]HBED, which is unique. Due to the good longevity and the high ortho-ortho level, the dose rates are also lower, compared to a standard Fe-EDDHA 4.8% ortho-ortho product (https://micronutrients.nouryon.com/products/fe/fe-hbed/ Accessed on February 23, 2020).

Some micronutrient fertilizer products are called "organic chelates" but actually are organically complexed micronutrient sources and formed by reacting metallic salts with various organic, industrial by-products (e.g., by-products of the wood pulp industry; sucrose-type materials like cane sugar molasses). The structure of these by-products is not well-defined (hence, the term "complexes"), and there is no evidence that the resulting product has true chelate structure or properties.

True chelate compounds might perform differently when bound with a particular micronutrient, like Fe, Zn, Cu, etc. For example, Zn-DTPA is slightly superior to Zn-EDTA in soils with pH $\geq$8.0; however, Zn-EDDHA is not as stable in the calcareous soil environment as Fe-EDDHA is. Similarly, Mn-EDTA is relatively a weak combination. Manganese might be lost in several soil types due to exchange with other metals like Fe or Cu. Soil application (with the right chelate) is preferred compared with foliar feeding in the case of Fe only. Especially, the Fe-HBED is extremely efficient in high pH soils, at very low dose rates. Similarly, in high pH, calcareous soils, Fe-EDDHA (6% Fe) (marketed under the trade name of *Sequestrene*-138 by BASF Europe) is more stable than Fe-EDTA and Fe-DTPA. In Israel, for alleviating Fe deficiency in mango orchards, Fe-*Sequestrene* is recommended through fertigation @ 1–5 mg L^{-1} (Kadman and Gazit, 1984).

Manufacturers of *chelate fertilizers* generally claim a 10:1 advantage compared to their inorganic sources. However, most of the research work has depicted that there is approximately a 3:1 to 5:1 advantage for Zn-EDTA, a "true" chelate. In most scientific trials, Zn-EDTA has been observed to be at least five times more effective than non-chelated or complexed sources of Zn.

7.3   SPECIFICATIONS FOR STANDARD MICRONUTRIENT FERTILIZERS

Specifications of the salient standard micronutrient fertilizer products, duly approved by the Government of Punjab (through Rapid Soil Fertility Survey and Soil Testing Institute (RSF&STI), Lahore), are given in **Annex 29**. Most of these comply with international standards set for micronutrient fertilizers. To ensure quality products in the market, the regulatory authorities randomly sample the marketed fertilizer products for laboratory analysis and then match the nutrient composition with the specified standards.

7.4 MICRONUTRIENT FERTILIZER PRODUCTS AVAILABLE IN PAKISTAN

In Pakistan, many micronutrient sources are produced/imported and/or marketed for agricultural use as well as for non-agriculture purposes; some of these products have been used in research studies as micronutrient nutrition of crop plants. For example, commercial-grade borax (11% B; mostly imported from Turkey and China) and boric acid (17% B; imported from Italy and Chile) have proven to be good sources of B for cotton (Rashid, 1996b; Rashid and Rafique, 1997). Much earlier, both sources of B also used to be imported from the US. Whereas borax is used by the ceramics and pesticides industries, boric acid is primarily consumed by the pharmaceutical industry. Similarly, commercial-grade ZnO (a by-product of a local galvanizing industry), $ZnSO_4$ (mainly imported from China), copper sulfate, and manganese sulfate, imported or produced for other purposes, have been used in rice and wheat field experiments.

Because of the multi-purpose usage of the micronutrient products, i.e., agricultural as well as non-agricultural, it is difficult to assess the correct volume of such products being imported for fertilizer purpose alone. Moreover, importers also mis-declare true usage of the micronutrient products, to save customs duties/taxes. As per the data available from the Pakistan Customs Department, 4274 tonnes of borax and 5094 tonnes of boric acid were imported during the fiscal year 2020–21. However, it is difficult to ascertain how much of these products were marketed by the fertilizer players, since these B chemicals are also used in ceramics, pesticides, and the pharmaceutical industry. During the same period, 6453 tonnes of zinc sulfate were imported which, alongside its usage as a fertilizer, were also used in animal feed. Such import data pertains to the ports established at Karachi, Lahore, Islamabad, and Peshawar; however, such data from Pak-China Border Port (Sost Port) is not available. The import tonnage for ferrous sulfate and manganese sulfate stood at 2714, and 2361, respectively, for the financial year 2020–21.

Considering both granular and liquid Zn formulations, total annual Zn fertilizer availability in the local market is crudely estimated at 7300 tonnes (of $ZnSO_4$. H_2O equivalent). Similarly, the annual B fertilizer market is crudely estimated at 907 tonnes (of various products of B). The major chunk of micronutrient fertilizers are marketed in Punjab province, followed by Sindh, KP, and Balochistan. This pattern is well correlated with the agricultural development across the country as the Punjab province is the lead. It is estimated that about 50% of the country's Zn fertilizer market (i.e., 3565 tonnes) is occupied by two fertilizer giants, i.e., Fauji Fertilizer Company Limited and Engro Fertilizers. Out of the 3565 tonnes marketed during the year 2015 by the two fertilizer giants, 63% was sold in Punjab, 29% in Sindh, 6% in KP, and 2% in Balochistan. As 66% of the total market is in Punjab, this province has developed strict regulatory measures for manufacturers, importers, and distributors of micronutrient fertilizer products. Compared to Punjab, other provinces are far behind in the implementation of regulatory measures to ensure good quality of the marketed micronutrient fertilizers. District-wise sale/offtake of Zn fertilizers made by the two fertilizer giants for the year 2015 has been depicted in the country map (Figure 7.1).

Crude estimates for various kinds of micronutrient fertilizer products available in Pakistan have been based on our discussions with individuals of different companies working in the fertilizer arena. These estimates for non-chelated $ZnSO_4$ are given in **Annex 19**, for chelated $ZnSO_4$ in **Annex 20**, for B fertilizer products in **Annex 21**, and for micronutrient complexes in **Annex 22**.

For Punjab, detailed structured datasets regarding micronutrient fertilizer products are available (online) with RSF&STI, Ayub Agricultural Research Institute (AARI), Ministry of Agriculture, Government of Punjab. The listed companies are either manufacturers or importers/distributors of micronutrient fertilizers. All these companies are allowed to market their products in Punjab as well as in other provinces by the respective Provincial Agriculture Departments. In the following sub-sections, summarized information is given about fertilizer products registered with RSF &STI, category-wise.

7.4.1 LOCALLY PRODUCED ZINC SULFATE FROM ZINC DUST

A by-product of galvanizing industry, i.e., Zn dust, is the main raw material for the local manufacture of zinc sulfate fertilizer ($ZnSO_4.7H_2O$, Zn 21%) in Pakistan. Zinc ash (also called zinc dross) is mainly collected by the industry from galvanizing industry, dry cell industry, and die casting industry. The contribution of the Gujranwala industry toward the Zn ash market in the country is about 40%, whereas the remaining 60% Zn ash supply comes from the related industry in Karachi. There are about 15 bulk suppliers of Zn ash in Gujranwala, each supplying 4500–5000 tonnes of Zn ash, every year. Thus, the annual Zn ash supply from Gujranwala alone is around 75,000 tonnes. Theoretically, 65 kg of 100% pure zinc ash reacted with 98 liters of standard sulfuric acid would yield about 161 kg of zinc sulfate ($ZnSO_4.7H_2O$) fertilizer. In India, the requirement of Zn ash (65–70% Zn) and sulfuric acid (specific gravity 1.85) for producing one tonne of zinc sulfate is 360 kg each (Singh, 1980). However, the actual volume of sulfuric acid required for the manufacture of commercial-grade zinc sulfate fertilizer (Zn 21%) depends upon the purity of locally available Zn-ash as well as the impurities (like Fe and Cu) it contains. To utilize local ash in an efficient way, the fertilizer industry may consider the feasibility of high-analysis zinc fertilizer ($ZnSO_4.H_2O$ ~ Zn 33%) manufacture after the extraction of zinc metal from the ash while getting rid of other metal impurities. At the same time, to minimize the zinc ash import bill, the government in support of fertilizer and galvanizing industry may also consider beneficiation from lead (galena)-zinc (sphalerite) ores in Besham, Chuddar, and Dudder districts of KP and Balochistan.

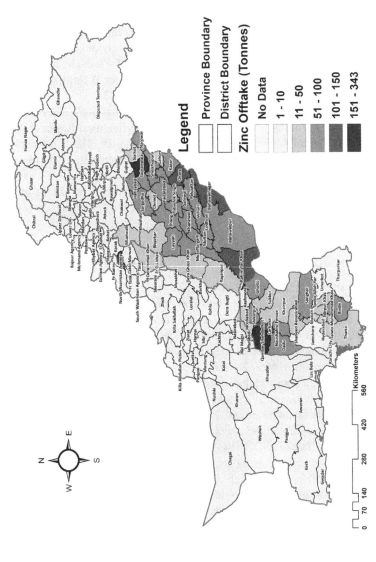

FIGURE 7.1 District-wise sale/offtake (tonnes per annum) of zinc fertilizer ($ZnSO_4.H_2O$, 33% Zn) in Pakistan. (Based on the FFC and Engro Fertilizers data, which cumulatively hold ~50% Zn fertilizer market share in the country; Courtesy of M. Iqbal, SUPARCO Pakistan.)

Elemental Zn concentration in the Zn ash, available as waste from local industry, varies from 40 to 75%. As per the 2018 assessment, for the Zn ash containing up to 40% Zn, the bulk sale price per kg of ash is PKR 1.00 per each %age of Zn, i.e., for the ash having 40% Zn, the price is PKR 40 per kg. However, for the Zn ash having more than 40% Zn, the price has no linear relationship with its Zn %age. For example, for the 60% Zn containing ash, the price is PKR 70 per kg, and for the 65% Zn containing ash, the price is PKR 75 per kg. If Zn concentration is more than 80% Zn, then the selling price of the Zn ash is approximately in the range of PKR 100–125 per kg. Due to economic reasons, medium-quality Zn ash, with a Zn concentration of 55–60%, is used for the local manufacture of zinc sulfate fertilizer; and such fertilizer is mainly manufactured in the Punjab province. To meet the analysis requirement, the manufacturers generally mix a certain volume of imported zinc ash (containing 90% Zn) with that of locally available medium-quality ash (containing 55–60% Zn).

Lack of finances, non-synchronization of the time of availability of Zn ash with need-based seasonal manufacture of $ZnSO_4$ (needed more during *Kharif* season), poor promotional activities, and unfair trade practices by a few (e.g., diversion of Zn ash meant for the production of $ZnSO_4$ for extraction of remunerative Zn metal) all collectively call for review of the Government of Pakistan fertilizer policy; assigning of zinc sulfate production to the organized sector would be a corrective measure.

7.4.2 MANUFACTURERS-CUM-DISTRIBUTORS OF MICRONUTRIENT FERTILIZERS

Under this category, a total of 114 companies manufacture a total of 312 products (Zn and B fertilizers or multi-micronutrient complexes). The multi-micronutrient complexes contain Zn and B as well as Fe, Cu, Mn, Mo, and Cl. All the micronutrient fertilizers products registered with Soil Fertility Research Institute, Government of Punjab, are listed in **Annex 30**. However, none of the 114 companies registered with Soil Fertility Research Institute, manufacture exclusive Fe, Cu, Mn, Mo, or Cl fertilizers in Pakistan. Out of the 114 companies, only three companies manufacture chelated micronutrient (Zn only) fertilizers. Two of these, i.e., LT Enterprises, Hattar, KP, and Kanzo Ag-Multan, Punjab, locally manufacture chelated Zn (Zn 5%) in solid form, whereas Pak-China Chemicals, Lahore, locally manufactures chelated Zn (Zn 6%) in liquid form.

7.4.3 IMPORTERS-CUM-DISTRIBUTORS OF MICRONUTRIENTS

Under this category, a total of 54 companies and importers of a total of 79 products are registered with Soil Fertility Research Institute, Punjab (**Annex 31**). Among the import list are Zn, B, and multi-micronutrient complexes that also include Fe, Cu, Mn, Mo, and Cl. None of the companies import exclusively Fe, Cu, Mn, Mo, or Cl fertilizer products in Pakistan. There are only four brands of

FIGURE 7.2   Zinc fertilizer being mixed by a farmer with N, P, and K fertilizers. (Photos courtesy of FFC; and Engro Fertilizers.)

multi-micronutrient complexes registered under the import category. Only five companies are importing chelated micronutrient fertilizers, as detailed below:

1. *Ch. Khair Din and Sons*: Fe= 4%, Mn= 2% w/w multi-micronutrient chelated solid
2. *R.B. Avari Enterprises (Pvt.) Ltd.*: Chelated Zn = 5% (solid form)
3. *Helb Agro Sciences Pakistan (Pvt.) Ltd.*: Chelated Zn = 5% (solid form)
4. *United Distributors Pakistan Ltd.*: Chelated Zn as Zn-EDTA= 5% (solid form)
5. *Swat Agro Chemicals*: Chelated Zn = 5% (solid form)

Based on the data of two major players in the fertilizer industry, district-wise zinc product (zinc sulfate, Zn 33%) offtake for Pakistan has been assessed in Figure 7.1. The mapping depicts that the highest consumption of zinc fertilizers is in the Punjab and Sindh provinces only. In the recent past, Engro Fertilizers has secured marketing rights for a chelated solid Zn fertilizer product in Pakistan (earlier marketed by Swat Agro Chemicals) under the brand name of *Librel Zinc* (chelated zinc as Zn-EDTA= 5%, manufactured by BASF, Germany; Figure 7.2). It has come to the knowledge of the authors of this book that this micronutrient product contains 14% Zn, but due to regulatory issues with Soil Fertility Research Institute, Government of Punjab, the product has been registered as Zn-EDTA, 5% Zn. This strange situation, if correct, merits reconsideration by the regulatory authorities.

7.4.4   DISTRIBUTORS OF MICRONUTRIENTS

A total of 272 distributor companies under this category purchase their fertilizer products either from registered local manufacturers-cum-distributors or from the registered importers-cum-distributors. Total fertilizer brands being marketed by these distributors are 1440 in number, with a lot of duplication, i.e., a single fertilizer brand is being marketed under different trade names by a number of distributors. Further details about this category of micronutrient distributors can be accessed at the official web page of Soil Fertility Research Institute, Government of Punjab –https://aari.punjab.gov.pk/rsfri_lhr.

7.5   FERTILIZER POLICY AND MICRONUTRIENTS IN PAKISTAN

In Pakistan, a significant yield gap exists between the potential and the actual yield of crops. In addition to other management lapses, inadequate and imbalanced use of fertilizers is responsible of this gap in yields. The inadequate availability of fertilizers and their high cost are the determinants of the imbalanced use of fertilizers. Thus, a sound farmer-friendly fertilizer policy can play a vital role in sustaining agricultural production. Currently, the Fertilizer Policy of 2001 is under implementation in Pakistan.

Since their introduction in Pakistan, fertilizers were heavily subsidized; marketing, imports, and prices were controlled. The public sector played a major role in the production and marketing of fertilizers, and a number of institutions were established for the import and distribution of this crucial farm input. In 1986, the Government of Pakistan initiated the process of market liberalization and deregulation on the one hand and divestiture of public corporations on the other. The subsidy and price controls on N-based fertilizers were abolished in 1986, on P-based fertilizers in 1993, and on K fertilizers in 1995. The fertilizer policy announced in 1989 gave more incentives to the private sectors such as gas supply for fertilizer production on priority/feedstock 10 years price freeze, location-related tax holidays, and machinery/equipment/raw material exempted from duty/sales tax. For phosphates, tariff adjustments were made to ensure that imported Di-ammonium phosphate (DAP) fertilizer cost is below US$ 250 per tonne, in bags.

The 1986 to 2000 period was a success story of deregulation and privatization. The Government of Pakistan pulled out of imports, price control, marketing, and manufacture of fertilizers. The public sector fertilizer manufacturing units were privatized. The first DAP plant was established by Fauji Fertilizer Company (FFC) with an annual capacity of 675,000 tonnes. There were balancing, *modernization*, and revamping (BMR) and additional capacities leading to expansion in urea production capacity. The production capacity reached the level of 5.6 million Mt and the consumption reached 6.2 million Mt. However, the picture regarding the balanced and efficient use of fertilizers remained bleak. Overall, the consumption of fertilizers increased but remained dominated by urea.

Based on the projected demand for fertilizers to increase crop production for attaining food security and economic growth, the Government of Pakistan announced a new Fertilizer Policy in 2001. Most of the incentives of the 1989 Fertilizer Policy were retained. The major incentive in the 2001 policy (for existing and new urea plants) was an adjustment of feedstock prices on concessional rates comparable to the Middle East countries to enable local fertilizer prices to stay below the prices of imported fertilizers.

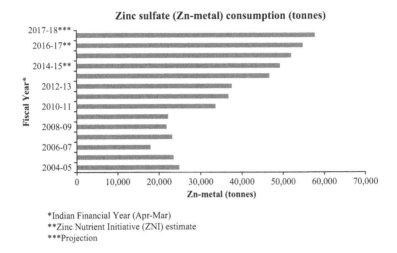

*Indian Financial Year (Apr-Mar)
**Zinc Nutrient Initiative (ZNI) estimate
***Projection

FIGURE 7.3   Impact of the Government of India subsidy on zinc fertilizer consumption in India. (Data source: Fertilizer Association of India)

Field-scale micronutrient deficiencies of economic significance have been established in the case of Zn, B, and Fe in Pakistan (Rashid et al., 2005, Rashid, 2006) but the Government of Pakistan's fertilizer policy still lacks emphasis on the use of these micronutrient fertilizers. Considerable positive crop responses to micronutrient fertilizers in terms of improvements in crop productivity, produce quality, and economic returns are now well-established in the Indo-Pak region. In India, government fixes minimum support prices for the main crops, controls the farm price of urea, and issues indicative selling prices of other fertilizers. The prices of fertilizers in India are heavily subsidized. The increased subsidy on micronutrient fertilizers, particularly on Zn fertilizer, has been instrumental in the accelerated use of Zn (Figure 7.3) across India. Consequently, Indian farmers pay subsidized prices for fertilizers and also receive reasonable prices for their crop produce, which make the use of fertilizers attractive and remunerative.

Contrary to India, in Pakistan no such incentives are being offered to farmers. Unfortunately, currently, there is no policy regarding promoting the use of micronutrients. Nevertheless, given the potential impact of Zn fertilizer use on public health at the macro-economic scale, including reduced healthcare costs and improved crop productivity, especially of the staple cereal (wheat), there seems to be a strong case for the Government of Pakistan to consider a subsidy program for Zn fertilizers (Joy et al., 2017) across the country. The government's interventions through subsidy and/or other means in promoting the use of other micronutrients, especially B and Fe, across different crop production regions would also certainly help in getting attractive economic returns.

In short, micronutrient fertilizers' quality and availability in time and place warrant improvements. For instance, the price per unit of micronutrient in commercial-grade products meant for non-agriculture uses is relatively high because of the Customs Duty. As mentioned above, currently the Fertilizer Policy of 2001 is under implementation in Pakistan. It lacks farmers' perspective and any mechanism for promoting the use of micronutrient fertilizers. As several variables have changed since 2001, there is a strong case for a review of the policy. At the time of writing this book, the Punjab Government is working on revising its Agriculture Policy for the province. The first draft of the policy, available with the authors, suggests that the Government of Punjab has recommended extending N, P, and K fertilizers' subsidies to micronutrient fertilizers as well. However, no such progress has been observed till the publication of this book, except for the rice crop under a five-year rice productivity enhancement project.

7.6 MICRONUTRIENT FERTILIZER QUALITY ISSUES

7.6.1 MICRONUTRIENT FERTILIZER QUALITY

The poor quality of fertilizers in Pakistan is not restricted to major nutrient fertilizers, like phosphatic and potassic fertilizer products. In the absence of adequate supply/marketing of a wide range of micronutrient fertilizers by the major players in the fertilizer industry (i.e., Fauji Fertilizer Company Limited, Engro Fertilizers, and Fatima Fertilizer Company), micronutrients are predominantly marketed by small vendors including the generic pesticide companies to fill in the demand gap. As a result, the quality issue persists in the case of micronutrients.

For one thing, the regulatory agencies in the provinces have registered too many micronutrient fertilizer products, irrespective of the requisite capability of the manufacturing and/or marketing entities. Consequently, a number of micronutrient fertilizers keep appearing in the local market from time to time. These include multi-nutrient formulations, inorganic salts as well as chelates, materials meant for soil application and foliar spray, and very well-described products to shabbily-presented ones. In packing, they come from a few gram pouches to a few kg bags or small plastic bottles to jerry cans. A few manufacturers/sellers provide reasonable technical information on the use of their products, while others hardly appreciate even the need, functions, and consequences of indiscriminate use of their products. Surprisingly, many such products are marketed as "chelated micronutrients" to fleece innocent farmers without consideration to the need or advantage of using these expensive products rather than inorganic micronutrient products which can bring the same or even better returns at a much lower cost. During the 1990s, it was observed that while a few micronutrient products contained the claimed composition, many of such products were, in fact, adulterated or totally fakes (NFDC, 1998). For instance, most samples of $ZnSO_4$ collected from the rice-growing areas of the Punjab province during 1990s were either adulterated or totally fake. Also, samples of $ZnSO_4$ collected from the rice-growing areas of Sindh during the 1992 rice season were either fake (containing 0.011.0% Zn) or badly adulterated (**Annex 23**; NFDC, 1998).

The classical example of adulteration in micronutrient fertilizers pertains to two samples collected in 1987 from the rice-growing area of Pasrur, Sialkot.

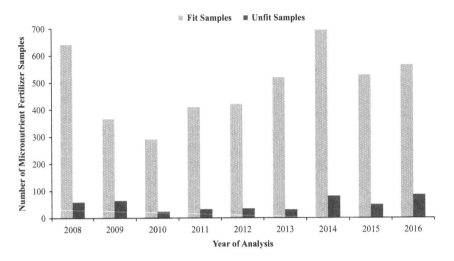

FIGURE 7.4   Micronutrient fertilizers analysis (quality) data. (data source: Soil Fertility Research Institute (SFRI), Government of Punjab.)

One of these so-called "micronutrient fertilizers", named *"Heera* Fertilizer", contained 0.01% Zn, 0.08% Cu, 0.03% Fe, and 0.73% Mn. Similarly, the other product *"Tiger* Micronutrient" contained 0.75% Zn, 028% Cu, 0.27% Fe, and 0.75% Mn. Many other "micronutrient fertilizers", like "Plantex", also contained negligible micronutrient contents. In research trials, the so-called Miracle Fertilizers claimed to contain micronutrients, yet they generally proved ineffective in addressing the micronutrient deficiencies. For example, Khattak (1988) observed that the micronutrient *"Zarzameen"* fertilizer product applied to potato @ 2.0 kg ha⁻¹ at Nowshehra proved totally ineffective.

The current situation regarding micronutrient quality issue is reported to have improved. For example, in the Punjab province, over a nine-year period (2008–16) a total of 4892 samples of micronutrient fertilizers were analyzed at the central laboratory of Soil Fertility Research Institute, Lahore. Only 9.3% of the total samples (i.e., 455 samples) were declared unfit (Figure 7.4), a percentage that is far lower than the one observed during an earlier period of 1990–98 (NFDC, 1998). This indicates strict regulatory control of the Punjab Government's Agriculture Extension Department regarding fertilizer quality over the reported period. In recent times, the Agriculture Extension Department in the Punjab province has increased the targets for fertilizer samples collection. This may also be a contributing factor in ensuring better quality of the fertilizer products including micronutrients being marketed. Nevertheless, the situation of micronutrient fertilizer quality in Sindh province is deplorable. It is no secret that small vendors marketing micronutrient fertilizer products in this province are involved in malpractices without punishment from regulatory authorities. The situation is equally worrisome in the KP and Balochistan provinces.

Although Fertilizer Control Order is also under implementation in KP and Balochistan, at least on paper, no fertilizer quality control data are available for these provinces. A fertilizer dealer in Quetta revealed that Fe-*Sequestrene* smuggling from neighboring Iran had ceased since 2014 due to tight border control measures, and, thus, at present, almost all of the Fe-*Sequestrene* being marketed in Balochistan is a counterfeit. This situation indicates the non-existence of a fertilizer quality control mechanism in Balochistan province.

The laboratory methodology adopted for analysis of certain solid micronutrient fertilizers, especially for Zn analysis, is sometimes challenged by the fertilizer companies. For instance, this complaint is quite common in the case of imported $ZnSO_4.H_2O$ fertilizer marketed by many multi-national companies like ICI, Syngenta, and others. The samples of imported $ZnSO_4.H_2O$ fertilizer collected by Agricultural Extension staff and analyzed in the laboratories of SFRI are reported to be unfit as these contain ~25–29% water-soluble Zn. However, after detailed meetings of the fertilizer industry officials with those of SFRI Lahore-based officials, it was agreed that while analyzing a zinc sulfate fertilizer sample, a standard zinc sulfate monohydrate (AR grade salt) will also be analyzed, simultaneously in addition to standard solutions of the salt. In case, standard zinc salt depicts 34% Zn in it compared to the expected value of 35.6%, this deficient value advantage would also be credited to the fertilizer test sample under analysis. Due to such complexities, many companies were forced to market their $ZnSO_4.H_2O$ fertilizer (containing 33% Zn) as $ZnSO_4$ fertilizer with 21% Zn. To resolve the situation, the Agriculture Department, Government of Punjab, has notified a new grade of zinc sulfate (Zn 27%) so that a cushion could be provided to fertilizer companies for the marketing of un-fit $ZnSO_4.H_2O$ fertilizer (containing 33% Zn) as $ZnSO_4$ fertilizer with 27% Zn. At the same time, the fertilizer marketing companies in the country also need to remain cautious while importing $ZnSO_4.H_2O$ fertilizer (containing 33% Zn) from the countries where a few exporters mix zinc oxysulfates (a mixture of $ZnSO_4$ and ZnO produced after partial acidification of ZnO with sulfuric acid) into this standard fertilizer product. The solubility of Zn-oxysulfates largely depends on the proportion of $ZnSO_4$ (highly water-soluble) and ZnO (almost water-insoluble). Thus, such adulteration by the exporters at the time of bulk shipment translates into lower water solubility of the zinc sulfate fertilizer product at fertilizer quality testing labs in the country.

It is also to be emphasized that the provincial governments' soil, water, plant, and fertilizer analysis laboratories capable of analyzing micronutrients need to upgrade their laboratory procedures for attaining quality analysis. A co-author of this book submitted two soil samples (after dividing the same well-mixed soil into two halves) to two government laboratories of a province for Zn and Fe analysis. The DTPA-extractable Zn reported by the two laboratories varied drastically. The concentration of Zn in the first soil sample was reported to be 0.64 mg kg^{-1} by Lab no. 1 and 1.2 mg kg^{-1} by Lab no. 2. Concentration of Zn in the second soil sample was reported to be 0.59 mg kg^{-1} by Lab no. 1 and 1.2 mg kg^{-1} by Lab no. 2. This depicts almost 100% variation in Zn analysis for the same soil sample by two laboratories located in different cities but working under the same authority. For information of the readers, for DTPA-extractable Fe, the values were 0.78 mg kg^{-1} and 3.0 mg kg^{-1} for the first soil sample and 1.2 mg kg^{-1} and 4.4 mg kg^{-1} for the second soil sample. Thus, the reported Fe values depicted >300% variation for the same soil sample that was divided into two halves, intentionally. However, a significant improvement in analytical quality is expected across the SFRI labs in Punjab after the completion of ISO standardization in the recent past and following of standardized test method protocols.

Similarly, when a co-author of this book submitted wheat grain samples from three locations for Zn analysis to a federal ISO-certified laboratory, the total Zn was about 18 mg kg^{-1}. Upon suspicion, the same wheat grain samples were dispatched to a UK Government laboratory where total Zn was found to be 34 mg kg^{-1}. The UK lab used a reference plant material (wheat flour) in all batches for analysis, which is not in practice in the laboratories in Pakistan. The analytical laboratories in Pakistan hardly use reference materials to assure the quality of the analytical results, obviously because of the high cost of such imported materials. However, to counter higher import costs, local materials can be developed very easily.

In the Punjab province, the staff of the Agricultural Extension Department collects fertilizer samples from the dealer's shop as a routine measure to check adulteration practices. However, analysis of the collected samples at the government laboratory may take up to several weeks time. By the time the laboratory report declares the sample "unfit", the dealer might have already sold out all the fertilizer stock from where the sample was collected. Therefore, there ought to be a regulation by which the dealer must be bound to keep a record of its customers in case it sells out its already-sampled fertilizer product and for which laboratory results are pending so that the customers could be compensated later on in case the product is found to be sub-standard. In the relevant law, there is an imprisonment possibility ranging from three months to a few years for the culprits who market adulterated fertilizers. However, in almost all cases the culprits are slapped only with a fine ranging from PKR 5000 to PKR 20,000. Due to the strict vigilance of Agriculture Extension officials in the Punjab province, the business of fake fertilizers in this province appears to be on the decline. In Balochistan and KP provinces, fertilizer analysis/quality data is non-existent, whereas in Sindh such data record was not shared by the regulators.

At present, the micronutrient fertilizer testing laboratories are functional at divisional headquarter level in the Punjab province. This analytical facility should be extended to the district level. The analysis facility in Sindh is limited to only two laboratories across the province, and regulators are lenient to fertilizer dealers, so samples are rarely collected and analyzed. In Balochistan and KP provinces, equipment for the laboratories are not available to monitor the quality of the micronutrient fertilizers.

Another aspect regarding the marketing of quality micronutrient fertilizers is the low literacy level among the farming community which is exploited by fertilizer dealers. A co-author of this book experienced this in a central Punjab market when he demanded a chelated Zn (EDTA-zinc) product. Instead, the dealer first presented some product which was labeled "Sea Weed Extract – Bio-stimulant". Upon pointing this out, the dealer, while apologizing, gave another product that too was not chelated Zn; rather, it was some organic chelate (not a true chelate). Upon pointing out this fact, the dealer realized that he was facing a literate customer; so, he had no option but to present a genuine product in the third instance. Experience indicates that the fertilizer dealers get lucrative commissions from generic companies for selling out their sub-standard products, and deceive their customers for financial greed.

REFERENCES

Basiouny, F. M., C. D. Leonard, and R. H. Biggs. 1970. Comparison of different iron formulations for effectiveness in correcting iron chlorosis in citrus. *Proceedings of the Florida State Horticultural Society* 83:1–6.

Correia, M. A. R., R. D. Prado, L. S. Collier, D. E. Rosane, and L. M. Romualdo. 2008. Zinc forms of application in the nutrition and the initial growth of the culture of the rice. *Bioscience Journal* 24:1–7.

Hsu, H.H. 1986. The absorption and distribution of metalosates from foliar fertilization. In: Ashmead, H. De W (Ed.). *Foliar Feeding of Plants with Amino Acid Chelates*. Noyes Publications, Park Ridge, NJ. p. 236–354.

Joy, E. J. M., W. Ahmad, M. H. Zia, et al. 2017. Valuing increased zinc (Zn) fertilizer-use in Pakistan. *Plant and Soil* 411:139–150.

Kadman, A., and S. Gazit. 1984. The problem of iron deficiency in mango trees and experiments to cure it in Israel. *Journal of Plant Nutrition* 7:283–290.

Karak, T., D. K. Das, and D. Maiti. 2006. Yield and zinc uptake in rice (*Oryza sativa*) as influenced by sources and times of zinc application. *Indian Journal of Agricultural Sciences* 76:346–348.

Khattak, J. K. 1988. Cooperative Research Program on Micronutrient Status of Pakistan Soils and Its Role in Crop Production. Annual Report, 1987–88. NWFP Agricultural University, Peshawar.

NFDC. 1998. *Micronutrients in Agriculture: Pakistan perspective*. National Fertilizer Development Center, Islamabad, 51 pp.

Rashid, A. 1996a. Secondary and micronutrients. p. 341–385, Chapter 12, In: *Soil Science*, Rashid, A., and Memon, K. S. (Managing Authors). National Book Foundation, Islamabad, Pakistan.

Rashid, A. 1996b. Nutrient Indexing of Cotton in Multan District and Boron and Zinc Nutrition of Cotton. Micronutrients Project, Annual Report, 1994–95. National Agricultural Research Center, Islamabad. 76 pp.

Rashid, A. 2006. Boron Deficiency in Soils and Crops of Pakistan: Diagnosis and Management. Pakistan Agricultural Research Council, Islamabad, Pakistan, viii+34 pp. ISBN: 969-409-184-5.

Rashid, A., S. Muhammad, and E. Rafique. 2005. Rice and wheat genotypic variation in boron use efficiency. Soil and Environment 24:98–102.

Rashid, A., and E. Rafique. 1997. Nutrient Indexing and Micronutrient Nutrition of Cotton and Genetic Variability in Rapeseed-Mustard and Rice to Boron Deficiency. Micronutrient/ Nutrient Management in cotton in Relation to CLCV. Annual Project Report 1995–96. NARC, Islamabad. 111 pp.

Shazly, S. A. 1986. The effect of amino acid chelated minerals in correcting mineral deficiencies and increasing fruit production in Egypt. In: Foliar Feeding of Plants with Amino Acid Chelates. In: Ashmead, H. De W. (Ed). Noyes Pub., Park Ridge, NJ. p. 236–254.

Singh, R. 1980. In: *Technical Papers Seminar on Zinc Wastes and their Utilization*. Indian Lead Zinc Information Centre and FAI, New Delhi, p. 1.46–1.58.

8 Micronutrient Fertilizer Use Efficiency, Residual Effect, and Soil Balances

8.1 INTRODUCTION

Micronutrients in the soil are present in several forms. The fractions in the soil solution are immediately available to plant roots (Tandon, 2013). The fractions occupying the soils' exchange complex and adsorbed on surfaces of soil constituents constitute the "available pool" which replenishes the soil solution. The fractions present in inorganic compounds, on dissolution, become plant available. Many micronutrients can become occluded in the mineral form and tie-up fertilizer micronutrients more so than soil organic matter (SOM). In general, healthy soils with high levels of SOM are not deficient in most micronutrients. The fractions which are present in SOM and in living organisms also become plant available on decomposition and death of microbes. The fractions present in secondary and primary minerals serve as potential medium- to long-term sources of available forms (Fageria et al., 2002; Ahmad et al., 2012; Ryan et al., 2013). When added to soil, inorganic micronutrient salts dissolve in the soil solution and react with the organic and mineral soil constituents. As a consequence, only a small fraction of the added micronutrients remain in the soil solution under most conditions, but these do not have to undergo any major chemical changes to furnish the forms in which the plant roots absorb them. The more mobile ions, such as sulfate, borate, and chloride, can also be leached down with drainage water.

8.2 MICRONUTRIENT FERTILIZER USE EFFICIENCY

Fertilizer use efficiency (FUE) is defined in many different ways. From an agronomic point of view, FUE means the amount of crop produce increased by the application of one unit of fertilizer nutrient/micronutrient. Soil scientists consider FUE as the %age recovery of the applied fertilizer nutrient/micronutrient in the harvested crop produce. And economists define FUE in terms of rupees returned for each rupee spent on fertilizer, i.e., in terms of value:cost ratio (VCR) or benefit:cost ratio (BCR). For the purpose of discussing residual and cumulative effects of soil-applied fertilizer micronutrients, in this section FUE will be dealt with from soil scientists' perspective, i.e., %age recovery of the applied fertilizer nutrient/micronutrient in the harvested crop produce.

DOI: 10.1201/9781003314226-8

Plants meet most of their micronutrient needs by absorption through roots from the soil solution. The absorption of micronutrients by crop roots depends on soil properties (governing soil solution concentrations of micronutrients), the crop and its variety, crop growth conditions, yield levels, etc.

Researchers in Pakistan have rarely reported on micronutrient FUE by various crops/cropping systems. To our knowledge, the only local research information on this aspect pertains to cotton–wheat and rice–wheat cropping systems generated by the NARC scientists. According to Rafique et al. (2012), in a five-year field study on two benchmark soils of the cotton belt in Punjab, fertilizer Zn use efficiency of the cotton–wheat system was 1.78–2.36% of the annually applied 5.0 kg Znha^{-1}. In cotton–wheat system, Abid et al. (2013) observed a similar residual and cumulative effect of Zn fertilizer. In the rice–wheat system, Hussain (2006) observed fertilizer Zn recovery efficiency of 1.72–2.40% for rice and 1.74–2.68% for wheat, depending on the dose of fertilizer Zn ranging from 2.5 to 5 kg Zn ha^{-1}; at 5 kg Zn ha^{-1}, dose recoveries efficiencies were around 2.5% both for rice and for wheat. In the rice–wheat system of Pakistan as well, residual Zn is applied to wheat crop, crop productivity, and grain quality of rice (Nadeem et al., 2020).

In the rice–wheat system, Hussain (2006) observed a B FUE of 1.78–2.9% for rice crop and much higher use efficiency for wheat crop, i.e., 2.59–4.49%. The B recovery efficiency was inversely proportional to the rate of fertilizer B ranging from 0.5 to 2 kg B ha^{-1}. At 1.0 kg B ha^{-1}, B FUE was 2.88% for rice crop and 4.44% for wheat crop. In numerous field experiments in farmers' fields within rice–wheat system of Punjab, Rashid et al. (2007) observed that the use efficiency of B fertilizer for rice crop varied from 1.78% to 3.40% of the applied B dose (@ 0.75 kg B ha^{-1}). The variation was largely attributed to the uptake efficiency of the rice cultivars; fine-grain basmati cultivars were less efficient in utilizing fertilizer B compared with coarse-grain IRRI-type cultivars. In a five-year permanent layout cotton–wheat study on two bench-mark/predominant soil series of the cotton belt of Pakistan, however, a much higher use efficiency of fertilizer B, i.e., 6.2–10.2% (1.8–2.8% for cotton), of the annually applied 1.0 kg B ha^{-1} was observed (Rashid et al. unpublished results).

International literature has listed quite high values for FUE for various micronutrients by various crops. For example, Katyal (1980) stated that fertilizer Zn uptake by the first crop seldom exceeds 10% of the applied fertilizer dose. However, data available in the literature indicates extremely wide variations in uptake of micronutrients by the same crop. For example, B uptake per tonne of potato dry matter varied from 2.1 g (Sakal and Singh, 1995) to 50.0 g (Kanwar and Youngdahl, 1985). Likewise, the figures for B uptake by wheat ranged from 18.0 to 48.0 g per tonne of grain. The variation in the case of B uptake by maize was phenomenal, i.e., 4.0 g B to 129.0 g B per tonne of grain produced (Tandon, 2013). Obviously, the values for FUE varied accordingly. Therefore, the values for FUE reported in the literature must be used with caution.

8.3 RESIDUAL AND CUMULATIVE EFFECT OF SOIL-APPLIED MICRONUTRIENT FERTILIZERS

The utilization of the freshly applied fertilizer micronutrients by the first crop is very low. Therefore, soil-applied micronutrient fertilizers leave a marked beneficial residual effect on subsequent crops grown in the same field (Nadeem et al., 2021). Thus, in micronutrient deficient soil situations, it is not necessary to apply micronutrient fertilizers to every crop in the cropping system. Soil-applied fertilizers of Zn as well as of B leave appreciable residual effects (Rafique et al., 2012; Nadeem et al., 2020, 2021). Therefore, unwanted repeated annual applications of micronutrient fertilizers can result in the build-up of these micronutrients to toxic levels for some of the crop species which are susceptible to the toxicities; this is highly likely in the case of B. Therefore, periodic soil testing is suggested to monitor changes in micronutrient levels of field soils as a result of fertilization. The application rate of the micronutrient(s) should be decreased appropriately, or the application should be ceased till available levels decrease to the response range.

However, the total quantity of un-utilized micronutrient fraction is not fixed irreversibly in the soil. Rather, an appreciable fraction of the fixed portion may be utilized by the subsequent crop(s). As Zn and Cu are immobile in soil, their leaching losses are apparently negligible. Thus, the major part of the fertilizer Zn or Cu remains in the surface soil layer and, thus, produces a residual effect on the succeeding crop(s). For example, Khattak and Perveen (1988), in three field experiments conducted in the Charsadda district of KP, observed that soil-applied micronutrients to maize crop increased grain yield by 18% and of the subsequent wheat crop by 16% over the respective control yields. In a rice–wheat system field experiment conducted by NIAB scientists, 5 kg Zn ha^{-1} applied to rice crop increased its yield by22% and of the subsequent wheat crop by 17% over control. Recovery of the fertilizer Zn was about 9% of the applied dose for rice and about 13% for wheat (Kausar and Tahir, 1994). In India, residual effect of 5.9 kg Zn ha^{-1} applied to a silty-loam soil under rice–wheat system lasted for as long as five rice crops and six wheat crops (Takkar, 1996).

On average, a single application of 5–10 kg Zn ha^{-1} may last for several crop seasons. In Australia, residual effects of zinc fertilizers have been reported for upto five crops. Generally, the higher the dose, the longer the residual effect. Soil characteristics possibly modify the residual effect. On an average, an application of Zn @ 5–6 kg ha^{-1} may remain effective for two or more crop seasons. Long-term field experiments are warranted in major cropping systems and soil types of Pakistan for determining the dose and frequency of Zn application under local agronomic conditions.

For boric acid, the dominant form of B in soil solution, being non-charged is prone to leaching, especially in light-textured soils. Boron fertilizer leaves a longer residual effect in silty and clay soils compared with that on sandy soils. In a long-term field study with oilseed rape in China, Yang et al. (2000) observed a residual effect of 1.65 kg B ha^{-1} for two to three years. Lower solubility materials (like Na-borates) have a more residual effect. In one field experiment in KP, B applied to maize @ 1.1 kg B ha^{-1} increased grain yield of maize by 20% and of subsequent wheat by 6% only over control (Khattak and Perveen, 1988). Thus, B application in medium- to heavy-textured soils @ 1.0 kg ha^{-1} is expected to be effective for two to three crop seasons. Long-term field studies are warranted for determining residual and cumulative effects of B application in local soils.

Iron and Mn do not have significant residual effects in calcareous soil, because the micronutrients are easily oxidized to unavailable forms in such soils. Annual application of Fe and Mn are, therefore, required to correct deficiencies. Foliar sprays, particularly of Fe, generally are more effective than soil application in providing plant-available supplies of the micronutrients.

The first crop fertilized with micronutrients removes only a small fraction of the applied micronutrient dose. Therefore, soil-applied micronutrient fertilizers leave a beneficial residual effect on succeeding crop(s) grown in the same field. This is exemplified by the data pertaining to major cropping systems in the country, i.e., cotton–wheat, maize–potato, rice–wheat, and sugarcane cropping systems (Tables 8.1 and 8.2).

In multi-year, rice field experiments in the Punjab and Sindh provinces, total B uptake in harvested plant parts (i.e., grains and straw) of various prevalent cultivars of rice (i.e., Super Basmati and Basmati-385) was 17.2–35.4 g ha^{-1} without B fertilization and 34.0–69.6 g ha^{-1} with B fertilization (Rashid et al., 2007). In similar multi-year, cotton field experiments in the Punjab province, total B uptake in harvested plant parts of cotton (i.e., lint, cottonseed, bur, and stalks) was 81.0–114.0 g ha^{-1} without B fertilization and 113.0–160.0 g ha^{-1} with B fertilization (Rashid et al. unpublished results).Though the literature indicates that a much higher fraction of the applied B is taken up by the current crop, multi-year, multi-location field experiments in Pakistan in rice–wheat and cotton–wheat cropping

TABLE 8.1
Residual and Cumulative Effect of Zn and B Fertilization in Major Cropping Systems of Pakistan

Cropping System	No. of Field Expts. and Locations	Yield Increase over Control (%)			
		First Crop/ Cumulative		Residual/ Second Crop	
		+ Zn	+ B	+ Zn	+ B
Cotton–wheat	4 nos.; Multan division	~16–20	~19–24	~9–12	~8–14
Maize–potato	3 nos., Sahiwal distt.	~22–28	~12–22	~8–15	~10–14
Rice–wheat	5 nos., Gujranwala and Sheikhupura	5–17	5–14	4–9	3–11
Sugarcane	4 nos., Thatta, Hyderabad	4–25	6–19		

Adapted/modified from Rafique, 2011.

TABLE 8.2

Five-Year Apparent Soil Balances of Zn and B in Irrigated Cotton–wheat System in Punjab, with Farmers' Fertilizer Use Practice (i.e., N, and P Fertilizer Use only)

Soil Series	Soil Texture	Total Input[†] (kg ha⁻¹)	Plant Uptake[‡] (kg ha⁻¹)	Apparent Soil Balance (kg ha⁻¹)
Zinc				
Awagat series	Coarse loamy	1.26	0.61	0.65
Shahpur series	Fine silty	1.41	0.85	0.56
Boron				
Awagat series	Coarse loamy	1.59	0.52	1.07
Shahpur series	Fine silty	1.24	0.75	0.49

[†] Input from irrigation water, rainwater, and crop residue

[‡] Uptake by above-ground plant parts, i.e., cotton stalks, leaves, bur, seeds and lint, and wheat grain and straw

Adapted from Rafique et al., 2012, with permission; and Rashid et al. unpublished results.

systems have revealed that the actual fraction of fertilizer B removed by the harvested plant parts of the first crop was only 1.7–3.4.% of the annually applied 1.0 kg B ha⁻¹ dose to rice crop (Rashid et al., 2007) and 1.8–2.8% of the annually applied 1.0 kg B ha⁻¹ dose to cotton crop; wheat crop following rice or cotton in the respective cropping system was not applied B fertilizer (Rashid et al. unpublished data). Because of irreversible B losses due to soil adsorption/precipitation and/or leaching, however, all the residual fertilizer B does not remain available for utilization by the subsequent crop(s). But the enhanced soil labile B definitely benefits succeeding crop(s) in the rotation (Nadeem et al., 2021).

In rice–wheat system field experiments on four predominant soil series of the rice belt in Punjab province, 1.00 kg B ha⁻¹ applied to the 2003 rice crop enhanced paddy yield by 22% over control. The residual effectiveness of B fertilizer also increased grain/paddy yields of four succeeding wheat and rice crops: of the following wheat crop by 24% and of the next year's rice crop by 13%, over respective control yields. The cumulative effect of B applied to each rice and wheat crop over a period of three years was also encouraging, although soil B build-up was negligible (**Annex 24**). Therefore, after applying a relatively low and safe B dose to the first rice crop (i.e., 0.75 kg B ha⁻¹), the succeeding two to three crops in rice–wheat rotation may not need B fertilization.

In fact, the actual length and magnitude of the beneficial residual effect of soil-applied micronutrient fertilizers depend on many soil factors, notably the ones influencing their adsorption/desorption and leaching. For example, micronutrient adsorption by fine-textured soils can result in a significant carry-over beneficial effect, whereas micronutrient fertilizers applied to coarse-textured soils may get leached under excessive moisture regimes, like torrential rains.

Therefore, long-term field experimentation is warranted, in various cropping systems/soil types, for ascertaining the longevity of soil-applied micronutrient fertilizer doses recommended in the country.

8.4   APPARENT SOIL MICRONUTRIENT BALANCES

Except for a couple of reports from Pakistan, we are not aware of any other report in the literature about the impact of cropping on soil micronutrient balances. The NARC Micronutrients Research Group has reported on apparent soil B and Zn balances in irrigated cotton–wheat cropping system in the Punjab province. In a five-year field study in cotton–wheat system on two benchmark soils of cotton belt in the Punjab, i.e., Awagat and Shahpur series (both soils being hyperthermic Fluventic Camborthids), Rafique et al. (2012) observed that only 1.78–2.36% of the annually applied 5.0 kg Zn ha^{-1} was taken up by the first cotton crop (Table 8.2, **and Annex 25**). Thus, ~98% of the applied Zn was retained in the soils. As Zn uptake by the cropping system was quite low (i.e., 62–123 g Zn ha^{-1} by cotton crop; 74–170 g Zn ha^{-1} by wheat crop) compared with Zn inputs (i.e., 1.12–1.79 kg Zn ha^{-1} year^{-1}), all nutrient management treatments, including farmers' fertilizer use practice of applying N and P fertilizers only (no Zn or B), resulted in positive apparent Zn balances in both the benchmark soils (Table 8.2). However, a much higher use efficiency of B fertilizer in a similar five-year field study in cotton–wheat system, i.e., 6.2–10.2% (1.8–2.8% for cotton) of the annually applied 1.0 kg B ha^{-1}, was observed (**Annex 24**). Thus, ~90–93% of fertilizer B was retained (fixed) in the soil. As the total B uptake by the cropping system was much less (e.g., 0.52–1.16 kg B ha^{-1} by both crops, i.e., cotton and wheat, in coarse loamy soil) compared with B inputs (i.e., 1.6–7.5 kg B ha^{-1}), all nutrient management treatments resulted in positive apparent B balances in both soils (**Annex 24**).

Despite positive apparent soil balances for Zn and B, both cotton and wheat grown in both the soils suffer from Zn and B deficiency (Rafique et al., 2012; Rashid et al., unpublished results). Whereas apparent soil N, P, and K balances are considered useful for predicting fertilizer requirements, such information for Zn and B appears to be of little value in this regard. High soil Zn and sorption because of the calcareous nature of the soils, rather than low total Zn and B in the soil, appear to be responsible for this ambiguity. Therefore, soil testing and plant analysis remain necessary tools for diagnosing Zn and B deficiency in the cropping systems.

REFERENCES

Abid, M., N. Ahmed, M. F. Qayyum, M. Shabaan, and A. Rashid. 2013. Residual and cumulative effect of fertilizer zinc on cotton-wheat productivity in an irrigated Aridisol. *Plant, Soil and Environment* 59:505–510.

Ahmed, H., M. T. Siddique, S. Ali, A. Khalid, and N. A. Abbasi. 2012. Mapping of Fe and impact of selected physico-chemical properties on its bioavailability in the apple orchards of Murree region. *Soil and Environment* 31:100–107.

Fageria, N. K., V. C. Baligar, and R. B. Clark. 2002. Micronutrients in crop production. *Advances in Agronomy* 77:185–268.

Hussain, F. 2006. Soil Fertility Monitoring and Management in Rice-Wheat System. Agricultural Linkages Program (ALP) Project, Final Report, 2002–2006. Land Resources Research Program, National Agricultural Research Center, Islamabad. 83 pp.

Kanwar, J. S., and L. J. Youngdahl. 1985. Micronutrient needs of tropical food crops. *Fertility Research* 7:43–67.

Katyal, J. C. 1980. Relative efficiencies of micronutrient sources. *Fertilizer News* 23:39–48.

Kausar, M. A., and M. Tahir. 1994. Susceptibility of eleven wheat cultivars to zinc and copper deficiency. p. 157–160. In: *Efficient Use of Plant Nutrients. Proc. 4th National Congress of Soil Science*, Islamabad, 24–26 May, 1992. Soil Science Society of Pakistan.

Khattak, J. K., and S. Perveen. 1988. Micronutrient status of NWFP soils. p. 62–74, In: *Proceedings of National Seminar on Micronutrients in Soils and Crops in Pakistan*, 13–15 December 1987, NWFP Agricultural University, Peshawar.

Nadeem, F., M. Farooq, B. Mustafa and A. Nawaz. 2021. Influence of soil residual boron on the rice performance and soil properties under conventional and conservation rice-wheat cropping systems. *Crop and Pasture Science* 72:335–347.

Nadeem, F., M. Farooq, B. Mustafa, A. Rehman and A. Nawaz. 2020. Residual zinc improves soil health, productivity and grain quality of rice in conventional and conservation wheat-based systems. *Crop and Pasture Science* 71:322–333.

Rafique, E. 2011. Micronutrient Management for Sustaining Major Cropping Systems and Fruit Orchards. ASPL-II Coordinated Project, Final Report 2006–2011. Land Resources Research Institute, National Agricultural Research Center, Islamabad.

Rafique, E., A. Rashid, and M. Mahmood-ul-Hassan. 2012. Value of soil zinc balances in predicting fertilizer zinc requirement for cotton-wheat cropping system in irrigated Aridisols. *Plant and Soil* 361:43–55.

Rashid, A., M. Yasin, M. A. Ali, Z. Ahmad, and R. Ullah. 2007. An alarming boron deficiency in calcareous rice soils of Pakistan: Boron use improves yield and cooking quality. p. 103–116, In: Xu, F., et al. (Ed.) *Advances in Plant and Animal Boron Nutrition. Proc 3rdInt Symp on All Aspects of Plant and Animal Boron Nutrition*, Wuhan, China, 9–13 Sep 2005. Springer, Dordrecht.

Ryan, J., A. Rashid, J. Torrent, et al. 2013. Micronutrient constraints to crop production in the Middle East-West Asia region: Significance, research, and management. *Advances in Agronomy* 122:1–84.

Sakal, R. and A. P. Singh. 1995. Boron research and agricultural production. In: Tondon, H.L.S. (Ed.), *Micronutrient Research and Agricultural Production*. FDCO, New Delhi. p. 1–31.

Takkar, P. N. 1996. Micronutrient research and sustainable productivity in India. *Journal of the Indian Society of Soil Science* 44:562–581.

Tandon, H. L. S. 2013. *Micronutrient Handbook from research to application. Fertilizer Development and Consultation Organization*, New Delhi, India. 234 pp.

Yang, X., Y. G. Yu, Y. Yang, R. W. Bell, and Z. Q. Ye. 2000. Residual effectiveness of boron fertilizer for oilseed rape in intensively cropped rice-based rotations. *Nutrient Cycling in Agroecosystems* 57:171–181.

9 Micronutrient Fertilizer Use in Pakistan

Current Use Level and Potential Requirements

9.1 INTRODUCTION

Since the establishment of field-scale Zn deficiency in rice crop in Pakistan in the early 1970s, farmers are being advised to apply Zn fertilizer to the crop. With the subsequent establishment of B and Fe deficiencies in a number of crops, along with Zn deficiency, the use of fertilizers for the deficient micronutrients is being advised to growers. Though many progressive farmers are fertilizing their crops with deficient micronutrients, along with the deficient major nutrients, the current use level is generally inadequate to meet crop requirements. Therefore, this chapter discusses the current use level of micronutrient fertilizers in the context of the use recommendations for various crops, potential micronutrient fertilizer requirements in the country, and constraints to micronutrient fertilizer use.

9.2 MICRONUTRIENT USE RECOMMENDATIONS FOR CROPS

Formal micronutrient use recommendations by provincial Agriculture Departments have been made only for a few selected major crops. Whereas Zn use in rice was recommended in the late 1970s, in both the Punjab and Sindh provinces, all other micronutrient use recommendations were made much later. For example, after witnessing highly convincing yield increases with the use of B and Zn fertilizers in multi-year field trials conducted by NARC scientists throughout the Punjab cotton belt, the use of these micronutrients in cotton crop was initially endorsed by a national forum on cotton production, organized by NFDC in 1997. Subsequently, a formal recommendation for using B and Zn fertilizers in cotton crop was made by the Punjab Agriculture Department in 1998, followed by the Sindh Agriculture Department in 1999. As wide-scale field testing of Zn-enriched rice nursery technology had proven to be as effective as field broadcast application of a much higher dose of Zn fertilizer, Zn enriching of rice nursery was highly economical and much easier to practice (Rashid et al., 2000). Later, in 2003 Punjab Agriculture Department recommended the use of Zn-enriched nursery technology for rice growers. As consequence of spectacular rice yield and grain quality improvements with B fertilization in the NARC-conducted field demonstration

DOI: 10.1201/9781003314226-9

trials, the recommendation for B use in rice by the Punjab Agriculture Department was made in 2005. During the same time, a realization of the beneficial impact of micronutrient use in many other crops like potato and fruits like citrus, apple, lemon, and pomegranate was also gained. Consequently, the progressive growers started adopting the use of Zn, B, and Fe in specific crops/fruits.

9.3 CURRENT USE OF MICRONUTRIENT FERTILIZERS

At present, wide-scale use of micronutrient fertilizer in Pakistan is more or less restricted to the application of zinc sulfate and borax to some specific crops. The major crops known to receive Zn fertilizer include rice, maize, potato, and citrus. In rice, zinc sulfate is used through broadcast soil application, which also provides benefit to the following wheat crop in the rice–wheat cropping system. As the availability of good-quality Zn fertilizer in the rice-growing areas, particularly during peak demand periods, remains a common problem, the actual use of Zn in rice is far less than the crop requirement. However, rice growers have an option to enrich the rice nursery by applying a higher dose of Zn fertilizer to the nursery beds. This option is much more economical and easier to practice (Rashid et al., 2000, 2002). Therefore, many rice growers are using this technology to take care of Zn deficiency in rice crop. Recent research has revealed that Zn use in wheat is also highly cost-effective (Zou et al., 2012; Rashid et al., 2014, 2016; Ram et al., 2016). This staple cereal is grown on about 9.0 million ha annually, and Zn use enhances crop productivity and enriches wheat grains with Zn, which in turn proves to be much better food as well as seed for the next crop. The promotion of Zn fertilizer use in this crop merits urgent attention by all concerned. The major recipients of B fertilizer in Pakistan include cotton (grown on about 3 million ha annually) and rice (grown on more than 2 million ha annually). Many other crops also receive B application. Crude estimates of the current use of fertilizer Zn in the country are given in Table 9.1 and **Annex 19** and **Annex 20**.

Because of obvious reasons, in cotton, the growers prefer using B through foliar sprays as it can be safely mixed with most pesticide spray solutions. However, B fertilizer availability in time and space is a problem; thus, only a few progressive farmers can obtain and use B, presently. Due to limited business volume and marginal profits, no major fertilizer player, except for the FFC, is marketing this fertilizer in Pakistan. In fact, during the early 2000s, Engro Fertilizers pioneered the marketing of B fertilizer products of Borax Europe (i.e., *Solubor* and *Granubor*) in Pakistan; however, this was abandoned after a few years. During the writing of this book, we have learned that Engro Fertilizers is now considering the marketing of micronutrients, preferably by way of fortification of NP and NPK products. The recent launching of zincated urea by Engro Fertilizers is an appreciable step in the right direction. Crude estimates of fertilizer B use in the country are given in **Annex 21**.

Iron chlorosis in fruit orchards is another area. Some fruit growers in Balochistan are applying Fe-*Sequestrene* to cure Fe chlorosis. Historically, Fe-*Sequestrene* smuggled from Iran used to be much cheaper than the one

TABLE 9.1
Current Zinc Fertilizer Market in Pakistan – Summary Calculations

Type of Zinc Fertilizer	Fertilizer Product (Tonnes)	$ZnSO_4.H_2O$ (33%) Equivalent (Tonnes)	Annex Source
FFC $ZnSO_4$ (33%) solid	2100	2100	Annex 19 – granular/ powder formulations
Engro $ZnSO_4$ (33%) solid	1547	1547	Annex 19 – granular/ powder formulations
Other Companies $ZnSO_4$ (21%) solid	4157	2643	Annex 19 – granular/ powder formulations
Others $ZnSO_4$ liquid (5%)	8272	1253	Annex 19 – liquid formulations
Annex 20			
Others $ZnSO_4$ chelated solid (5%)	1049	159	Annex 20– chelated solid formulations
Others $ZnSO_4$ chelated liquid (5%)	375	57	Annex 20– chelated liquid formulations
Total ($ZnSO_4.H_2O$) (33% equivalent)		7759	Annex 19 and Annex 20
Total elemental Zn		**2560 tonnes elemental Zn**	

marketed by CIBA-Geigi, and thus was a favorite choice with the orchard growers. It is learned that currently, Iranian Fe-*Sequestrene* is hardly available even in Balochistan. Consequently, the orchard growers are facing difficulty in rectifying the Fe chlorosis problem.

Another reason for the un-judicious use of micronutrients in Pakistan is the lack of good soil advisory services to the farmers. Despite the extensive network of soil testing laboratories, farmers can hardly get their soils and plants analyzed for micronutrients. The foremost handicap in this respect appears to be ignorance on the part of farmers about the availability of this important facility at nominal cost. As micronutrient analysis facility is available only at divisional-level soil and water testing laboratories, most farmers, especially the small landholders, can hardly avail this facility. Therefore, improvements regarding awareness and upgradation of soil advisory services are warranted. The communication barrier between the farmers and the labs can be improved through the provision of greater mobility to extension staff across the country.

In short, despite the convincing experimental evidence about the prevalence of field-scale micronutrient deficiencies and highly cost-effective yield increases with micronutrient fertilizer applications in a number of crops, actual micronutrient fertilizer use in the country remains much less compared with the actual requirement. Crude estimates of the current use of micronutrient fertilizers in the country are given in Annexures; **Annex 19** pertains to non-chelated Zn fertilizer products,

Annex 20 to chelated Zn fertilizer products, **Annex 21** to B fertilizer products, and **Annex 22** to micronutrient complexes. Table 9.1, presenting summary calculations regarding the current use of Zn fertilizers in Pakistan (based on the information given in **Annex 19 and Annex 20**), reveals that about 2560 tonnes (7759 product tonnes) of elemental Zn are used as fertilizer in all crops in the country every year.

- For conversion of 33% concentration product into 100% pure elemental Zn, product volume is multiplied by 0.33
- For conversion of 21% concentration product into 100% pure elemental Zn, product volume is multiplied by 0.21
- For conversion of 5% concentration product into 100% pure elemental Zn, product volume is multiplied by 0.05
- For conversion of elemental Zn (100% pure product) into 33% concentration, product volume is multiplied by 3.03

9.4 POTENTIAL REQUIREMENT OF MICRONUTRIENT FERTILIZERS

It is quite obvious that the current use of micronutrient fertilizers in the country is far less compared the potential requirements. Consequently, despite having adequate scientific evidence that the widespread deficiencies of certain micronutrients (like Zn, B, and Fe) are hampering crop productivity, impairing produce quality, and reducing farm income, there is inadequate commitment on the part of concerned decision-makers to improve upon this persistent adverse situation. In this regard, it is important to compare and contrast current micronutrient fertilizer use and potential requirement of micronutrient fertilizers in the country.

It is a known fact that estimates of potential fertilizer requirements will be hardly useful to individual farmers. Still, we have attempted to estimate the potential micronutrient fertilizer requirements in the country because:

a) Estimates of fertilizer needs are attention-getters. They emphasize the enormity of the need to manage fertilizer inputs, especially at the end of fertilizer marketing companies.
b) Estimates of fertilizer needs help identify the more pressing problems; thus, they are useful to emphasize the need to advance relevant research and development programs.
c) Reasonable estimates of potential requirements for fertilizer provide essential benchmark material for sensible resource development, resource management, and resource conservation.

Several different approaches could be used for estimating national fertilizer requirements, four of which are as follows:

i) **Socio-economic approach**: This applies past trends and market conditions to predict future fertilizer requirements.

ii) **Balance sheet approach**: It is based on nutrient removal and seeks to restore to the soil the nutrients removed by cropping (and, perhaps, also removed by soil erosion).

iii) **Soil test approach**: It is based on the fertilizer requirements to bring soils to a certain prescribed soil test nutrient levels.

iv) **Recommended fertilizer rate approach**: This targets the nutrient-deficient areas or fractions of total areas under major crops.

All of these approaches involve a series of bold assumptions. In the background of minimal use of micronutrient fertilizers in Pakistan till now, the socio-economic approach is hardly worth consideration. Because of the enormous complexity involved in the balance sheet approach, this approach is easier but not valid in the case of micronutrients. For one thing, information about soil micronutrient balances is hardly available in the literature. On top of this, according to a five-year field study in cotton–wheat system of Punjab province, despite estimated slightly positive soil Zn balances without Zn fertilization, both cotton and wheat crops are widely affected by Zn deficiency in the cotton belt (Rafique et al., 2012). The situation is similar with respect to soil B balances and the prevalence of B deficiency in cotton and wheat crops (Rashid et al. unpublished results). The soil test approach is fraught with practical limitations especially in the case of micronutrients – because it is virtually impossible to carry out extensive soil testing for micronutrients in a country like Pakistan, which has very limited requisite facilities and resources for the purpose. Therefore, for this exercise, the potential requirements of Zn, B, and Fe fertilizers in Pakistan have been calculated using the recommended fertilizer rates approach.

9.4.1 POTENTIAL REQUIREMENT OF FERTILIZER ZINC

In Pakistan, the total area under field crops (i.e., wheat, cotton, rice, maize, sugarcane, rapeseed (canola), pulses, tobacco, and peanut), which is deficient in Zn – and, thus, in need of using Zn fertilizer – is around 6.68 million ha (Table 4.3; **Annex 3)**. In an earlier similar exercise, the total Zn-deficient area under field crops in the country was estimated to be much less, i.e., 3.27 million ha (NFDC, 1998). The predominant reasons being that till that time Zn deficiency was not well established in certain crops like sugarcane and that the percentage areas known to be affected by Zn deficiency under some major crops (like wheat and maize) were much lesser compared with the more widespread Zn deficiency known now. Also, the field areas under vegetables (potato, tomato, and onion) and fruits (citrus, apple, plum, peach, pear, pomegranate, grape, mango, and banana) requiring Zn fertilization are almost double now compared with the areas under such vegetables and fruits requiring Zn fertilizer use in the late 1990s (NFDC, 1998). Obviously, the micronutrient research during the time period, carried out by various agricultural institutions in the country, has added to our knowledge about the extent of micronutrient problems in all sorts of agricultural crops.

According to our estimates, the annual potential fertilizer Zn requirement for all crops (i.e., field crops, vegetables, and fruits) is around 12,163 tonnes of elemental Zn (Table 9.2) against the current market volume of 2560 tonnes. The total fertilizer Zn requirement for the whole deficient field areas sown with these crops is 30,902 tonnes of elemental Zn. As the application of Zn fertilizer leaves a beneficial residual effect on many subsequent crops in the same field, Zn fertilizer need not be applied to every crop in the rotation. The relevant literature, as well as local research, tells that an application of 5 kg Zn ha^{-1} remains effective for about six crop seasons in all cropping systems, except for two crop seasons in rice–wheat system. Thus, the annual potential fertilizer Zn requirement for field crops has been calculated accordingly (Table 9.2). For cotton crop, farmers prefer to add micronutrient fertilizer in the pesticide spray solutions. Therefore, for 50% of the Zn-deficient cotton area, fertilizer Zn requirement has been estimated for foliar feeding (i.e., three foliar sprays of 0.1% elemental Zn concentration @ 300 L per foliar spray per ha).

For the susceptible fruit orchard areas, fertilizer Zn requirement has been estimated to cater to Zn by foliar fertilization only; thus, there would not be any beneficial residual effect. Therefore, total and annual fertilizer Zn requirements for fruit orchards are the same, i.e., 228 tonnes of elemental Zn.

Because of much greater extent of Zn deficiency in crops known now, compared with two decades ago, the estimated annual fertilizer Zn requirement in the country (i.e.,12,537 tonnes of elemental Zn) is much greater than the Zn requirement estimated in 1998 (i.e., 5729 tonnes of elemental Zn; NFDC, 1998).This much higher potential fertilizer Zn requirement estimation has resulted despite considering much longer beneficial residual effect of soil-applied Zn fertilizer in all cropping systems (i.e., six crop seasons), except for the rice–wheat system in which every rice crop in the rotation requires Zn fertilization. Earlier, in the 1990s, the general perception was that soil-applied Zn fertilizer leaves a beneficial residual effect on three subsequent crops only in all cropping systems, and, hence, this consideration was used in estimating annual Zn requirements for the 1998 report (NFDC, 1998).

9.4.1.1 Fertilizer Zinc Requirement for Enriching Wheat Grains with Zinc

A recent realization regarding micronutrient nutrition of staple food crops, like wheat and rice, is the predominant contribution of low Zn-containing cereal grains in the incidence of alarming Zn malnutrition (i.e., *"hidden hunger"*) in the human population in Pakistan. Global as well as local research has established that wheat grains can be enriched with Zn by applying two foliar sprays of zinc sulfate solution (@ 0.32% $ZnSO_4.H_2O$ or 0.1% elemental Zn) to wheat, i.e., first spray one week prior to heading and second spray one week after heading. Whereas soil Zn application is necessary for optimizing wheat grain productivity in soils deficient in Zn, foliar sprays of Zn are essentially needed to enrich wheat grains with Zn. Therefore, an additional fertilizer Zn requirement has been calculated for 50% of the wheat area which is deficient in Zn. As wheat is grown on a 9.02 million ha

TABLE 9.2
Potential Fertilizer Zn (Elemental) Requirement in Pakistan

Crop	Crop Area[1] (million ha)	Extent of Deficiency (%)	Zn-deficient Crop Area (million ha)	Recommended Zn Dose (kg ha^{-1})	Total Zn Requirement (tonnes)	Annual Zn Requirement[3] (tonnes)
Field crops	**18.175**	**20–76**	**6.681**	**5**	**30,902**	**12,163**
Wheat	9.022	56	1.847[2]	5	9233.6	1538.9
+ *Wheat grain enrichment with Zn* (2 foliar Zn sprays (on 50% of Zn-deficient wheat area): 9.022 Mha x 56% = **5.052 Mha** x 50% area = 2.52 Mha @ 2 sprays of 0.5% ZnSO$_4$.7H$_2$O solution (i.e., 3 kg ZnSO$_4$.7H$_2$O ha^{-1} = **0.63 kg Zn ha^{-1}**) = **3183 tonnes**					3183.0	3183.0
Cotton	2.882	41	50%: 0.591	Soil application: 5	2953.5	492.3
			50%: 0.591	3 foliar sprays: 0.9[4]	531.6	531.6
Rice	2.663	76	2.024	5	10,118.5	5059.3
Maize	1.124	60	0.674	5	3370.5	561.8
Sugarcane	114.7	57	0.654	5	3269.5	545.0
Rapeseed (canola)	0.015	80	0.012	5	60.5	10.1
All pulses	1.181	20	0.236	5	1181.5	196.9
Tobacco	0.051	50	0.026	5	127.5	21.3
Peanut	0.091	60 (30%*)	0.027	5	136.0	22.7
Vegetables	**0.361**	**40–60**	**0.171**	**5**	**855.0**	**142.6**
Potato	0.167	40	0.067	5	333.5	55.6
Tomato	0.061	40	0.024	5	121.5	20.3
Onion	0.133	60	0.080	5	400.0	66.7
Fruits	**0.544**	**29–80**	**0.318**	**0.6–0.9**	**231.3**	**231.3**
Citrus	0.194	74	0.143	0.90[4]	129.1	129.1
Apple	0.103	80	0.082	0.60[5]	49.4	49.4
Plum	0.007	70	0.005	0.60	2.9	2.9
Peach	0.014	70	0.010	0.60	5.9	5.9
Pear	0.002	70	0.001	0.60	0.8	0.8
Pomegranate	0.011	50	0.005	0.60	3.2	3.2
Mango	0.171	29	0.044	0.60	26.6	26.6
Banana	0.028	57	0.016	0.60	9.5	9.5
Grapes	0.015	70	0.011	0.36[6]	3.9	3.9
				Total	31,988	9354
				Grand Total	35,171	12,537

[1] Crop area: Mean of three crop years, 2012–13, 2013–14, 2014–15 (Agricultural Statistics of Pakistan, Government of Pakistan, 2014–15 (GoP, 2016))

[2] Net Zn-deficient wheat area: (Total Zn-deficient wheat area, 5,052,200 ha) – Zn-deficient cotton area, 1,221,900 ha + Zn-deficient rice area, 2,023,700 ha) = 1,806,600 ha

[3] Assuming that soil-applied 5 kg Zn ha^{-1} lasts for six crops in all cropping systems, except for rice–wheat, where it lasts for 2 crop seasons.

[4] Three foliar sprays of 0.1% Zn @ 300 L ha^{-1} each spray

[5] Two foliar sprays of 0.1% Zn @ 300 L ha^{-1} each spray

[6] Two foliar sprays of 0.06% Zn @ 300 L ha^{-1} each spray

* The percentage area assumed to be receiving Zn fertilizer

field area and 56% of the total wheat area is reported to be deficient in Zn, fertilizer Zn requirement to make foliar sprays on one-half of the Zn-deficient wheat area (i.e., 2.52 million ha) came out to be 3183 tonnes of elemental Zn. By adding up this Zn requirement, the grand total requirement of fertilizer Zn in the country comes out to be 12,537 tonnes of Zn per annum (Table 9.2).

9.4.1.2 Potential Requirement and Actual Use of Fertilizer Zinc

Despite decades-old convincing evidences about the prevalence of Zn deficiency in a number of crops and the highly cost-effective impact of its fertilizer use, the current use of Zn fertilizer in the country remains much less than its actual requirement. According to our estimates, the potential annual fertilizer micronutrient requirement in the country is around 12,537 tonnes of elemental Zn (Table 9.2). As the current use of Zn fertilizer in the country for all crops is around 2560 tonnes of elemental Zn per annum, the potential requirement of fertilizer Zn for all crops is about five-times the current use level. According to our estimates, the actual use of fertilizer Zn in the country for Zn-deficient field areas under all crops (i.e., field crops including rice, vegetables including potato, and fruits including citrus) is around 5703 elemental Zn per annum (assuming 33% elemental Zn in all Zn products; **Annex 19** and **Annex 20**). Thus, the potential fertilizer Zn requirement is 2.2 times higher compared with its actual use. Consequently, even rice, having the oldest and well-known history of Zn deficiency, remains widely affected by this micronutrient disorder.

9.4.2 Anticipated Enhancement in Use of Zinc Fertilizer

Similar to the slow adoption of any other farmer-friendly new technologies by the stakeholders, the adoption of micronutrient fertilizer use by the growers to the desired level cannot be expected instantly. The experience so far, especially since the compilation of a comprehensive status report on micronutrients some 20 years ago (NFDC, 1998), has demonstrated that micronutrient fertilizer use adoption by the growers is much slower than the expectation of the concerned agricultural researchers. To our understanding, this slower-than-desired pace of adoption is the consequence of some genuine constraints on the part of growers. In addition to inadequate awareness about the needs and benefits of using micronutrient fertilizers, such constraints also include quality and availability issues concerning micronutrient fertilizer products. With enhanced awareness and advancement in time, adoption of Zn fertilizer use in the country within the next 10 years may be enhanced to 70% of the total crop areas needing Zn fertilization. The anticipated fertilizer Zn requirements with progressively enhanced use adoption over the next 10 years are given in Table 9.3.

Zinc sulfate is the suggested Zn source; depending upon hydration of the $ZnSO_4$ molecule, Zn concentration in zinc sulfate fertilizer varies from 21% (in $ZnSO_4.7H_2O$) to 33% (in $ZnSO_4.H_2O$). Currently, the predominant Zn fertilizer product (powder, crystal, and granular forms) available in the country is $ZnSO_4.H_2O$ (33% Zn).

TABLE 9.3
Anticipated Enhancement in Zn Fertilizer Use and Fertilizer Requirement over the Next Ten-Year Period

Crop Year	Expected Zn use Adoptability	Elemental Zn Requirement (tonnes per Annum)[1]			
		Field Crops	Vegetables	Fruits	Total
2nd	30% of deficient area	3649	43	69	3760
4th	40% of deficient area	4865	57	91	5014
6th	50% of deficient area	6081	71	114	6268
8th	60% of deficient area	7299	86	136	7521
10th	70% of deficient area	8516	100	159	8774

[1] Potential annual requirement: Field crops, 8980 tonnes Zn; vegetables, 143 tonnes Zn; fruits, 231 tonnes Zn.

9.4.3 POTENTIAL REQUIREMENT OF FERTILIZER BORON

According to our estimates, the total potential fertilizer B requirement for all crops (i.e., field crops, vegetables, and fruits) is 2396 tonnes of elemental B per annum (Table 9.4). The total B-deficient agricultural area in the country under field crops, vegetables, and fruits, requiring B fertilization, is around 5.27 million ha – of which 4.8 million ha are under field crops, 0.20 million ha under vegetables, and 0.27 million ha under fruit orchards. The total area under field crops affected by B deficiency (i.e., wheat, cotton, rice, maize, sugarcane, rapeseed (canola), certain pulse crops, tobacco, and peanut) that requires the use of B fertilizer is around 4.80 million ha. In an earlier similar exercise as well, the total B-deficient area under field crops was estimated to be close to the present estimate, i.e., 4.54 million ha (NFDC, 1998). The difference in these two estimates is that sugarbeet and sunflower crops included in the earlier estimate have been dropped as their acreage in the country is negligible and sugarcane crop has been added to the list because of relatively recent experimental evidence of B deficiency in this crop in Punjab as well as in Sindh province (Table 6.1).

Information gathered from the relevant literature (published as well as unpublished), especially considering the results of a few permanent layout field studies in cotton–wheat and rice–wheat systems in the Punjab province, indicates that the residual effect of soil-applied B fertilizer (@ ~1.0 kg B ha^{-1}) lasts for one succeeding crop in cotton–wheat rotation and maize–potato rotation and for three succeeding crops in rice–wheat rotation. Thus, soil application of B fertilizer is needed for every cotton crop in cotton–wheat rotation, for every maize crop in maize–potato rotation, for every alternate crop in all other upland cropping systems, but for every second rice crop in the rotation in the rice–wheat cropping system. Using this analogy, the potential annual fertilizer B requirement for all crops is around 2396 tonnes of elemental B. In consideration of using a lower B dosage (i.e., 0.75–1.0 kg B ha^{-1}) for all field crops and the residual effect of soil B

TABLE 9.4
Potential Fertilizer B (Elemental) Requirement in Pakistan

Crop	Crop Area[1] (Million ha)	Extent of Deficiency (%)	B-deficient Crop Area (million ha)	Recommended B Dose (kg ha⁻¹)	Total B Requirement (Tonnes)	Annual B Requirement[4] (Tonnes)
Field crops	**18.175**	**10–68**	**4.800**	**0.75–1.0**	**4188**	**2122**
Wheat	9.022	46	1.748[2]	0.75	1311.2	655.6**
Cotton[3]	2.882	51	50%: 0.735	1.00 kg B ha⁻¹	734.9	367.5**
			50%: 0.735	0.55 kg B ha⁻¹	404.2	404.2
Rice	2.663	35	0.932	0.75	699	174.8
Maize	1.124	50	0.562	0.75	421.4	210.7
Sugarcane	1.147	56	0.642	0.75	481.8	241.0**
Rapeseed (canola)	0.015	68	0.010	0.75	7.7	3.9
All pulses	1.181	50 (10%*)	0.118	0.75	88.6	44.3
Tobacco	0.051	50	0.026	0.75	19.1	9.6
Peanut	0.091	50 (30%*)	0.027	0.75	20.4	10.2
Vegetables	**0.454**	**20–55**	**0.203**	**0.75**	**152.1**	**38.0**
Potato	0.167	55	0.092	0.75	68.8	17.20
Tomato	0.061	50	0.030	0.75	22.8	5.70
Onion	0.133	40	0.053	0.75	40.0	10.01
Cauliflower	0.010	50	0.005	0.75	3.62	0.90
Turnip and Radish	0.020	50	0.010	0.75	7.50	1.88
Chili	0.063	30 (20%*)	0.013	0.75	9.5	2.37
Fruits	**0.544**	**45–61**	**0.268**	**0.16–0.90**	**236**	**236**
Citrus	0.194	45	0.087	0.90[5]	78.49	78.49
Apple	0.103	61	0.063	0.90	56.55	56.55
Plum	0.007	50	0.004	0.90	3.11	3.11
Peach	0.014	50	0.007	0.90	6.35	6.35
Pear	0.002	50	0.001	0.90	0.86	0.86
Pomegranate	0.011	50	0.005	0.90	4.82	4.82
Grapes	0.015	50	0.008	0.16[6]	1.22	1.22
Mango	0.171	45	0.077	0.90	69.17	69.17
Banana	0.02	60	0.017	0.90	15.01	15.01
		Total	**5.271**		**4576**	**2396**

[1] Crop area: Mean of three crop years, 2012–13, 2013–14, 2014–15 (Agricultural Statistics of Pakistan, Government of Pakistan, 2014–15 (GoP, 2016))

[2] Net B-deficient wheat area: (Total B-deficient wheat area, 4,149,982 ha) – B-deficient Cotton area, 1,469,800 ha + B-deficient rice area, 1,065,100 = 1,615,000 ha

[3] Soil application @ 0.75–1.0 kg B ha⁻¹ *or* three foliar sprays @ 0.55 kg B ha⁻¹; mean of both methods, 1.03 kg B ha⁻¹

[4] Assuming that soil-applied 0.75–1.0 kg B ha⁻¹ lasts for four crop seasons, in general.

[5] Two foliar sprays of 0.15% B @ 300 L ha⁻¹ each spray

[6] Two foliar sprays of 0.05% B @ 300 L ha⁻¹ each spray

* The percentage area assumed to be receiving B fertilizer

** Assuming that soil-applied 0.75–1.0 kg B ha⁻¹ lasts for two crop seasons in cotton–wheat system, two sugarcane crops – the latter being an annual crop

fertilization on subsequent crops grown in the same field, the currently estimated annual fertilizer B requirements for field crops only (i.e., 2122 tonnes of B per annum) is much less compared with the earlier estimation of 3075 tonnes of B per annum made in 1998 (NFDC, 1998).

The currently known B-deficient field area under vegetables (potato, tomato, cauliflower, turnip, radish, and onion) is 0.20 million ha and under fruits (citrus, apple, plum, peach, pear, pomegranate, grape, mango, and banana) is 0.27 million ha. Both B-deficient areas requiring B fertilization are multiple times higher compared with the B-deficient areas under vegetables and fruits estimated in the late 1990s (NFDC, 1998). Based on the experimental evidences, the number of B fertilizer-requiring vegetables increased to six now, compared with only cauliflower and turnip included in the 1998 status report (NFDC, 1998).

For the B-deficiency-susceptible fruit orchard areas, the B fertilizer requirement has been estimated to cater to foliar fertilization needs only. As cotton farmers prefer to mix micronutrients with pesticide spray solutions, fertilizer B requirement estimates for cotton crop have been made for both modes of fertilizer application, i.e., for 50% of deficient area by soil application and the rest 50% of deficient area by foliar feeding. As foliar feeding hardly leaves any residual effect, total and annual B requirements for fruit orchards are the same, i.e., 236 tonnes of elemental B. According to the 1998 estimates, the potential fertilizer B requirement in the country at that time was 3100 tonnes of elemental B per annum (NFDC, 1998).

9.4.3.1 Actual Fertilizer B Use and Potential Fertilizer B Requirement in Pakistan

Though the magnitude of B deficiency in crops is almost equivalent to Zn deficiency, its current use level is much lower than that of Zn fertilizer. In our estimation, the potential requirement of fertilizer B for all crops (i.e., field crops, vegetables, and fruits) is around 2396 tonnes of elemental B per annum (Table 9.4). However, the current use of fertilizer B in Pakistan is very low, i.e., 101.31 tonnes of elemental B (assuming 11% elemental B in 921 tonne B fertilizer products; **Annex 21**). Currently, the use of B fertilizer is limited to cotton, rice, and some fruit crops by progressive growers, Thus, the potential requirement of fertilizer B in the country is about 23-times higher than its current use level in the country.

9.4.4 ANTICIPATED ENHANCEMENT IN USE OF BORON FERTILIZER

As the current use of B fertilizer (101 tonnes of elemental B) is dismally low compared with its potential requirement (2396 tonnes of elemental B) in the country, enhancement of its use by the growers will be an uphill task, compared with the enhancement of Zn fertilizer. To start with, within the next two-year period, by enhancing stakeholders' awareness about the need and benefits of B fertilization, the use of B fertilizer may increase to 10% of the total deficient crop areas, i.e., from the current use level of 101 tonnes of elemental B to 240 tonnes of elemental

TABLE 9.5
Anticipated Enhancement in B Fertilizer Use and Fertilizer Requirement Over the Next Ten-Year Period

Crop Year	Expected B Use Adoptability	Elemental B Requirement (Tonnes per Annum)[1]			
		Field Crops	Vegetables	Fruits	Total
2nd	10% of deficient area	212.2	3.8	23.6	239.6
4th	15% of deficient area	424.4	7.6	47.2	359.4
6th	20% of deficient area	636.6	11.4	70.8	479.2
8th	25% of deficient area	848.8	15.2	94.4	599.0
10th	30% of deficient area	1061.0	19.0	118.0	718.8

[1] Potential annual requirement: Field crops, 2122 tonnes B; Vegetables, 38.03 tonnes B; Fruits, 236 tonnes elemental B (Table 9.4).

B per annum. This substantial enhancement by 2.4 times within a short span of two years will require a lot of strenuous efforts. With continued efforts, thereafter, B fertilizer use may increase progressively to additional crop areas every year, and within the next 10 years, its use adoption may reach 30% of the deficient areas under different field crops, vegetables, and fruits. The estimated fertilizer B requirement with the anticipated increase in adoption of B fertilizer use over the next 10-year period is given in Table 9.5. The anticipated seven-time enhancement in use adoption of fertilizer B (from the current use of 101 tonnes of B to 719 tonnes per annum) will be a challenging task. Of course, the rate of use adoption will vary in different cropping systems, vegetables, and fruit orchards. We antici-pate that because of more attractive economic returns, among field crops better adoptability of B fertilizer use is expected in rice and cotton crops.

9.4.5 POTENTIAL FERTILIZER IRON REQUIREMENT

Determining the potential requirement of fertilizer Fe is much more complex, compared with those of Zn and B, due to the following reasons:

i) Unlike Zn and B, reliable estimates of the extent of Fe deficiency are not available. This is so because of the ineffectiveness of soil testing as well as plant tissue analysis for the total Fe concentration. The data on the reliable index of Fe status in crop plants, i.e., Fe^{2+} concentration in fresh tissue, are almost lacking – not only in Pakistan but also the world over. Visual symptoms of Fe deficiency in plants, i.e., Fe chlorosis, are very peculiar and conspicuous, but quantitative estimates cannot be based on visual observations.

ii) Frequently, soil application of inorganic Fe fertilizer products (like $FeSO_4$) proves ineffective in curing the deficiency. On the other hand, the effective Fe-chelate, i.e., *Sequestrene* (Fe-EDDHA), is quite expensive and hence cannot be used for soil application, especially for low-value field crops. Even its foliar sprays cannot be recommended indiscriminately. Therefore, the use of Fe fertilizer cannot be anticipated in all the field crops and vegetables in showing response to its application. There is a good possibility of Fe fertilization of fruit orchards, ornamentals, and some vegetable crops (like chili) which are susceptible to Fe chlorosis, primarily through foliar sprays. Iron fertilization in agronomic crops would be rather very limited, in the foreseeable future.

As foliar feeding of Fe is more effective than soil application of inorganic Fe salt (i.e., $FeSO_4$), the potential Fe requirement for the susceptible fruits, chili, and peanut has been made to cater to foliar sprays only. In our estimation, the total $FeSO_4$ requirement for the purpose will be around 1298 tonnes per annum (Table 9.6). The use of Fe fertilizer in other fruits prone to Fe chlorosis, like walnut, strawberry, etc., will require additional fertilizer. Estimates of fertilizer Fe requirement for other field crops, vegetables, and ornamentals are more difficult to attempt. Though Fe-*Sequestrene* is more effective, particularly for soil application in calcareous environments, its high cost is prohibitive for its use in most crops. Therefore, its use can be considered for high-value plants, like ornamentals, only.

It must be realized that, despite basing the estimates on sound scientific information, the future micronutrient fertilizer demand will largely depend on the

TABLE 9.6
Potential Fertilizer Fe Requirement in Pakistan

Crop	Crop Area[1] (million ha)	Extent of Deficiency (%)	Fe-Deficient Crop Area (million ha)	Annual FeSO₄ Requirement[2] (tonnes)	Recommended Dose(kg ha⁻¹)
Field crops					All crops, vegetables,
Peanut	0.091	50 (30*)	0.027	122	and fruits:
Vegetables					a) *Foliar feeding*: 4.5 kg
Chili	0.063	35	0.022	99	FeSO₄ (3 sprays of
Fruits					0.5% FeSO₄ solution
Citrus	0.194	71	0.138	619	or b) *Soil application*:
Apple	0.103	80	0.082	371	50 kg FeSO₄ ha⁻¹
Plum	0.007	60	0.004	18	Fe-*Sequestrene* is
Peach	0.014	60	0.009	38	more effective, but is
Pear	0.002	60	0.001	55	extremely expensive;
Grapes	0.015	50	0.008	34	thus, it can be used for
Pomegranate	0.011	45	0.005	22	high-value plants only.
	Total		**0.296**	**1298**	

[1] Crop area: Mean of three crop years, 2012–13, 2013–14, and 2014–15 (GoP, 2016)

[2] Three foliar sprays of 0.5% $FeSO_4$; 900 L solution @ 300 L ha⁻¹ for each spray

* The percentage area assumed to be receiving Fe fertilizer

availability of micronutrients to the farmers, prices of fertilizer products, the marketing infrastructure, and the promotional efforts.

9.5 CONSTRAINTS TO MICRONUTRIENT FERTILIZER USE

Despite the establishment of field-scale deficiencies of Zn, B, and Fe in many crops and fruit orchards, micronutrient fertilizer use in Pakistan is almost restricted to Zn application in rice only. The factors constraining the use of micronutrients in Pakistan are as follows:

   i) Inadequate awareness about the nature, extent, and severity of micronutrient deficiencies in various regions and cropping systems in Pakistan. Soil fertility research should give more emphasis on micronutrient nutrition studies. Soil testing laboratories should also be adequately equipped for routine micronutrient analysis.
   ii) Non-availability of the needed micronutrient fertilizer at appropriate places at right time. Micronutrient fertilizer availability should be improved substantially.
   iii) Lack of awareness among farmers regarding the need for use of micronutrients. An effective transfer of technology campaign is suggested through print and electronic media along with extensive field demonstrations on the benefit of micronutrient use.
   iv) Lack of information on the diversity of available micronutrient fertilizer products.
   v) Availability of poor-quality micronutrient products. The lack of serious implementation of fertilizer quality control law is a major factor, and, thus, its immediate promulgation is warranted.
   vi) High price of micronutrient fertilizer. The market price of micronutrient sources, particularly of Fe-*Sequestrene* and *Solubor*, is on the high side. Suggested measures include i) import of micronutrient products as fertilizer and ii) availability of alternate micronutrient sources for fair market competition.
   vii) Inability of farmers to thoroughly mix micronutrients with major nutrient fertilizers in correct proportion. Commercial availability of appropriate micronutrient-fortified fertilizer suitable for various crop/orchards is suggested.

There are genuine reasons for the non-adoption of such a beneficial technology by the growers. Salient constraints responsible for the non-adoption of micronutrient use are as follows:

   i) *Ignorance about the Need to Fertilize*: Ignorance on the part of stakeholders (i.e., policymakers, technology transfer/extension staff, fertilizer industry, and growers) is perhaps the most crucial constraint in the adoption of micronutrient use in the country. Though most of the research information was documented and publicized through scientific conferences/seminars, its transmission to the stakeholders is rather slow and ineffective. In fact, it is a recent development that the agricultural extension service and the fertilizer industry in the country have got convinced about the use of micronutrients; still, the masses remain ignorant. Thus, a vigorous technology transfer campaign is warranted to enhance the awareness among all stakeholders, substantially. Print and electronic media as well as extensive field demonstrations and farmers' field days on major crops and fruits may be employed.

ii) *Inadequate Fertilizer Availability*: Difficulty in access to micronutrient fertilizers is one of the major constraints in the adoption of micronutrient use by the growers. Out of the numerous micronutrient sources being marketed around the world, only a few are available as fertilizer in Pakistan. In fact, the micronutrient research carried out in Pakistan has predominantly relied upon the use of analytical reagent-grade (AR-grade) and commercial-grade zinc sulfate, borax, boric acid, etc., imported and marketed in the country for non-agricultural uses. The only micronutrient fertilizer marketed widely in the country used to be zinc sulfate in the rice-growing areas. However, with the enhanced realization of the benefits of micronutrients in a number of crops, since the year 2001, major fertilizer companies in the country (like Engro Fertilizers and Fauji Fertilizer Company Limited) initiated the marketing of Zn and B fertilizer products. Concurrently, numerous small vendors, including pesticide firms, are also in the race to capture the fast-growing micronutrient market. Thus, the concerned government agencies need to be proactive for quality assurance.

iii) *Dubious Fertilizer Quality*: Dubious quality of the available micronutrient products in the local market is also a serious factor in the non-adoption of micronutrient use. Because of enhanced awareness about incidence of micronutrient deficiencies and the cost-effectiveness of their use, currently, there is a strong urge for their use, especially in crops like cotton and fruits, etc. Consequently, a number of "micronutrient" products introduced by the non-reputed small vendors keep floating in the market, which invariably are not true-to-label contents, and even some are totally fake, and hence phase out with time. This, serious quality problem with the micronutrient sources has, in fact, led to the erosion of growers' confidence in investing in micronutrients. However, as a consequence of the enhanced realization of the need and benefits of micronutrient use, major fertilizer industries in the country (like Engro Fertilizer Company and Fauji Fertilizer Company Limited), finally, launched the marketing of some micronutrient products. Consequently, the standard micronutrient products (like *Solubor*, *Granubor*, and zinc sulfate), used elsewhere in the world, are now becoming available to growers in Pakistan as well. Even some of the companies involved in the agro-chemicals business, like pesticides, have also started the import of chelated micronutrient fertilizers manufactured in Europe. Still, improved availability of good-quality micronutrient fertilizers is highly warranted.

iv) *Growers' Application Problems*: The difficulty in field application of micronutrients and preparation of correct spray solution is another serious constraint. As the dosage of micronutrients is small, unless they are blended with major nutrient granular fertilizers, their uniform distribution across the field area is not an easy task. Various studies have found that as Zn concentration and granule size decreased, the distribution of Zn increased in the soil, making it more available to plant roots. For example, if zinc sulfate (Zn 33%) fertilizer is used @ 6 kg per acre, particle number per square foot field area would be 33 for 2 mm, 265 for 1 mm, and 2116 for 0.5 mm size. This would translate into 1.44, 11.52, and 92.19 million particles per acre for 2, 1, and 0.5 mm size, respectively. In case recommended dose (9.4 kg per acre) of zinc sulfate (Zn 21%) fertilizer is used by farmers, particle number per square foot area would be 52 for 2 mm, 414 for 1 mm, and 3316 for 0.5 mm size. This would translate into 2.26, 18.05, and 144.43 million particles per acre for 2, 1, and 0.5 mm sizes, respectively. Consequently, as particle diameter is decreased from 2 to 0.5 mm, the number of particles per square foot field area increases by a factor of 64. Similarly, when a low concentration of zinc sulfate (Zn 21%) is used, particles per square foot area increase by a factor of 57.

Similarly, the preparation of spray solutions of correct concentration is also troublesome for an average farmer in the country. Only easy-to-use formulations, like micronutrient-fortified fertilizers and ready-to-use spray solutions, can help promote their use.

Micronutrient fertilizer use experience in other countries of the world reflects that most convenient formulations are micronutrient-fortified fertilizers, i.e., fertilizers blends/mixtures containing micronutrients. Micronutrients could be added to major nutrient fertilizers prior to granulation (e.g., during the formulation of compound fertilizers) or could be added during the dry blending process. Thus, the fertilizer industry in the country has an obligation to cater to high-quality, easy-to-use, crop-specific micronutrient-fortified fertilizers. To start with, micronutrient-fortified formulations may be prepared for a few specific crops with good economic returns (e.g., cotton, rice, potato, and fruits).

In short, unless genuine constraints on the part of resource-poor farmers are addressed adequately, wide-scale adoption of micronutrient use is not expected in Pakistan.

REFERENCES

GoP (Government of Pakistan). 2016. Agricultural Statistics of Pakistan 2014–15. Ministry of Finance, Economic Affairs, Revenue, Statistics & Privatization, Statistics Division, Islamabad. Available at www.statistics.gov.pk [Accessed on Jan 01, 2022.]

NFDC. 1998. *Micronutrients in Agriculture: Pakistan perspective.* National Fertilizer Development Center, Islamabad, 51 pp.

Rafique, E., A. Rashid, and M. Mahmood-ul-Hassan. 2012. Value of soil zinc balances in predicting fertilizer zinc requirement for cotton-wheat cropping system in irrigated Aridisols. *Plant and Soil* 361:43–55.

Ram, H., A. Rashid, W. Zhang, et al. 2016. Biofortification of wheat, rice and common bean by applying foliar zinc fertilizer along with pesticides in seven countries. *Plant and Soil* 403:389–401.

Rashid, A., M. A. Kausar, F. Hussain, and M. Tahir. 2000. Managing zinc deficiency in transplanted flooded rice grown in alkaline soils by nursery enrichment. *Tropical Agriculture* 77:156–162.

Rashid, A., K. Mahmood, Mahmood-ul-Hassan, M. Rizwan, and Z. Iqbal. 2014. *HarvestPlus Zinc Fertilizer Project, Phase-II Country Report, 2011–2014.* Nuclear Institute for Agriculture and Biology, Faisalabad.

Rashid, A., K. Mahmood, M. Rizwan, Z. Iqbal, and A. Naeem. 2016. *HarvestPlus Zinc Fertilizer Project, Country Report, July 2016.* Nuclear Institute for Agriculture and Biology, Faisalabad.

Rashid, A., M. Yasin, and M. Ashraf. 2002. Mat-type nursery enrichment for correcting zinc deficiency in rice. *International Rice Research Notes* 27:32–33.

Zou, C. Q., Y. Q. Zhang, A. Rashid, et al. 2012. Biofortification of wheat with zinc through zinc fertilization in seven countries. *Plant and Soil* 361:43–55.

10 Strategies for Improving Micronutrient Management

10.1 INTRODUCTION

The information narrated in preceding chapters of this book indicates that we have reasonably adequate information regarding the nature and extent of micronutrient deficiency problems as well as the benefits of micronutrient fertilizer use in terms of increases in crops yields, farm income, and produce quality improvements. However, the current use of micronutrient fertilizers is far less compared with the crop requirements. Though some promotional activities have also been undertaken to enhance awareness regarding the need and benefits of using micronutrients, more concerted efforts are required to promote micronutrient fertilizer use. Obviously, there are genuine constraints on the part of growers which are prohibitive to the adoption of farmer-friendly technologies regarding micronutrient fertilizer use. In the following paragraphs, we endeavor to elaborate on the strategies to promote the use of micronutrient fertilizers, which include the following:

  i) Awareness enhancement
 ii) Subsidy on micronutrient fertilizers at the country scale
iii) Micronutrient-fortified fertilizers
 iv) Enhanced availability in time and space
  v) Micronutrient fertilizer quality control measures

10.2 AWARENESS ENHANCEMENT

The first and foremost task is to enhance awareness on the part of the farming community and the decision-makers on the need and benefits of using micronutrient fertilizers. The following steps are suggested in this regard:

  i) *Field Demonstrations*: As seeing is believing, more field demonstrations on important crops are needed by involving the stakeholders, i.e., farmers, agricultural extension staff, fertilizer industry, and decision-makers in the government.
 ii) *Technology Transfer Materials*: More technology transfer materials (i.e., local language brochures and popular articles) on proven technologies must be prepared and disseminated.

iii) *Electronic and Print Media*: Electronic media (i.e., ratio, PTV, and private TV channels like *Sohni Dharti*) must be used for talk shows on eve of major crop seasons as well as during the early phase of the major crops. Similarly, local language articles must be published in national and regional newspapers prior to the commencement of crop seasons.

10.3   SUBSIDY ON MICRONUTRIENT FERTILIZERS

It is always difficult to convince farmers to adopt new technologies, particularly when the technology will need some additional expense (Ahmad et al., 2021). This has been the case with the introduction of fertilizers in the 1950s and 1960s. Thus, in the interest of inducing the farmers to use micronutrient fertilizers, this farm input (micronutrient fertilizers) must be subsidized like nitrogenous, phosphatic, and sometimes potassic fertilizers. Considering the enormous impact of micronutrient fertilizer application on the quality of produce; there seems to be a strong case for the Government of Pakistan to consider a subsidy program at the national scale. Recently, the Government of Punjab, in coordination with the federal government, has announced a provision of micronutrient fertilizers on 50% cost sharing basis to rice-growing farmers for 181,300 acres during 2019–20 to 2023–24 period (Project ref: National Program for Enhancing Profitability through Increasing Productivity of Rice", see http://ext.agripunjab.gov.pk/rice_project).

10.4   MICRONUTRIENT-FORTIFIED FERTILIZER PRODUCTS

World over, especially in developed countries of the world, micronutrient fertilizers are used by way of micronutrient-fortified NPK/NP fertilizers or as complex fertilizers containing the required micronutrient in the fertilizer granules. Both granular and fluid NPK/NP fertilizers are used as carriers of micronutrients. Fortified fertilizers containing micronutrients can be prepared by:

   i) Incorporation during the manufacture of granular fertilizers
  ii) Bulk blending with granular fertilizers
 iii) Coating onto granular fertilizer
  iv) By mixing with fluid fertilizers just before application to soils or making foliar sprays

Micronutrient mixtures for a particular cropping system/area are prepared in accordance with the reported micronutrient deficiencies. Such fertilizer products, if water-soluble, can also be used for foliar sprays of fruit orchards and field crops like cotton.

In Pakistan, one practical problem with micronutrient fertilization is the uniform application of the small quantity of micronutrient fertilizer over a large area. Uniform field application is essential because, apart from the decreased beneficial impact of crop yield, uneven distribution (particularly of B) could prove toxic. It is to be emphasized that while correction of a micronutrient deficiency is easy and cost-effective, amelioration of B toxicity is far more difficult and expensive. In addition, the availability of micronutrient-fortified fertilizers will help eliminate the cost of separate applications of micronutrients.

10.5   ADEQUATE FERTILIZER AVAILABILITY IN TIME AND SPACE

Adequate availability of quality micronutrient fertilizer products in time and space at affordable prices is a prerequisite for its use by the growers. Therefore, this practical aspect merits serious consideration by decision-makers in the government as well as the fertilizer industry.

10.6   MICRONUTRIENT FERTILIZER QUALITY CONTROL

The poor quality of micronutrient fertilizer products (solid as well as liquid) being marketed unchecked is not only causing financial loss to the farmers but also eroding confidence of the growers in the usefulness of fertilizers. Therefore, the concerned regulatory agencies must play a more effective role to improve the situation. The situation in the Punjab province has started improving visibly during the recent years (~9% of the total samples were unfit out of ~4892 samples analyzed at the central laboratory of Soil Fertility Research Institute, Lahore – a percentage that is far lower than that reported by NFDC (1998) for an earlier period). However, the current national scenario demands special attention from the regulators, especially in Sindh, KP, and Balochistan provinces.

REFERENCES

Ahmad, W., N. Ullah, L. Xu, and A. EL Sabagh. 2021. Global food and nutrition security under changing climate. Editorial: Frontiers in Agronomy. doi:10.3389/fagro.2021.799878.

NFDC. 1998. Micronutrients in Agriculture: Pakistan perspective. National Fertilizer Development Center, Islamabad, 51 pp.

11 Micronutrient Biofortification for Addressing Malnutrition in Humans

11.1 INTRODUCTION: SOIL–PLANT MICRONUTRIENT DEFICIENCIES AND HUMAN MALNUTRITION

The range of essential micronutrients for humans and animals is broader than that for plants and extends to a range of organic compounds such as Vitamin A (Welch and Graham, 2005). For instance, while selenium (Se) has been recognized as essential for humans and animals since 1957 (Hartikainen, 2005), it is not recognized as essential for plants. However, strong grain responses to Se have been reported for some flowering plants like *Astragulus* and *Arabidopsis* (Graham et al., 2005). Conversely, B is essential for plants but not yet recognized as essential for animals and humans. However, research is building evidence that B is essential for animals and humans (Nielsen, 2002). For example, B regulates the endocrine system, metabolism, immune response, and hematology parameters in humans and animals (Başoğlu et al., 2010; Kabu and Akosman, 2013). Therefore, any deficiency of B, in human diet, may affect the bone health and calcium (Ca) metabolism and may increase the incidences of prostate cancer (Gupta et al., 2011). Boron should, therefore, be included in human diet.

Billions of people in developing countries suffer from an insidious form of hunger known as micronutrient malnutrition or *hidden hunger* (Bouis et al., 2012). Deficiencies of Zn and Fe afflict more than one-third of the world population (Host and Brown, 2004; Stein, 2010); women and children in resource-poor families are the most affected (Welch, 2008). A close geographical overlap is observed between the global distribution of soil and human Zn deficiency (Alloway, 2008; Cakmak, 2008a), which tells that inadequate Zn supply through the food chain is hampering human health (Welch, 2008). Human Zn deficiency mainly occurs in areas where cereals are the staple food (Cakmak, 2008a; Bouis et al., 2012). This scenario is particularly true for resource-poor communities in developing countries, like Pakistan, dependent predominantly on staple cereals (i.e., wheat and rice) to meet their dietary requirements. Cereal grains are known to be inherently low in Zn density as well as bioavailability. This relatively recent realization of the pivotal role of low micronutrient content staple food grains as a

DOI: 10.1201/9781003314226-11

cause of micronutrient malnutrition in developing countries of the world (Graham and Welch, 2000) calls for better attention to adequate micronutrient nutrition of crop plants.

The incidence of widespread Zn and Fe malnutrition around the world is attributed largely to inadequate concentrations of these micronutrients in staple foods, like wheat and rice, grown on micronutrient-deficient soils. Micronutrient-deficient soils produce Zn- and Fe-deficient feed- and food-stuffs, which, in turn, cause animal and human malnutrition (Graham and Welch, 2000). Also, release of high-yielding cereal cultivars has contributed to high incidence of Zn deficiency in humans by reducing Zn concentration in grains because of the dilution effect (Cakmak, 2008b; Hussain et al., 2012b).

Two national nutrition surveys in Pakistan conducted in 2001, 2011 and 2018 have revealed an alarming magnitude of Zn malnutrition in humans in all provinces of the country, i.e., in 41% of mothers and 37% of children. Not surprisingly, the disorder is more prevalent among the rural poor (NNS, 2002, 2011). However, a recent survey (NNS, 2019) indicates a slight improvement in the population's Zn status. In Pakistan, Fe deficiency anemia is also a well-recognized common health disorder; it affects 49% of mothers and 29% of children. In recent times, public awareness about the serious gravity of micronutrient malnutrition has increased to some extent.

11.2 BIOFORTIFICATION: A STRATEGY TO ADDRESS MICRONUTRIENT MALNUTRITION

Inadequate concentration of micronutrients in staple foods is one of the causes of malnutrition affecting many people in the developing world. There are many methods to increase the micronutrient quality of food. One of them is biofortification – a method to increase the nutritional value of crops.

In Pakistan, 50–70% of the total 22 million ha cultivated area is deficient in Zn, depending on the cropping system (Rashid, 2006). Likewise, its one-half population, mostly resource-poor women and children, suffer from Zn malnutrition (Table 11.1; NNS, 2002, 2011). However, the situation seems to have improved during National Nutrition Survey carried out in 2018. Mild to acute Zn deficiency has also been observed in cattle and buffaloes in the country, which is hampering growth, reproduction, and/or lactation (Rashid, 2006). Cereal grains are inherently very low both in concentration and in bioavailability of Zn, particularly when grown on potentially Zn-deficient soils (Cakmak et al., 2010; Hussain et al., 2012a). Thus, of the plant micronutrients, only Zn is directly linked in the food chain such that its deficiency is extensive in both humans and their food crops.

The unpublished data of Munir Zia (per. com.) pertaining to micronutrient concentrations of wheat grains collected from 75 field locations across Pakistan, including Pakistan Administered Kashmir, reveals wide variations as per the following values: total Fe (min 30, max 233, median 42 mg kg^{-1}); Cu (min 3.0, max 7.7, median 4.3 mg kg^{-1}); Mn (min 16.6, max 60.3, median 34.3 mg kg^{-1}); Zn (min 15.1, max 39.7, median 24.5 mg kg^{-1}); Mo (min 0.17, max 1.1, median

TABLE 11.1

Alarming Magnitude of Human Zinc Deficiency in Pakistan

	National Nutrition Surveys		
Age Group (Zn-Deficient*)	2018	2011	2001
Children (<5 yr)	19%	39%	37%
Women	22.1%		
Pregnant		48%	49%
Non-pregnant		41%	40%

Adapted from National Nutrition Survey -NNS, Planning Commission, Government of Pakistan (2002, 2011, and 2019).

* Serum Zn: <60 mcg/dL

0.63 mg kg^{-1}); and Ni (min 0.06, max 7.5, median 0.14 mg kg^{-1}). Based on per capita wheat flour consumption of 330 g a day by a healthy individual weighing 70 kg, this would ideally provide a daily intake of:

i) 8.6 mg Zn against recommended daily intake of 11 mg. Since the bio-availability of Zn consumed via grains is very low due to phytic acid contents, the majority of the population of this country is deficient in this essential mineral;

ii) 15.9 mg Fe against recommended daily intake of 8 mg. Since the bio-availability of Fe consumed via grains is very low, the majority of the population of this country is deficient in this essential mineral;

iii) 1.5 mg Cu against recommended daily intake of 0.9 mg;

iv) 12.2 mg Mn against recommended daily intake of 2.3 mg;

v) 0.054 mg Ni against the maximum allowable intake of 1.0 mg;

vi) 0.22 mg Mo against recommended daily intake of 0.05 mg.

This depicts that the population of Pakistan that meets more than 50% of its caloric needs from wheat flour is probably not deficient in Cu, Mn, Ni, and Mo minerals.

The best option for addressing micronutrient malnutrition is the enrichment of crop produce through biofortification, which can be accomplished by growing micronutrient-efficient crop genotypes coupled with micronutrient fertilizer use. Thus, adequate micronutrient nutrition of crop plants has gained enhanced emphasis because of the realization of their critical importance for catering to micronutrient requirements in the human diet (Welch and Graham, 2005). Thus, the necessity of promoting micronutrient fertilizer use in the country cannot be over-emphasized.

In short, while crop production in the past focused on higher productivity per unit of land area as its primary goal, the focus for the future will also include the produce quality to cater to human nutrition and health.

11.3   BIOFORTIFICATION OF STAPLE CEREAL GRAINS WITH ZINC IODINE, SELENIUM, AND IRON

Biofortification aims at increasing the density and balance of nutrients essential to humans in harvested plant produce, mainly by means of genetic manipulation and crop management practices. The global implications of enriching foodstuff from agriculture for human nutrition have been expanded forcefully in many countries of the world, including Pakistan, India, and China. Consequently, there is a substantial increase in research for enhancing micronutrient concentrations in staple foods (like wheat, rice, sorghum, cassava, and common bean). A salient outcome of this realization is the global effort by launching the HarvestPlus/HarvestZinc program.

The predominant biofortification approaches used to enrich micronutrients in the crop produce are genetic biofortification and agronomic biofortification

11.3.1   Genetic Biofortification with Zinc

Like other crop species, wheat genotypes possess variation in their capacity for Zn acquisition and utilization (Banziger and Long, 2000; Oikeh et al., 2003). In a pot culture study at University of Agriculture, Faisalabad, Maqsood et al. (2009) studied partitioning of Zn in grain and straw of 12 wheat genotypes and observed useful genetic variations among wheat genotypes for Zn acquisition and its utilization to produce biomass. In their study, three genotypes (*Sehar-2006*, *Shafaq-06*, and *Inqalab-91*) retained more Zn in grain compared to straw. Efficient cultivars translocated more Zn from the straw to grains. Thus, these wheat genotypes can be exploited to develop Zn-efficient wheat cultivars in Pakistan.

Genetic biofortification or breeding plants with traits that result in more efficient acquisition and accumulation of bioavailable nutrients in the edible portions of staple crops is a long-term approach to combat malnutrition. Opposed to this, agronomic biofortification refers to the enrichment of the crop produce by using micronutrient fertilizers. Agronomic biofortification is a rapid and cost-effective approach to enhancing micronutrient density in staple food grains grown in soils affected by micronutrient deficiencies. Genetic biofortification is relatively a time-consuming strategy for fortification but is potentially more cost-effective in the long run compared to agronomic biofortification.

Right from its inception during the early 2000s, Pakistan is an active partner in the HarvestPlus global program for enriching staple crop produce with micronutrients. Though the mandate of HarvestPlus includes a wide array of staple crops, including wheat and rice and micronutrients (i.e., Zn, Fe, I, vitamin A, etc.) and Pakistan was assigned to work on wheat enrichment with Zn only. Wheat cultivars grown in Pakistan are genetically less diverse with regard to grain micronutrient concentration, owing to the fact that the emphasis of breeding programs remained on the development of high-yielding varieties only. Therefore genetically diverse wheat germplasm with high endospermic Zn density and high grain yield potential should be used in the breeding programs, aiming at improving Zn bioavailability (Rehman et al., 2018). The HarvestPlus research efforts, so far, have resulted in the development of five Zn-efficient wheat lines (HarvestPlus, 2014). Three high-Zn wheat lines under test in Pakistan, *"Zincol-2016"*, *"Akbar-2019"*, and *"Nawab-2021"*, have been approved to-date as varieties by the Punjab Seed Council, Government of Punjab, Pakistan. Whereas the average concentration of Zn in Pakistani wheat varieties is around 25.0 mg kg^{-1} (average of 75 locations across Pakistan), the high-Zn wheat varieties developed by the National Agriculture Research System, with the financial support of HarvestPlus, contain around 35–39 mg Zn kg^{-1} (average of many field locations across the country).

To measure the Zn uptake potential of *Zincol-2016*, replicated field trials were carried out using soils with contrasted zinc (DTPA-extractable) status (Zia et al., 2020). The study confirmed that the performance of biofortified wheat (*Zincol-2016*) is site-specific when tested across zinc-deficient, zinc-medium, and zinc-rich soil environments. This means that variety alone does not determine the accumulation of zinc within grains. Therefore, to evaluate the potential nutritional impact of Zn-biofortified crops it is important to understand how varietal performance is influenced by environmental factors, including soil type and crop management. Designing studies to detect realistic effect sizes for new varieties and crop management strategies is, therefore, an important consideration. For example, Zia et al. (2020) indicated that 16 replicate plots would be needed to achieve 80% power to detect a 25% increase in grain Zn concentration at a zinc-deficient site. Recent pilot trials of feeding *"Zincol-2016"*-derived whole-wheat flour in Pakistan (Khan et al., 2017) have confirmed significant improvement in human plasma Zn concentration (PZC), i.e., from 680 µg L^{-1} to 794 µg L^{-1}. In the follow-up second trial (Lowe et al., 2022), 50 households were fed with Zn-biofortified and control wheat flower for a period of eight weeks. Significant increase in plasma Zn concentration was noted after four weeks (mean difference 41.5 µg L^{-1}); however, the difference was not present at the end of the intervention period, i.e., eight weeks. In a follow-up study (Gupta et al., 2022), Zn-biofortified flour was fed to adolescent girls of 10-16 years age ($n = 517$) for a period of 5.5 months. This had non-significant effect on PZC (control 641.6 ±95.3 µg/L vs. intervention 643.8 ± 106.2 µg/L; $p = 0.455$); however, there was an overall reduction in the rate of storage iron deficiency (SF < 15 µg/L; control 11.8% vs. 1.0% intervention). To validate the results, further feeding trials are ongoing under BiZiFED2 project using a larger set of population.

11.3.2 AGRONOMIC BIOFORTIFICATION OF STAPLE CEREALS WITH ZINC, IODINE, SELENIUM, AND IRON

Agronomic biofortification is the enrichment of the crop produce by using micronutrient fertilizers. Zinc fertilization is an effective agronomic approach for Zn, I, and Fe biofortification of staple food grains for overcoming human micronutrient malnutrition. There has been a research emphasis on micronutrient density in staple foods and feeds because of their critical importance for the provision of micronutrients for sustaining human and animal health. Initially, field studies in Central

Anatolia, Turkey, revealed that Zn fertilization of the severely Zn-deficient soils can enhance wheat grain Zn concentration up to three fold (Yilmaz et al., 1997).

Since the inception of HarvestZinc Fertilizer Project in 2008, under the leadership of Prof. Ismail Cakmak of Sabanci University, Istanbul, Turkey, NIAB, Faisalabad, under the technical guidance of the lead author of this book, Dr. Abdul Rashid, is an active partner in this global effort for agronomic biofortification of staple cereals (i.e., wheat and rice in case of Pakistan). This collaborative project aims at enhancing Zn, iodine (I), selenium (Se), and Fe density in wheat and rice grains by using their fertilizer products to address micronutrient malnutrition in resource-poor segments of the society. Following are salient achievements of this multi-country, multi-year project:

i) Extensive multi-year field research in many countries, including Pakistan, has established that foliar feeding of fertilizer Zn is more effective in enhancing Zn density in wheat and rice grains, compared with soil-applied Zn fertilizer (Phattarakul et al., 2012; Zou et al., 2012). Foliar Zn application alone or in combination with soil application increased wheat grain Zn concentration from 27 mg kg^{-1} to 48 mg kg^{-1} across all site-years in seven countries (Table 11.2). The best strategy for enhancing grain Zn density is a combination of soil application and foliar feeding of Zn at the booting stage (Cakmak, 2008a; Zou et al., 2012). Foliar Zn sprays are effective even when applied after mixing with pesticide spray solutions, like *Confidor* in the case of Pakistan – which is spray-applied to control aphids in wheat (Ram et al., 2016);

ii) Field-cum-laboratory research in this project has also established that simultaneous foliar sprays of Zn, I, and Se are effective for increasing their density in wheat and rice grains (Zou et al., 2019; Prom-u-thai et al., 2020). Foliar-applied cocktail of Zn, I, and Se also proved effective for simultaneous biofortification of these micronutrients in grains of diverse rice cultivars to address the *hidden hunger* problem in human populations (Naeem et al., 2021).

iii) The use of high-Zn wheat grains as seed for the next crop increases wheat yield because of better crop establishment (Rashid et al., 2019).

Mobility of Fe within the plant is restricted and foliar or soil application has very little or no effect on the amount of Fe accumulated in edible plant parts. On the other hand, foliar or combined soil plus foliar application of Zn fertilizer have been found to be effective in maximizing grain Zn density (Cakmak, 2008a, 2008b; Zou et al., 2012). In fact, foliar feeding of Zn in combination with pesticide spray solutions is also equally effective in enhancing cereal grain Zn density (Ram et al., 2016).

In a field experiment on a calcareous soil of Faisalabad containing 0.71 mg kg^{-1} DTPA-extractable Zn, Hussain et al. (2013) observed an increase in wheat grain Zn concentration from 21.6 mg kg^{-1} to 40.8 mg kg^{-1} with soil-applied 18 kg Zn ha^{-1}. The soil Zn fertilization also improved Zn bioavailability significantly by

TABLE 11.2

Grain Zn Concentration of Wheat Grown with Zn Fertilization in Seven Countries

Country	Location	Year	Grain Zn concentration (mg kg^{-1})				F test	LSD$_{0.05}$
			Nil	Soil Zn	Foliar Zn	Soil + foliar Zn		
China	Quzhou	2009	27.7	32.4	47.2	53.9	**	3.5
		2010	29.5	40.9	44.3	52.0	**	6.8
	Yongshou	2009	18.8	21.0	26.5	22.5	**	3.8
		2010	19.5	22.1	31.4	34.0	***	3.5
India	Varanasi	2008	29.0	32.0	44.0	47.0	**	12.0
	Kapurthala	2010	49.0	52.0	64.8	65.3	**	7.4
		2011	31.4	30.2	51.1	49.1	**	7.9
	Ludhiana	2010	25.5	30.3	61.0	60.8	**	7.5
		2011	27.3	36.7	58.3	57.0	**	6.4
Pakistan	AARI,	2009	27.0	25.3	48.2	44.6	*	4.3
	Faisalabad	2008	29.0	29.0	60.0	59.0	***	9.0
	Muridke-1	2009	45.3	38.5	52.6	53.6	*	9.3
		2010	28.0	33.2	44.1	42.7	*	4.6
	Muridke-2	2009	40.5	46.2	50.9	49.3	*	5.7
		2010	37.3	46.7	53.5	54.7	*	4.3
Turkey	Eskisehir	2009	21.5	21.4	43.9	41.9	**	4.3
		2010	30.1	30.6	43.0	44.8	**	4.4
	Konya	2009	11.6	11.7	19.6	22.1	**	3.2
		2010	14.1	14.1	31.9	32.6	**	4.0
Grand mean			27.4	30.5	48.0	49.0		
		Increase (%) in grain Zn concentration over nil Zn treatment						
Mean				12.3	83.5	89.7		

Adapted from Zou et al., 2012, with permission.
n.d., no data; *, **, and ***, significant at $P<0.05$, $P<0.01$, and $P<0.001$ level, respectively

increasing Zn concentration and decreasing phytate concentration. Recently, Rehman and Farooq (2016) have reported that coating of wheat seed with 1.25 g Zn kg^{-1} seed (either as $ZnSO_4$ or $ZnCl_2$) resulted in improving grain yield and grain Zn biofortification of wheat cvs. *Lasani-2008* and *Faisalabad-2008*.In a field study on rice at two sites in Pakistan having calcareous soils, Zn application significantly increased paddy yield as well as grain Zn concentration (Farooq et al., 2018). Recently, in a field study in Pakistan Zn seed treatment (i.e., seed priming, seed coating) of *desi* and *Kabuli* chickpea improved crop productivity as well as grain Zn density and grain bioavailable Zn (Ullah et al., 2020).

In flooded rice (cv. Basmati-515), grown in a calcareous soil containing 0.37 mg kg^{-1} DTPA-extractable Zn, seed priming (in 0.5% w/v Zn solution), root dipping (in 0.5% Zn solution for 2 h), soil application (20 kg Zn ha^{-1}), two foliar sprays of 0.25% Zn solution, and soil + foliar treatments significantly (P≤0.05) increased grain and straw yield. However, seed priming and root dipping increased paddy yield only by ≤5%. Brown rice Zn concentration increased from 22 (without Zn application) to 29 mg kg^{-1} (with soil + foliar Zn application). Zinc applications, especially soil + foliar, decreased grain [phytate]:[Zn] ratio and increased estimated human Zn bioavailability in rice grains based on a trivariate model of Zn absorption. These authors concluded that soil + foliar Zn application optimized paddy yield and agronomic Zn biofortification of rice. In view of a low increase in grain Zn enrichment (i.e., by 7 mg kg^{-1}) by Zn fertilization (soil + foliar combination), these researchers have suggested the exploitation of molecular and genetic approaches for Zn biofortification in rice (Imran et al., 2015).

In the case of maize grown in a calcareous soil containing 0.72 mg kg^{-1} DTPA Zn in Faisalabad, the maximum increase in grain yield was 21% for Hybrid (FHY-421) with 18 kg Zn ha^{-1} and 13% for synthetic variety (Golden) with 6 kg Zn ha^{-1}. The corresponding increases in grain Zn concentration were 22.0 mg kg^{-1} to 29.5 mg kg^{-1} in the case of Hybrid and 23.0 mg kg^{-1} to 26.0 mg kg^{-1} in the case of synthetic variety (Kanwal et al., 2010). In field-grown mungbean (cv. NM-92) in two calcareous soils of Pakistan as well, substantial increases in grain yield (up to 63%), as well as grain Zn density (up to 79%), were observed with soil+foliar Zn application (Haider et al., 2020).

11.4   BENEFITS OF HIGH-ZINC WHEAT GRAINS AS SEED

High-Zn wheat grains when used as seed for the next-year crop result in much better seedling density, especially in Zn-deficient soil situations. Extensive multi-year wheat field experiment conducted in Pakistan under HarvestZinc Fertilizer Project revealed that both high-Zn density seed and Zn-primed seed increased the density and height of wheat seedlings. Consequently, high-Zn wheat seed, as well as Zn-primed wheat seed, resulted in wheat yield increases at 67% site-years in India and 80% site-years in Pakistan (P≤0.05). Also, high-Zn wheat seed resulted in increased grain Zn density in 25% of site-years in China, 33% of site-years in India, and 40% of site-years in Pakistan (P≤0.05) (Rashid et al., 2019).

Thus, Zn biofortification not only results in high-Zn grain produce, but also the use of high-Zn grains used as seed leads to higher crop productivity because of denser crop stand establishment. In case high-Zn seed is not available, the same benefit can be realized by the use of Zn-primed wheat seed (i.e., by dipping the wheat seed in ZnSO$_4$ solution) (Rashid et al., 2019).

11.5   NEED FOR PREMIUM PRICE FOR ZN-ENRICHED WHEAT

The current use of micronutrient fertilizers in the country is much less compared with actual requirements. Apart from some other constraints, one factor inhibiting farmers to make investments in micronutrient fertilizers is the absence of incentive of better market price for the better-quality crop produce by using micronutrients. The case in point is the price of high-Zn wheat grains produced by making two foliar sprays of Zn fertilizer. The better-quality wheat grains so produced involve an expense (on Zn fertilizer product and labor cost in applying the sprays). However, there is no premium price mechanism for such a valuable staple food produce which is so badly needed in the country to combat Zn malnutrition in resource-poor segments of the society. Therefore, this is a strong case for the provision of premium prices to the growers in the national interest. A policy decision, as well as a workable strategy, is warranted in this regard.

REFERENCES

Alloway, B. J. 2008. *Micronutrient Deficiencies in Global Crop Production.* Springer, Dordrecht.

Banziger, M. and J. Long. 2000. The potential for increasing the iron and zinc density of maize through plant breeding. *Food Nutrition Bulletin* 21:397–400.

Başoglu, A., N. Baspinar, A. S. Ozturk, and P. P. Akalin. 2010. Effects of boron administration on hepatic steatosis, hematological and biochemical prolife les in obese rabbits. *Trace Elements and Electrolytes* 27:225–231.

Bouis, H., E. Boy-Gallego, and J. V. Meenakshi. 2012. Micronutrient malnutrition: Causes, consequences, and interventions. p. 29–64, In: Bruulsema, T. W., et al. (Ed.), *Fertilizing Crops to Improve Human Health: A Scientific Review; Vol. 1 Food and Nutrition Security.* International Plant Nutrition Institute, Norcross, GA, USA, and International Fertilizer Industry Association, Paris, France.

Cakmak, I. 2008a. Zinc deficiency in wheat in Turkey. p. 181–200, In: Alloway, B.J. (Ed.), *Micronutrient Deficiencies in Global Crop Production.* Springer, Dordrecht.

Cakmak, I. 2008b. Marschner review. Enrichment of cereal grains with zinc: Agronomic or genetic biofortification? *Plant and Soil* 302: 1–17.

Cakmak, I., W. H. Pfeiffer, and B. McClafferty. 2010. Biofortification of durum wheat with zinc and iron. *Cereal Chemistry* 87:10–20.

Farooq, M., A. Amanullah, A. Rehman, et al. 2018. Application of zinc improves the productivity and biofortification of fine grain aromatic rice grown in dry seeded and puddled transplanted production systems. *Field Crops Research* 216: 53–62.

Graham, R. D., and R. M. Welch. 2000. Plant food micronutrient composition and human nutrition. *Communications in Soil Science and Plant Analysis* 31:1627–1640.

Graham, R. D., J. C. R. Stangoulis,Y. Genc, and G. H. Lyons. 2005. Selenium can increase growth and fertility in vascular plants. p. 208–209. In: Li, C.J., et al. (Ed.), *Plant Nutrition for Food Security, Human Health and Environmental Protection. Proceedings of the XVth Int. Plant Nut.Colloq.* 9–13 September 2005, Beijing, China. Tsinghua University Press, Beijing.

Gupta, C. U., P. C. Srivastava, and S. C. Gupta. 2011. Role of micronutrients: boron and molybdenum in crops and in human health and nutrition. *Current Nutrition and Food Science* 7:126–136.

Gupta, S., M. Zaman, S. Fatima, et al. 2022. The Impact of consuming zinc-biofortified wheat flour on haematological indices of zinc and iron status in adolescent girls in ruralPakistan: A cluster-randomised, double-blind, controlled effectiveness trial. *Nutrients* 14, 1657. DOI:10.3390/nu14081657

Haider, M. U., M. Hussain, A. Nawaz and M. Farooq. 2020. Zinc nutrition for improving the productivity and grain biofortification of mungbean. *Journal of Soil Science and Plant Nutrition* 20:1321–1335.

Hartikainen, H. 2005. Biogeochemistry of selenium and its impact on food chain quality and human health. *Journal of Trace Elements in Medicine and Biology* 18:309–318.

HarvestPlus. 2014. Biofortification Progress Briefs. Available at www.HarvestPlus.org. (Accessed on Apr 26, 2022).

Host, C., and K. H. Brown. 2004. Assessment of the risk of zinc deficiency in populations for its control. International Zinc Nutrition Consultative Group (IZiNCG). *Food and Nutrition Bulletin* 25:S91–S204.

Hussain, S., M. A. Maqsood, Z. Rengel, and T. Aziz. 2012a. Biofortification and estimated human bioavailability of zinc in wheat grains as influenced by methods of zinc application. *Plant and Soil* 361:279–290.

Hussain, S., M. A. Maqsood, Z. Rengel, and M. K. Khan. 2012b. Mineral bioavailability in grains of Pakistani bread wheat declines from old to current cultivars. *Euphytica* 18:1–11.

Hussain, S., M. A. Maqsood, Z. Rengel, T. Aziz, and M. Abid. 2013. Estimated Zn bioavailability in milling fractions of biofortified wheat grains and flours of different extraction rates. *International Journal of Agriculture and Biology* 15:921–926.

Imran, M., S. Kanwal, S. Hussain, T. Aziz, and M. A. Maqsood. 2015. Efficacy of zinc application methods for concentration and estimated bioavailability of zinc in grains of rice grown on a calcareous soil. *Pakistan Journal of Agricultural Sciences* 52:169–175.

Kabu, M., and M. S. Akosman. 2013. Biological effects of boron. *Reviews of Environmental Contamination and Toxicology* 225:57–75.

Kanwal, S., Rahmatullah, A. M. Ranjha, and R. Ahmad. 2010. Zinc partitioning in maize grain after soil fertilization with zinc sulfate. *International Journal of Agriculture and Biology* 12:299–302.

Khan, M. J., U. Ullah, Usama, et al. 2017. Effect of agronomically biofortified zinc flour on zinc and selenium status in resource poor settings; a randomised control trial. *Proceedings of The Nutrition Society* 76(OCE4), DOI:10.1017/S0029665117003457.

Lowe, N. M., M. Z. Afridi, M. J. Khan, et al. 2022. Biofortified wheat increases dietary zinc intake: A randomised controlled efficacy study of Zincol-2016 in rural Pakistan. *Frontiers in Nutrition* 8:809783, DOI:10.3389/fnut.2021.809783.

Maqsood, M.A., S. Rahmatullah, T. Kanwal A. Aziz, and M. Ashraf. 2009. Evaluation of Zn distribution among grain and straw of twelve indigenous wheat (*Triticum aestivum* L.) genotypes. *Pakistan Journal of Botany* 41:225–231.

Naeem, A., Aslam, M., Ahmad, M., et al. 2021. Biofortification of diverse basmati rice cultivars with iodine, selenium and zinc by individual and cocktail spray of micronutrients. *Agronomy* 12, 49. DOI:10.3390/agronomy12010049.

Nielsen, F. H. 2002. The nutritional importance and pharmacological potential of boron for higher animals and human. p. 37–49. In: Goldbach, H.E., et al. (Ed.) *Boron in Plant and Animal Nutrition*. Kluwer Academic Publishers, Dordrecht.

NNS. 2002. National Nutrition Survey, 2001–2002. Planning Commission, Government of Pakistan, Islamabad, 69 pp.

NNS. 2011. National Nutrition Survey – Pakistan. Planning Commission, Government of Pakistan, Islamabad, 105 pp.

NNS. 2019. National Nutrition Survey, 2018: Key Findings Report. Ministry of National Health Services, Regulation and Coordination, Government of Pakistan, Islamabad, 52 pp.

Oikeh, S. O., A. Menkir, B. Maziya-Dixon, R. Welch, and R. P. Glahn. 2003. Genotypic differences in concentration and bioavailability of kernel-iron in tropical maize varieties grown under field conditions. *Journal of Plant Nutrition* 26:2307–2319.

Phattarakul, N., B. Rerkasem, L. Li, et al. 2012. Biofortification of rice grain with zinc through zinc fertilization in different countries. *Plant and Soil* 361:131–141. DOI:10.1007/s11104-012-1211-x.

Prom-u-thai, C. T., A. Rashid, H. Ram, et al. 2020. Simultaneous biofortification of rice with zinc, iodine, iron and selenium through foliar treatment of a micronutrient cocktail in five countries. *Frontiers in Plant Science* 11:589835.

Ram, H., A. Rashid, W. Zhang, et al. 2016. Biofortification of wheat, rice and common bean by applying foliar zinc fertilizer along with pesticides in seven countries. *Plant and Soil* 403:389–401.

Rashid, A. 2006. Incidence, diagnosis and management of micronutrient deficiencies in crops: Success stories and limitations in Pakistan. In: *Optimizing Resource Use Efficiency for Sustainable Intensification of Agriculture. IFA Agriculture Conference,* Kunming, PR China. 27 February–2 March, 2006.

Rashid, A., H. Ram, C. Zou, et al. 2019. Effect of zinc-biofortified seeds on grain yield of wheat, rice, and common bean grown in six countries. *Journal of Plant Nutrition and Soil Science* 182:791–804.

Rehman, A., and M. Farooq. 2016. Zinc seed coating improves the growth, grain yield and grain biofortification of bread wheat. *Acta Physiologiae Plantarum* 38:238.

Rehman, A., M. Farooq, A. Nawaz, et al. 2018. Characterizing bread wheat genotypes of Pakistani origin for grain zinc biofortification potential. *Journal of the Science of Food and Agriculture* 98:4824–4836.

Stein, A. J. 2010. Global impacts of human mineral malnutrition. *Plant and Soil* 335:133–154.

Ullah, A., M. Farooq, F. Nadeem, et al. 2020. Zinc seed treatments improve productivity, quality and grain biofortification of desi and kabuli chickpea (*Cicerarietinum*). *Crop and Pasture Science* 71:668–678.

Welch, R. M. 2008. Linkages between trace elements in food crops and human health. In: Alloway, B.J. (Ed), *Micronutrient Deficiencies in Global Crop Production.* Springer, The Netherlands, p. 287–309.

Welch, R. M., and R. D. Graham. 2005. Agriculture: the real nexus for enhancing bio-available micronutrients in food crops. *Journal of Trace Elements in Medicine and Biology* 18:299–307.

Yilmaz, A., H. Ekiz, B. Torun, et al. 1997. Different zinc application methods on grain yield and zinc concentrations in wheat grown on zinc deficient calcareous soils in Central Anatolia. *Journal of Plant Nutrition* 20: 461–471.

Zia, M. H., I. Ahmed, E. H. Bailey, et al. 2020. Site-specific factors influence the field performance of a Zn-biofortified wheat variety. *Frontiers in Sustainable Food Systems* Special Issue. DOI:10.3389/fsufs.2020.00135.

Zou, C. Q., Y. Q. Zhang, A. Rashid, et al. 2012. Biofortification of wheat with zinc through zinc fertilization in seven countries. *Plant and Soil* 361:43–55.

Zou, C., Y. Du, A. Rashid, et al. 2019. Simultaneous biofortification of wheat with zinc, iodine, selenium and iron through foliar treatment of a micronutrient cocktailin six countries. *Journal of Agricultural and Food Chemistry* 67:8096–8106.

12 Some Misconceptions about Micronutrients

12.1 INTRODUCTION

Micronutrient fertilizer use is a specialized area of crop nutrition. Because of the lesser R&D as well as its lesser duration, compared with major nutrients like N, P, and K, many aspects of micronutrient management in crops remain blurred to growers and even to the agricultural extension personnel. In specific, some well-known misconceptions, like B toxicity hazard in crops, P–Zn antagonism, and superiority of high-cost chelated micronutrient products, are prohibitive in enhancing micronutrient fertilizer use in the country. Therefore, such misconceptions are discussed in this chapter.

12.2 BORON TOXICITY APPREHENSION IN SOILS OF PAKISTAN

The range between deficiency and toxicity of B is quite narrow. Therefore, great care is warranted in applying high doses or inadvertent uneven distribution of soil-applied B fertilizer. In case of repeated fertilizer application to the same field(s), periodic soil testing is strongly suggested for monitoring soil B levels.

In a global study sponsored by FAO, Sillanpää (1982) observed that 11% of the 242 soil samples provided by Pakistan fell in the high B range. Many of the high values were observed in soil from the Multan district of Punjab and from Sindh province. In a subsequent study in 1993, HWE B content in 164 soil samples collected from throughout the country ranged from 0.10 to 3.16 mg kg^{-1}, indicating the prevalence of B deficiency as well as high soil B contents (NFDC, per. com). Subsequent extensive nutrient indexing studies, however, have failed to verify such *high soil B levels in Multan or elsewhere in the Punjab cotton belt*. Contrarily, widespread occurrence of B deficiency has been observed in the soils and cotton plants of five districts of the cotton belt including Multan (Rafique et al., 2002).

Salt-affected soils (i.e., saline and sodic soils) and the fields irrigated with the B-contaminated irrigation water are prone to B toxicity. However, extensive data are not available about the B status of salt-affected soils and irrigation waters in Pakistan. The rates of B fertilizer suggested for soil application in Pakistan (i.e., 0.75–1.0 kg B ha^{-1}) are not expected to create B toxicity in alkaline/calcareous soils prevalent in the country. In a field experiment at Faisalabad, 4 kg B ha^{-1} increased wheat grain yield by 8%, and fertilizer B dose of up to 10 kg B ha^{-1} had no adverse effect on crop yield (Kausar et al., 1988). Similarly, 10 kg B ha^{-1} applied to rice at Faisalabad did not cause any toxicity in the crop (NIAB, 1997).

DOI: 10.1201/9781003314226-12

In a field trial, Hayat et al. (2008) recorded 10.8% increased yield of cured tobacco leaf with 10 kg B ha^{-1} but also observed maximum reducing sugars and nicotine with this B treatment. At Islamabad, application of 4.0 kg B ha^{-1} to a B-deficient soil increased the grain yield of wheat by 11 %; however, 8 and 16 kg B ha^{-1} proved toxic for wheat; the decrease in grain yield with the highest dose, i.e., 16 kg B ha^{-1} was 14% over control yield. The requirement of B fertilizer for wheat estimated in Islamabad was only 1.2 kg B ha^{-1} (Rashid and Qayyum, 1991; Rashid et al., 2002; Rashid et al., 2011). In cotton, the B fertilizer requirement is about 1.0 kg B ha^{-1}. In extensive field experiments conducted throughout the Punjab cotton belt, B applied up to 3.0 kg B ha^{-1} did not prove toxic (Rashid, 1996; Rashid and Rafique, 2002). Cotton is known to be tolerant to B toxicity (Keren and Bingham, 1985); and on average, at least two crops per year are grown in all cropping systems of Pakistan. The residual effect of B applied to cotton crop is expected to benefit the following one crop (i.e., wheat, in the case of cotton–wheat system). Thus, there appears no possibility of toxicity hazard at the suggested rate of B application, i.e., 0.75–1.0 kg B ha^{-1}. Very high rates of micronutrient fertilizer applications are not expected to be practiced by farmers due to economic reasons, and uneven field distribution of micronutrients can be avoided by using micronutrient-fortified fertilizers or even by pre-mixing of the small fertilizer dose with well-pulverized soil of the same field.

Plant species vary with respect to their tolerance to B toxicity. Thus, in case of soil B toxicity, the crop genotypes tolerant to B toxicity can be grown in such situations. Keren and Bingham (1985) have categorized crop species into sensitive, semi-tolerant, and tolerant and have listed the species in order of increasing tolerance to B within each of the three tolerance classes. In Pakistan, only wheat and barley cultivars have been screened for their tolerance to B toxicity (Kausar et al., 1988, 1991).

High soil levels of other micronutrients, like Zn, Cu, Mn, and Fe, do not prove toxic to plants per se. A major fraction of the soil-applied micronutrient cations becomes inactive because of fixation reactions. If at all, high levels of these micronutrients become harmful, it could be through their antagonistic effect on the uptake of other micronutrients, like Zn-Cu antagonism. In these sorts of situations, the induced problem can be corrected by the application of the affected micronutrient.

Thus, contrary to the B toxicity apprehension in cotton-growing soils of Pakistan, expressed by the FAO global study report on micronutrients (Sillanpää, 1982), subsequent extensive B research carried out by local scientists throughout the country (comprising nutrient indexing of farmer-grown crops and field experimentation) has rather identified and established B deficiency as a widespread problem in the cotton soils (Rafique et al., 2002; Rashid and Rafique, 2002). Contrary to the general perception of high B content in salt-affected soils, Yasin et al. (2002) rather observed B deficiency even in the saline and saline-sodic soils of the cotton-belt in Pakistan (EC of up to 16.3 dS m^{-1}) as maximum HWE B in these soils was 2.2 mg B kg^{-1} soil only. Also, simultaneously, Aslam et al. (2002) observed appreciable increases in paddy yield with B fertilizer application to salt-affected soils of the rice-belt in Punjab province. Thus, so far, B toxicity has not been observed anywhere in the country.

12.3   CAN ZINC AND PHOSPHORUS FERTILIZERS BE APPLIED TOGETHER?

In consideration of widespread deficiencies of Zn and P and their likely build-up in soils upon year after year of fertilization, the interaction between these two nutrients has been studied quite extensively. Usually, this interaction has been perceived as P-induced Zn deficiency. Zinc concentration in plants, in general, has been reported to decrease with P fertilization even at P fertilizer rates which did not affect plant growth adversely (Reddy et al., 1973; Verma and Minhas, 1987). Thus, historically P–Zn interaction has been perceived as being antagonistic. If the interaction is antagonistic, crop yield must decrease due to the application of either of the two nutrients. Contrarily, many experiments have revealed that with their normal rates of application, P–Zn interaction was rather synergistic and optimum crop yields were obtained with balanced application rates of Zn and P fertilizers (Reddy et al., 1973; Takkar et al., 1976; Verma and Minhas, 1987). This synergistic effect, however, can become antagonistic if the soil nutrient availability or fertilizer rates are biased in favor of one of these nutrients (Nayyar and Chhibba, 1992). For instance, in a field experiment on pigeonpea and blackgram in Tamil Nadu, India, maximum grain yields of both legume crops were obtained with the combined application of 25 kg P_2O_5 ha^{-1} and 2.5 kg Zn ha^{-1}. However, a much higher rate of P fertilizer, i.e., 50 kg P_2O_5 ha^{-1}, and the same rate of fertilizer Zn, i.e., 2.5 kg Zn ha^{-1}, resulted in decreased yields of both crops (Devarajan et al., 1980). In another field experiment on maize in a calcareous soil of Bihar, India, 60 kg P_2O_5 ha^{-1} applied along with 5 kg Zn ha^{-1} resulted in maximum grain yield, but the interaction between these two nutrients turned antagonistic at very high rates of both the nutrients, i.e., 240 kg P_2O_5 ha^{-1} and 20 kg Zn ha^{-1} (Sakal and Singh, 1995). In yet another field study on maize crop, a progressive decrease in Zn concentration in leaves, stems, and roots was observed with increasing rates of fertilizer P; and maize plants exhibited Zn deficiency symptoms where Zn fertilizer was not applied. This P-induced Zn deficiency in maize was corrected by an increased supply of Zn (Takkar et al., 1976). Hence, P-induced Zn deficiency can occur in case soil P levels are high or a high rate of P fertilizer is applied but, Zn despite being deficient in the soil, Zn fertilizer is not applied. For better uptake of both the nutrients by field crops, recommended ratio of Zn to P is 1:10.

Historically, the decrease in soil Zn availability due to high rates of P was attributed to the formation of relatively insoluble zinc phosphate in the soil resulting in reduced Zn concentration in the soil solution. This hypothesis was, however, not accepted because several studies have revealed that finely ground zinc phosphate, being soluble in water to some extent could also be a good source of Zn (Takkar et al., 1989). For soil and foliar applications however, $ZnSO_4$ is the most commonly used in alkaline, calcareous soils. Sparingly soluble zinc-fertilizer sources, e.g., ZnO (67–80% w/w Zn), $ZnCO_3$ (56% w/w Zn), and zinc frits (4–16% w/w Zn), are also commonly used in fine-powdered form(s) (Alloway, 2008).

It is worth mentioning here that the water solubility of Zn in ZnO depends on the physical and chemical nature of the product, which is influenced by manufacturing processes, i.e., the French process, the Direct/American process, and the Chemical process. In an interesting study, (McBeath and McLaughlin, 2014) there were no differences in Zn fertilizer recovery between Zn sources (ZnO manufactured through three different processes – all differing in morphology and water solubility, and standard zinc sulfate) and Zn fertilizer recovery was in the order of 1.5–2 % for both soil types, i.e., calcareous and acidic soils. The concentration of Zn in plants increased with the addition of Zn in both types of soils and was above the critical value of 13 mg Zn kg^{-1} plant tissue (in both the soil types). Despite the huge differences in water solubility of Zn between the ZnO products and $ZnSO_4$, the authors (McBeath and McLaughlin, 2014) observed little difference in their bioavailability to plants. In some cases, ZnO products showed slightly higher Zn supply to plants than $ZnSO_4$ when uniformly mixed with soil. The reactions driving dissolution to completion do not occur in a simple laboratory test for water solubility, and hence these tests grossly underestimate the solubility of ZnO in soil systems when the product is mixed throughout the soil, the authors concluded. Hence, the choice of Zn fertilizer should be driven not only by considerations of water solubility of the product measured in the laboratory, but also by the intended application method or placement/mixing in soil.

A co-author of this book was told by many officers of the Punjab and Sindh Agriculture Departments that Zn and phosphatic fertilizers should not be mixed together or applied together, as after mixing these two fertilizers transform into an "insoluble stone" in the soil through some reaction that makes Zn fertilizer unavailable to crops. This perception is clearly in ignorance of the fact that in neighboring India, Zincated-DAP (Paras Fertilizers – Tata Chemicals Limited) and Zincated-SSP (containing 0.5–1% Zn) fertilizers (Rama Phosphates Ltd.; M/s Patel Phoschem Ltd.; and CIL Ranipet; Khaitan Chemicals and Fertilizers Limited, etc.) are being manufactured on a commercial scale, and even zinc phosphate itself is being listed as a Zn fertilizer (Alloway, 2008). Lindsay (1972) reviewed that there are no phosphate compounds of Zn which are sufficiently insoluble to account for the observed correlation of zinc deficiency problems in crops which have received heavy applications of phosphatic fertilizers through the formation of an insoluble zinc phosphate complex.

Single superphosphate (SSP) fertilizer used in Australia contains about 600 mg Zn kg^{-1} fertilizer that is considered sufficient enough to meet the long-term Zn requirement of wheat crop (Brennan, 2001). Similarly, the use of compound NPK fertilizers containing 1% Zn is being promoted on a wide-scale across Turkey. Also, a long-term 25-year study on maize crop grown on a calcareous soil concluded that there was no induction of Zn deficiency by long-term high P application (Mallarino and Webb, 1995). Recently, improved wheat and corn growth response to P was realized when the Zn supply was sufficient or when Zn was co-granulated with P fertilizer by the scientists (A. Yazici and I. Cakmak) at Sabanci University, Turkey. Thus, despite the fact that international literature has refuted the old myth of P–Zn antagonism or P-induced Zn deficiency, this perception still prevails in Pakistan. A recent review of 96 peer-reviewed publications by the International Fertilizer Development Center (IFDC) (Rietra et al., 2017) suggested that two of the reviewed studies concluded synergism between Zn and P; four studies concluded Liebig-synergism; while a single study noted zero-interaction between the two nutrients (Figure 12.1). Thus, this review by IFDC clearly negates the misconception that Zn and P fertilizers should not be applied together to crops. Thus, there is a need to update the knowledge level of local Agricultural Extension staff about micronutrient fertilizers.

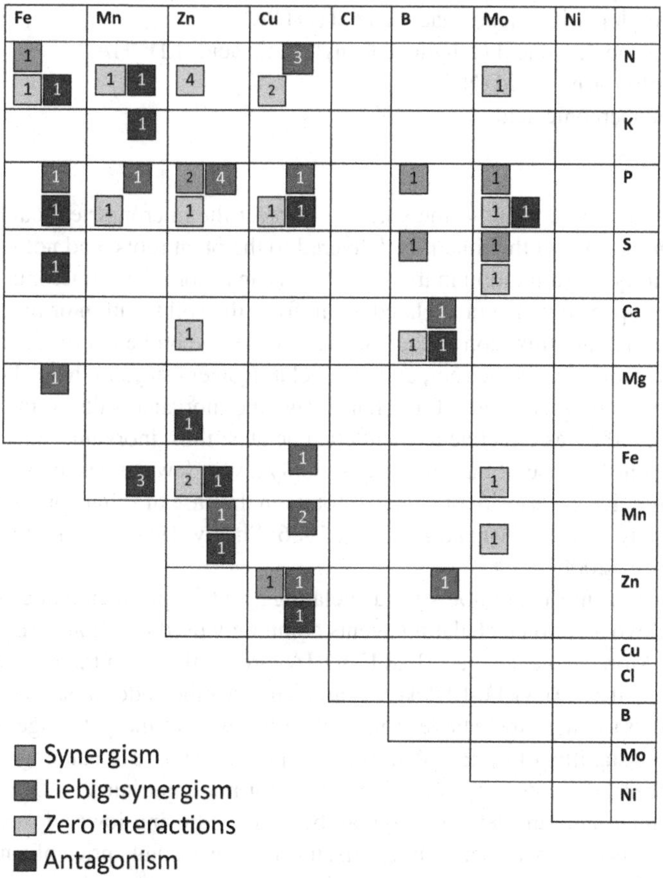

	Fe	Mn	Zn	Cu	Cl	B	Mo	Ni	
	1 / 1 1	1 1	4	3 / 2			1		N
			1						K
	1 / 1	1	2 4 / 1	1 / 1 1		1	1 / 1 1		P
	1					1	1 / 1		S
			1			1 / 1 1			Ca
	1		1						Mg
		3 / 2 1		1			1		Fe
			1 / 1		2		1		Mn
				1 1 / 1		1			Zn
									Cu
									Cl
									B
									Mo
									Ni

■ Synergism
■ Liebig-synergism
□ Zero interactions
■ Antagonism

FIGURE 12.1 A review of nutrient interactions (numbers in squares refer to the number of studies). (Rietra et al., 2017, adapted with permission.)

12.4   ARE CHELATED MICRONUTRIENTS DEFINITELY MORE EFFECTIVE?

The terms "chelate", "chelated complexes", or "chelated compounds" refer to organo-metallic molecules of varying size and shapes in which the organic part binds the metal cation (like Zn^{2+}) in a ring-like structure. The organic chelating agent (e.g., EDTA) is known as the "ligand" which is negatively charged and forms covalent bonds with the positively charged metal cation (like Zn^{2+}). Only micronutrient cations are chelated. Other non-metal micronutrients (i.e., B, Cl, and Mo) are not chelated.

Some salient synthetic chelating agents used for the production of micronutrient chelates are as follows:

- Ethylene diaminetetraacetic acid (EDTA)
- Hydroxy ethylene diaminetriacetic acid (HEDTA)
- Diethylenetriaminepentaacetic acid (DTPA)
- Ethylene diamine di(o-hydroxphenyl acetic acid (EDDHA)
- Nitrilotriacetic acid (NTA)
- Glucoheptonic acid
- Citric acid

The job of a good chelate in the soil is to protect the micronutrient cation from fixation in the soil till the time it is delivered to the plant roots, and not to release it prematurely for re-fixation in the soil. Plants can absorb these micronutrients as cations, and possibly also in chelated form, from the soil solution or the leaf surface. For efficient utilization of the chelate, it has to be broken at the root surface so that the cation enters the root, and the chelating agent stays in the soil (Mengel and Kirkby, 1987). However, for foliar sprays the inorganic salts of micronutrients are generally as effective as synthetic chelates; thus, inorganic salts are preferred options because of their lower cost. Only a very few studies, however, have reported relative superiority of chelated forms in the case of foliar sprays, e.g., for Fe (Basiouny et al., 1970) and Zn (Hsu, 1986; Shazly, 1986; Karak et al., 2006; Correia et al., 2008).

EDTA is the most commonly used chelating agent for producing micronutrient fertilizer. Two additional chelating agents commonly used for Fe are HEDTA and EDDHA. More stable chelates (like EDDHA) are costlier than the relatively less stable ones (like EDTA). Fe-EDDHA (marketed under the trade names *Sequestrene* and *Feriplex*) is the most stable iron chelate throughout the pH range of 4–10, whereas the stability of Fe-EDTA falls above pH 7.0. Thus, even among chelating sources, there is a wide range of effectiveness for micronutrients.

Since micronutrient deficiencies may be corrected using lower doses of chelated products than that of inorganic salts, it is not the absolute price of a micronutrient fertilizer product alone that is important but rather the input–output relationship is of prime interest to the farmer. In general, inorganic salts are considered suitable for "low-value" agronomic crops and the chelated micronutrients may find acceptance in high-value cash crops like cotton, fruits, vegetables, and ornamentals – particularly for foliar sprays. However, opinions vary on the agronomic superiority of chelated products over inorganic salts. According to Tandon (2013), the higher efficiency of chelated micronutrients has not been well proven, except for Fe. Research on Zn nutrition of rice and wheat in Pakistan and several other countries has also revealed that chelated Zn products were not superior to inorganic $ZnSO_4$ (Rashid et al., 2014; Ismail Cakmak, per. com). In any case, prior to recommending the use of any chelated micronutrient product, that particular chelated micronutrient fertilizer must be extensively evaluated, in comparison with the predominant inorganic fertilizer of the same micronutrient, in field conditions in terms of their application rates and the net returns they can produce.

REFERENCES

Alloway, B. J. (Ed.). 2008. *Micronutrient Deficiencies in Global Crop Production.* Springer, Dordrecht.

Aslam, M., I. H. Mahmood, R. H. Qureshi, S. Nawaz, and J. Akhtar. 2002. Salinity tolerance of rice as affected by boron nutrition. *Pakistan Journal of Soil Science* 21:110–118.

Basiouny, F. M., C. D. Leonard, and R. H. Biggs. 1970. Comparison of different iron formulations for effectiveness in correcting iron chlorosis in citrus. *Proceedings of the Florida State Horticultural Society* 83:1–6.

Brennan, R. F. 2001. Residual value of zinc fertilizer for production of wheat. *Australian Journal of Experimental Agriculture* 41:541–547.

Correia, M. A. R., R. D. Prado, L. S. Collier, D. E. Rosane, and L. M. Romualdo. 2008. Zinc forms of application in the nutrition and the initial growth of the culture of the rice. *Bioscience Journal* 24:1–7.

Devarajan, R., M. M. Sheriff, G. Ramanathan, and G. Selvakumari. 1980. Effect of phosphorus and zinc fertilization on yield, content and its uptake by pulse crops. *Indian Journal of Agricultural Research* 14:47–52.

Hayat, Z., H. Gul, H. Akbar, et al. 2008. Yield and quality of flue-cured Virginia tobacco, Nicotianatobacum L. as affected by different levels of fico-micron and boron. *Sarhad Journal of Agriculture* 24:211–216.

Hsu, H. H. 1986. The absorption and distribution of metalosates from foliar fertilization. p. 236–354, In: Ashmead, H. De W. (Ed.). *Foliar Feeding of Plants with Amino Acid Chelates.* Noyes Publications, Park Ridge, NJ.

Karak, T., D. K. Das, and D. Maiti. 2006. Yield and zinc uptake in rice (*Oryza sativa*) as influenced by sources and times of zinc application. *Indian Journal of Agricultural Science* 76:346–348.

Kausar, M. A., M. Tahir, and M. Sharif. 1988. Wheat response to field application of boron and evaluation of various methods for the estimation of soil boron, p. 132–138. *Proc. National Seminar on Micronutrients in Soils and Crops in Pakistan*, December 13–15, 1988. NWFP Agriculture University, Peshawar.

Kausar, M. A., M. Tahir, R. Ahmed, and A. Hamid. 1991. Boron tolerance in five barley varieties. *Pakistan Journal of Soil Science* 6:1–4.

Keren, R., and F. T. Bingham. 1985. Boron in waters, soils, and plants. *Advances in Soil Science* 1:229–276.

Lindsay, W. L. 1972. Zinc in soil and plant nutrition. *Advances in Agronomy* 24:147–186.

Mallarino, A. P., and J. R. Webb. 1995. Long-term evaluation of phosphorus and zinc interactions in corn. *Journal of Production Agriculture* 8:52–55.

McBeath, T. M., and M. McLaughlin. 2014. Efficacy of zinc oxides as fertilisers. *Plant and Soil* 374:843–855.

Mengel, K. and E. A. Kirkby. 1987. *Principles of Plant Nutrition*. International Potash Institute, Worblaufen-Bern, Switzerland.

Nayyar, V. K., and I. M. Chhibba. 1992. Interactions of zinc and other plant nutrients in soils and crops. p. 116–142, In: *Management of Nutrient Interactions in Agriculture. Fertilizer Development and Consultation Organization*, New Delhi, India.

NIAB. 1997. Silver Jubilee of NIAB: 25 Year Report. Nuclear Institute for Agriculture and Biology, Faisalabad.

Rafique, E., A. Rashid, A. U. Bhatti, G. Rasool, and N. Bughio. 2002. Boron deficiency in cotton grown in calcareous soils of Pakistan. I. Distribution of B availability and comparison of soil testing methods. p. 349–356, In: Goldbach, H. E., et al. (Ed.), *Boron in Plant and Animal Nutrition*. Kluwer Academic Publishers, New York.

Rashid, A. 1996. Nutrient Indexing of Cotton in Multan District and Boron and Zinc Nutrition of Cotton. Micronutrients Project, Annual Report, 1994–95. National Agricultural Research Center, Islamabad. 76 pp.

Rashid, A., and E. Rafique. 2002. Boron deficiency in cotton grown in calcareous soils of Pakistan. II. Correction and criteria for foliar diagnosis: p. 357–362, In: Goldbach, H. E., et al. (Ed.), *Boron in Plant and Animal Nutrition*. Kluwer Academic Publishers, New York.

Rashid, A., and F. Qayyum. 1991. Micronutrient Status of Pakistan Soils and Their Role in Crop Production, Cooperative Program. Final Report, 1983–1990. National Agricultural Research Center, Islamabad. 84 pp.

Rashid, A., E. Rafique, A. U. Bhatti, et al. 2011. Boron deficiency in rainfed wheat in Pakistan: Incidence, spatial variability, and management strategies. *Journal of Plant Nutrition* 34:600–613.

Rashid, A., E. Rafique, and N. Bughio. 2002. Boron deficiency in rainfed alkaline soils of Pakistan: Incidence and boron requirement of wheat. p. 371–379, In: Goldbach, H.E., et al. (Ed.), *Boron in Plant and Animal Nutrition*. Kluwer Academic Publishers, New York.

Rashid, A., K. Mahmood, Mahmood-ul-Hassan, M., Rizwan, and Z. Iqbal. 2014. HarvestPlus Zinc Fertilizer Project, Phase-II Country Report, 2011–2014. Nuclear Institute for Agriculture and Biology, Faisalabad.

Reddy, G. D., V. Vankatasubbaiah, and J. Venkateswarlu. 1973. Zinc-phosphate interaction in maize. *Journal of the Indian Society of Soil Science* 21:433–445.

Rietra, R. P. J. J., M. Heinen, C. O. Dimkpa, and P. S. Bindraban. 2017. Effects of nutrient antagonism and synergism on yield and fertilizer use efficiency. *Communications in Soil Science and Plant Analysis* 48:1895–1920.

Sakal, R. and A. P. Singh. 1995. Boron research and agricultural production. p. 1–31, In: Tondon, H.L.S. (Ed.), *Micronutrient Research and Agricultural Production*. Fertilizer Development and Consultation Organization, New Delhi.

Shazly, S. A. 1986. The effect of amino acid chelated minerals in correcting mineral deficiencies and increasing fruit production in Egypt. p. 236–254, In: Ashmead, H. De W. (Ed.), *Foliar Feeding of Plants with Amino Acid Chelates*. Noyes Pub., Park Ridge, NJ.

Sillanpää, M. 1982. Micronutrients and the Nutrient Status of Soils: A Global Study. FAO Soils Bulletin 48, FAO, Rome.

Takkar, P. N., I. M. Chibba, and S. K. Mehta. 1989. Twenty Years of Coordinated Research on Micronutrients in Soils and Plants. Bulletin 314, Indian Institute of Soil Science, Bhopal.

Takkar, P. N., M. S. Mann, R. L. Bansal, N. S. Randhawa, H. Singh. 1976. Yield and uptake response of corn to zinc as influenced by phosphorus fertilization. *Agronomy Journal* 68:942–946.

Tandon, H. L. S. 2013. *Micronutrient Handbook from Research to Application.* Fertilizer Development and Consultation Organization, New Delhi, India. 234 pp.

Verma, T. S., and R. S. Minhas. 1987. Zinc and phosphorus interaction in wheat grown in limed and unlimed acid Alfisol. *Fertility Research* 13:77–86.

Yasin, M., A. Rashid, and E. Rafique. 2002. Boron status of cotton soils: Relationship with salinity. In: *Abstracts, 9th Congress of Soil Sciences*, 18–20 March, 2002, Faisalabad. Soil Science Society of Pakistan, Islamabad, p. 25.

13 Micronutrient Research and Development Needs

13.1 INTRODUCTION

Research and development (R&D) is a systematic activity that combines basic and applied research to explore simple, appropriate, cost-effective, and feasible solutions to new or existing problems. Collaborative R&D, however, could be a potentially important contributor to facilitating the transfer and uptake of technologies by the farming community, in developing countries (Ockwell et al., 2015). As elaborated in the earlier chapters, relevant R&D in the country, so far, has adequately established the prevalence of certain micronutrient problems and the cost-effectiveness of using micronutrient fertilizers for specific crops (Rashid, 2006; Rashid and Rafique, 2017). Positive impacts of using micronutrient fertilizer on grain quality of rice (Rashid et al., 2004, 2007; Prom-u-thai et al., 2020; Naeem et al., 2021) and wheat (Zou et al., 2012, 2019) have also been well established. However, a lot remains to be explored and propagated to optimize and sustain crop productivity and produce quality in the country. The earlier-suggested R&D needed in the NFDC (1998) status report on micronutrients is outdated now. Therefore, this chapter suggests micronutrient R&D needs in the country in consideration of the gaps observed in earlier chapters of this book.

13.2 RESEARCH NEEDS

- Nutrient indexing of farmer-grown crops in the left-over cropping systems and geographic regions of the country for precise diagnosis of the nature, extent, and severity of micronutrient disorders.
- Determining micronutrient management strategies with an emphasis on Soil Order-specific critical concentrations, major cropping systems, and fruit orchards.
- Investigating residual/build-up effect of micronutrient use in major cropping systems. Long-term micronutrient balance sheets in major cropping systems must also be determined.
- Introducing fertigation in fruits, vegetables, and other high-value crops.
- Developing and/or screening micronutrient efficient crop varieties.
- Investigating micronutrient role in animal and human health.

DOI: 10.1201/9781003314226-13

13.3 DEVELOPMENT NEEDS

- Improvement of micronutrient fertilizer availability in quantity, quality, time and place and import of micronutrients as fertilizers.
- Development and production of good-quality micronutrient products for soil application, foliar feeding, and fertigation.
- Commercial-scale production and marketing of micronutrient(s) coated major/popular fertilizers like Urea, DAP, etc.
- Effective soil–plant–fertilizer micronutrient advisory services and staff training.
- Formulation of micronutrient management recommendations for fruit orchards.
- As "seeing is believing", the utility of field demonstrations in popularizing the use of micronutrients cannot be overemphasized. Thus, extensive field demonstrations on micronutrient use are warranted for educating the stakeholders on their use benefits.
- Media campaigns (print and electronic) emphasizing the need of using micronutrients.
- Promulgation and effective enforcement of a fertilizer quality control law at the national level.

REFERENCES

NFDC. 1998. *Micronutrients in Agriculture: Pakistan perspective*. National Fertilizer Development Center, Islamabad, 51 pp.

Naeem, A., M. Aslam, M. Ahmad, et al. 2021. Biofortification of diverse Basmati rice cultivars with iodine, selenium and zinc by individual and cocktail spray of micronutrients. *Agronomy* 2021(12):49. doi:doi.org/10.3390/agronomy12010049.

Ockwell, D., A. Sagar, and H. de Coninck. 2015. Collaborative research and development (R&D) for climate technology transfer and uptake in developing countries: towards a needs driven approach. *Climatic Change* 131:401–415. doi:doi.org/10.1007/s10584-014-1123-2.

Prom-u-thai, C. T., A. Rashid, H. Ram, et al. 2020. Simultaneous biofortification of rice with zinc, iodine, iron and selenium through foliar treatment of a micronutrient cocktail in five countries. *Frontiers in Plant Science* 11:589835. doi:10.3389/fpls.2020.589835.

Rashid, A. 2006. Incidence, diagnosis and management of micronutrient deficiencies in crops: Success stories and limitations in Pakistan. In: *Optimizing Resource Use Efficiency for Sustainable Intensification of Agriculture. IFA Agriculture Conference*, 27 Feb–2 Mar, 2006, Kunming, P. R. China.

Rashid, A., and E. Rafique. 2017. Boron deficiency diagnosis and management in field crops in calcareous soils of Pakistan: A mini review. *BORON* 2:142–152.

Rashid, A., M. Yasin, M. Ashraf, and R. A. Mann. 2004. Boron deficiency in calcareous soils reduces rice yield and impairs grain quality. *International Rice Research Notes* 29:58–60.

Rashid, A., M. Yasin, M. A. Ali, Z. Ahmad, and R. Ullah. 2007. An alarming boron deficiency in calcareous rice soils of Pakistan: Boron use improves yield and cooking quality. p. 103–116. In: Xu, F., et al. (Ed.) *Advances in Plant and Animal Boron Nutrition. Proc 3rd Int Symp on All Aspects of Plant and Animal Boron Nutrition*, Wuhan, China, 9–13 Sep 2005. Springer, Dordrecht.

Zou, C., Y. Du, A. Rashid, et al. 2019. Simultaneous biofortification of wheat with zinc, iodine, selenium and iron through foliar treatment of a micronutrient cocktail in six countries. *Journal of Agricultural and Food Chemistry* 67:8096–8106. doi:doi. org/10.1021/acs.jafc.9b01829.

Zou, C. Q., Y. Q. Zhang, A. Rashid, et al. 2012. Biofortification of wheat with zinc through zinc fertilization in seven countries. *Plant and Soil* 361:43–55.

14 Managing Micronutrient Deficiencies in the Context of 4R Nutrient Stewardship

14.1 INTRODUCTION: THE 4R NUTRIENT MANAGEMENT STEWARDSHIP

The 4R nutrient management stewardship refers to using the *right source* and the *right rate* of fertilizer at the *right time* and the *right place* (IFA, 2009; IPNI, 2012). Extensive research on major nutrient management, around the world, has demonstrated that the application of fertilizers by following this scientifically sound strategy results in the most effective and cost-effective use of these expensive and crucially needed farm inputs (Johnston and Bruulsema, 2014; Sally, 2020). As the basic principles for managing the deficient macronutrients and micronutrients in crop plants remain the same, this chapter is devoted to the use of micronutrient fertilizers in accordance with the 4R stewardship.

14.2 CROPS REQUIRING MICRONUTRIENT FERTILIZER USE IN PAKISTAN

Formal micronutrient use recommendations in the country pertain only to a few major crops, fruits, and vegetables. Whereas Zn use in rice was recommended in the 1970s, both in Punjab and Sindh provinces, all other micronutrient use recommendations were made in subsequent years. For example, B and Zn use in cotton was initially endorsed by a national forum on cotton production, organized by the National Fertilizer Development Center (NFDC) during 1997 in Islamabad. However, its formal recommendation by the Punjab Agriculture Department was made later, in 1998, followed by the Sindh Agriculture Department in 1999. In extensive field trials in major rice-growing areas of the Punjab province, Zn-enriched rice nursery proved to be as effective as field application of Zn fertilizer, but quite economical and much easier to practice (Rashid et al., 2000). In the year 2003, the use of Zn-enriched rice nursery was recommended by the Punjab

Agriculture Department. Later in 2005, the recommendation for using B fertilizer in rice was made by the Punjab Agriculture Department. During these years, the realization of the beneficial impact of micronutrient use in many other crops (like maize, sugarcane, sorghum, and tobacco), vegetables (like potato and chili), and fruits (like citrus, apple, lemon, and pomegranate) was also gained, and the progressive growers adopted the use of Zn, B, and Fe fertilizers for many crops. The field crops, vegetables, fruits, and ornamentals requiring micronutrient fertilizer use in Pakistan are listed in Table 2.1 and **Annex 15**.

14.3   MICRONUTRIENT FERTILIZER USE RECOMMENDATIONS

Micronutrient deficiencies in crops can be managed by applying fertilizers to soil or through foliar sprays. For instance, deficiencies of Zn and B can be corrected effectively by applying low doses of their fertilizers by soil broadcast or by repeated foliar sprays. Iron deficiency is the most difficult micronutrient disorder to be corrected in the field. Soil application of $FeSO_4$ is generally ineffective; and the high cost of the chelated compound, which is effective by soil application, i.e., Fe-*Sequestrene*, is prohibitive in its use for field crops, etc. Repeated foliar sprays are needed for effective control of micronutrient disorders.

Fertilizer dosage for deficient soils largely depends on crop sensitivity and crop micronutrient requirements rather than the soil test micronutrient status *per se*. Great care should be exercised in recommending fertilizer rates to avoid toxicity (especially in case of B) or excessive levels which may induce deficiency of other micronutrient(s). Generalized recommendations for micronutrient fertilizer use in the country have been summarized in Table 14.1.

TABLE 14.1
Generalized Micronutrient Fertilizer Use Recommendations in Pakistan

Crops	*General Application Rate and Method	
	Soil Application	**Foliar Feeding**
Zinc		
Field crops (excluding rice) and vegetables	2–5 kg elemental Zn ha^{-1}	0.1% elemental Zn solution; 2–3 sprays per season
Rice	20 kg elemental Zn ha^{-1} nursery bed area OR 5 kg elemental Zn ha^{-1} field area	
Wheat, cotton, maize, sugarcane	5 kg elemental Zn ha^{-1} field area	0.1% elemental Zn solution 0.32% solution of $ZnSO_4.H_2O$ 0.5% solution of $ZnSO_4.7H_2O$
Citrus, apple, grape, apricot, peach, plum, pear		Foliar sprays of 0.1% elemental Zn solution, 3 times a year

TABLE 14.1   (CONTINUED)
Generalized Micronutrient Fertilizer Use Recommendations in Pakistan

	*General Application Rate and Method	
Crops	Soil Application	Foliar Feeding
Boron		
Cotton	1.0 kg elemental B ha^{-1}	3 sprays of 0.1% elemental B solution, 45, 60, and 90 days after sowing
Wheat, rice, rapeseed, cauliflower, tobacco, sunflower, sugarbeet	0.75 kg elemental B ha^{-1}	
Peanut, turnip	0.6 kg elemental B ha^{-1}	
Potato, tomato	0.75 kg elemental B ha^{-1}	2 sprays of 0.05% elemental B solution per season
Apple, grape, apricot, peach, plum, pear		2 sprays of 0.1% elemental B solution, every year
Rose and other ornamentals		2 foliar sprays of 0.05% elemental B solution, during a year
Iron		
Crops and fruits		0.5% ferrous sulfate solution or 1% Fe-*Sequestrene* solution; 3–4 sprays per season/per year
Copper		
All crops	5 kg elemental Cu ha^{-1}	0.5% copper sulfate solution; 2–3 sprays per season/per year
Manganese		
All crops	5–10 kg elemental Mn ha^{-1}	0.2–0.5% manganese sulfate; 2–3 sprays per season/per year

Adapted/modified from Rashid and Rafique, 1998, 2017; Rashid et al., 2000, 2002b; Rashid, 2006, 2006b.

* Soil application rates are for the broadcast method; band placement rates are substantially lesser.

For 0.1% foliar elemental Zn spray, dissolve 303 g of $ZnSO_4.H_2O$ 33% Zn *or* 476 g of $ZnSO_4.7H_2O$ 21% Zn in 100 liters of water (approx. five spray tanks, each of 20 liters) and spray over 1-acre crop area.

For 0.1% foliar elemental B spray, dissolve 909 g of borax 11% B *or* 588 g of boric acid 17% B *or* 478 g of *Solubor* 20.9% B in 100 liters of water (approx. five spray tanks, each of 20 liters) and spray over 1-acre crop area.

For 0.05% foliar elemental B spray, dissolve 455 g of borax 11% B *or* 294 g of boric acid 17% B *or* 239 g of *Solubor* 20.9% B in 100 liters of water (approx. five spray tanks, each of 20 liters) and spray over 1-acre crop area.

1) For assuring uniform field broadcast, mix the fertilizer dose with 4–5 times the volume of pulverized soil before broadcast.
2) For avoiding leaf injury, where necessary, neutralize the spray solution with lime or urea.
3) In foliar spray solution, add 0.05% detergent powder, like *Surf*, as surfactant. Depending on crop/fruit age, the required volume of the spray solution will vary from 100 to 400 L ha^{-1}.

Generalized guidelines for soil application of micronutrient fertilizer are as follows:

- Care must be exercised in using the right dose of micronutrient (especially boron) fertilizer, because of the narrow range between their deficient and toxic level.
- For soil application to agronomic crops, broadcast the fertilizer and incorporate it before crop sowing.
- For soil application to fruit orchards, apply the fertilizer under the tree canopy, 50 cm away from the tree trunk.

The soil application rates are for the broadcast method; band placement rates are substantially lower. Depending on crop/fruit age, the required spray solution volume will vary from 100 to 300 L ha^{-1}.

For most field crops, the optimum dosage of micronutrient fertilizers is 2.0–5.0 kg Zn ha^{-1} and 0.75–1.0 kg B ha^{-1} by soil application or 2–3 foliar sprays of 0.1% Zn solution/0.32% solution of $ZnSO_4.H_2O$/0.5% solution of $ZnSO_4.7H_2O$ and 0.05–0.1% solution of B (see **Annex 16** for crop-specific recommendations for using micronutrient fertilizers). A good soil application of Zn and B is adequate for 2–3 crop seasons, at least. Soil application of inorganic Fe sources is generally ineffective; the only effective Fe source for soil application in calcareous soil is Fe-*Sequestrene*, which is quite expensive and, thus, can be used only for high-value plants, like ornamental flowers. Therefore, a general recommendation to cure Fe chlorosis in plants is repeated foliar sprays of 0.5% $FeSO_4$ solution or 1.0% Fe-*Sequestrene* solution.

Also, the growers must observe the following guidelines for attaining cost-effective and safe use of micronutrient fertilizers:

i) Micronutrient use may be based on soil test results and/or experience of positive crop response to micronutrient use on similar soils in the area. As soil testing of each field every year is not practically possible, soil testing for micronutrients of interest must be performed every three to four years. This is extremely important to monitor the residual and cumulative effect of micronutrient fertilizer use, and especially to monitor B build-up following repeated application of B fertilizer.

ii) Soil application of micronutrients may be made pre-planting along with other basal fertilizers.

iii) For assuring uniform field broadcast, mix the fertilizer dose with 4–5 times the volume of pulverized soil or with major nutrient fertilizers, before broadcast.

iv) As the recommended dosage may remain effective at least for 2–3 crop seasons, soil application may be made only to one crop per annum; e.g., to rice crop in the rice–wheat system and to cotton in the cotton–wheat system.

v) Extreme care must be exercised in avoiding an overdose of B fertilizer. Instead of using B fertilizer on every crop or every year, the recommended dose may be used after two to three crop seasons. To avoid the cumulative toxic effect of B fertilizer use, the B status of the soil must be monitored, periodically, to tailor its dosage or application frequency, as dictated by the soil test result.

vi) Wherever feasible, foliar sprays may be practiced. Foliar solutions must be of correct concentration, must contain surfactant (*Surf* or any other detergent powder @ 0.05% in solution) at the recommended concentration, and, if needed, must be properly neutralized. Repeated sprays will be needed to rectify micronutrient deficiencies.

vii) In deciduous fruits (apple, peach, plum, pear, apricot, grape), foliar sprays may be carried out annually at petal fall and one week later.

14.4 COMPATIBILITY OF MICRONUTRIENTS WITH PESTICIDES AND HERBICIDES

Chemical mixing of micronutrient fertilizers with pesticides or herbicides should be approached with caution as some chemicals might be incompatible. Compatible chemicals can be mixed without adverse reactions. The known reactions of the chemicals most commonly used for field crops and orchards are listed in **Annex 17**. This information has been extracted from literature for the guidance of growers to check the compatibility of pesticides and weedicides before mixing with micronutrient fertilizers. Where no information is given, it is advisable to consult the material safety data sheet (MSDS) or check with the chemical manufacturer before trying combinations. For example, "*Solubor*" may be sprayed together with most of the crop protection chemicals as well as with urea. *Solubor* is not compatible with $ZnSO_4$ and $MnSO_4$ (www.chempac.net/images/files/luqhjrsflwcecspg50ebedc23b1bb. pdf). Similarly, a widely used herbicide "Glyphosate" is not compatible with zinc fertilizers since it forms glyphosate chelation of metals such as Zn and Mn. Once molecularly locked up by glyphosate, Zn and Mn lose mobility and efficacy. The chelation process also negatively impacts the herbicidal activity of glyphosate. To address the issue, "System-Ready" has recently been developed by the Agro-K Corporation, USA (www.croplife.com/crop-inputs/ micronutrients/new-micronutrient-formulation-is-glyphosate-compatible/).

Until the development of "System-Ready", growers have had to wait for eight or more days for the residual glyphosate in crop tissue to dissipate before the nutrients (Zn and Mn) could provide benefit. Zinc when applied in liquid form, e.g., through fertigation, is not compatible with humic acid; however, it can be combined with fluvic acid (http://blog.nutri-tech.com.au/zinc-the-leaf-enlarger/). A broader advice regarding mixing compatibility is to "*Never mix the fertilizer solutions containing Ca with the solutions containing phosphates or sulfates*".

14.5 FACTORS AFFECTING AVAILABILITY OF MICRONUTRIENTS TO CROPS

Healthy soils are enriched with micronutrients. Ideal physical and chemical properties lead to ideal biological conditions, and the soil's nutrient cycling functionality is enhanced. Degraded soils are typically those in most need of micronutrients. Management that improves soil properties and overall soil health will lead to less micronutrient deficiency. If soil health is not improved or the soil is degraded further, there will be more need for micronutrients. The following factors regulate the availability of micronutrients to plants.

14.5.1 RESIDUAL VALUE OF PREVIOUSLY APPLIED MICRONUTRIENT FERTILIZER

Micronutrient cations are largely immobile in soil and only move short distances from the point of placement. Micronutrient movement is soil texture dependent, i.e., more in sandy soils than in heavy textured soils. Therefore, the inclusion of soil texture in subjective assessment for Zn availability is suggested (Norton and Roberts, 2012). Micronutrient fertilizers applied to soil have a residual effect on crop growth for several years depending on crop and soil type. For example, based on a review of literature, Lindsay (1972) concluded that soil-applied Zn fertilizers have a residual effect for two to eight years after application. However, this timeframe cannot be extrapolated to all soil-crop situations.

14.5.2 SOIL WATER

Soil moisture affects the availability of micronutrients to plants because most of the movement of micronutrients in the soil is by diffusion. The uptake of micronutrients by plants is reduced under drier conditions as the diffusion rate is slowed. For example, in the case of Zn, except for flooded rice, about 55% of its movement toward crops roots is through diffusion that would be severely affected under dry spells commonly observed in arid agriculture of Pakistan, as well as during the periods between irrigations. Dry soil also restricts root growth and, thus, the ability of the plant to seek out micronutrients in the soil (Marschner, 1993).

14.5.3 SOIL pH

Soil pH is critical to the solubility of micronutrients and the buffering capacity of the soil (Dang et al., 1994; Lindsay, 1972). Solubility of all micronutrient cations, except for Mo and B, decreases as soil pH increases. This explains why critical soil micronutrient values derived for one soil cannot be readily extrapolated to another soil. Reduced solubility of micronutrients in an alkaline soil environment helps explain why the incidence of micronutrient deficiencies is common in crops grown in alkaline soils of Pakistan.

Alkalinity is the main soil chemical constraint that directly reduces micronutrient uptake due to the dependence of micronutrient solubility on pH. Other physico-chemical constraints that reduce root growth through the soil reduce access to soil micronutrients and will, therefore, reduce the availability of soil micronutrients to the plant (Lindsay, 1972).

14.5.4 EFFECT OF OTHER NUTRIENTS ON RESPONSE TO A PARTICULAR MICRONUTRIENT FERTILIZER

Due to the limited number of studies, a systematic overview of interactions between all nutrients is not available. In a review, Rietra et al. (2017) used "yield" as the main parameter to assess the nutrient interactions, which can either be positive or beneficial (synergism), negative or adverse (antagonism), zero-interaction (additive, no interaction, neutral), or partly positive (Liebig-synergism). Plant roots must actively grow into soil layers that contain micronutrients. This requirement means that the limited availability of the major plant nutrients, like N and P, that restricts root growth also restricts micronutrients uptake. Conversely, crops exhibit a micronutrient deficiency due to nutrient dilution when the availability of that micronutrient is marginal and major nutrients, like P and N, are adequate (Loneragan and Webb, 1993). Plant uptake of one particular micronutrient also affects the uptake of other specific micronutrients and crop growth. For example, adequate soil availability of Zn coupled with inadequate or marginal availability of Cu in the soil leads to Cu deficiency in crop plants due to antagonism between the two (Chaudhry et al., 1973). Similar antagonistic interactions are also reported between other micronutrients (Loneragan and Webb, 1993; Rietra et al., 2017). Contrary to the general misconception that an antagonistic interaction exists between P and Zn, the review by Rietra et al. (2017) concluded three types of interactions between the two nutrients, i.e., synergism, Liebig-synergism, and zero-interaction. However, a clear antagonism has been reported between Fe and Mn in this review. For more information on this aspect, readers are referred to **Figure12.1** in Chapter 12.

14.5.5 GENOTYPIC VARIATION IN MICRONUTRIENT UPTAKE AND USE EFFICIENCY

Genetic variation in micronutrient uptake and use efficiency by different cultivars of the same crop species is well known. In Pakistan as well, many instances of diverse variability in susceptibility of different cultivars of the same crop to micronutrient deficiencies have been observed (Table 2.2). For example, in widescale field trials conducted in major rice-growing areas of the Punjab province, the prime long-grain aromatic rice cv. Basmati-370 was less susceptible to Zn deficiency compared with coarse-grain cv. IR-6 (Chaudhry et al., 1977). Similarly, in a study by Maqsood et al. (2009), it was concluded that wheat genotypes with better growth performance and Zn utilization efficiency, such as *Sehar-06* and *Auqab-2000*, could be recommended for Zn-deficient soils as well as for breeding purposes. In Australia, McDonald et al. (2001) observed that the Zn efficiency of 13 wheat cultivars ranged from 77% to 96% when grown in the Victorian Mallee (Birchip). In a pot culture investigation, Rashid et al. (2002a) observed that old cultivars of rice, i.e., Basmati-370 and IR-6, were relatively tolerant (less susceptible) to B deficiency compared with later-developed cvs. Super Basmati and Basmati-6129. In recent field trials on B nutrition of rice conducted in the Punjab and Sindh provinces, the severity of B deficiency was less in Basmati cultivars (Super Basmati and Basmati-385) compared with coarse-grain cvs. IR-6 and KSK-282 (Rashid et al., 2007). In two-year trials under rainfed conditions in Pakistan, B use efficiency of three peanut genotypes declined in the order BARD-479 > ICG-7326> ICGV-92023 (Nawaz et al., 2014). However, an increase in the yield of peanut noted in the study, i.e., up to 34%, seems exceptional. Differential susceptibility to B deficiency was also observed in various cultivars of wheat and rapeseed (Table 2.2). Similarly, chickpea cv. CM-72 was tolerant to Fe deficiency, whereas chickpea cvs. C-44 and CM-88 grown side by side, in the same field, were highly susceptible to Fe chlorosis (Rashid and Din, 1992). The severity of visual symptoms of micronutrient deficiencies is related to plant micronutrient efficiency in soils with low micronutrient status, with less-efficient crop cultivars exhibiting more severe symptoms than more efficient cultivars (Genc and McDonald, 2004).

14.6   MICRONUTRIENT DEFICIENCY MANAGEMENT IN THE CONTEXT OF 4R NUTRIENT STEWARDSHIP

The principles for effective use of any fertilizer (including micronutrient fertilizers) are to select the *right source* of nutrients, applied at the *right rate*, at the *right time*, and at the *right place*. These four rights (4Rs) of nutrient stewardship, principally developed by The Fertilizer Institute (TFI), International Plant Nutrition Institute, Canada (IPNI Canada), International Fertilizer Industry Association (IFA), and Canadian Fertilizer Institute (CFI), indicate that these four factors are all interlocking, and if one is altered, the whole approach should be reconsidered (Bruulsema et al., 2012; Norton and Roberts, 2012).

The 4R approach of nutrient management is linked with various aspects of agricultural production systems (Figure 14.1). The 4R nutrient stewardship is an integral part of the best management practices (BMPs). The *right source* matches nutrient type to crop needs; the *right rate* matches the amount of nutrients to the crop needs and depends on soil and water analyses, crop, and soil fertility status; the *right time* involves making nutrients available when the crop needs them and is correlated with crop-responsive growth stages; and the *right place* encompasses keeping nutrients where crops can use them and is associated with using the best method which shows higher nutrient use efficiency and/or minimal loss.

The 4Rs concerning the management of micronutrient deficiencies are described below.

14.6.1   RIGHT SOURCE

All of the common micronutrient fertilizer products listed in Table 7.1 are not used everywhere. Rather, their use in a particular country, cropping system, or situation depends on the consideration of factors like availability in time and place,

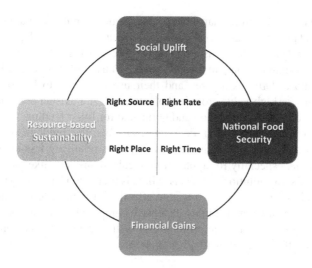

FIGURE 14.1   4R approach of nutrient management and its link with various aspects of agricultural production systems (FAO, 2017a, 2017b; Ahmad et al., 2018)

soil reaction (pH), solubility, ease of use, and economics. For example, globally, the predominant Zn sources are $ZnSO_4$, ZnO, and chelated-Zn. Zinc oxide has low solubility, especially in alkaline and calcareous soils, and chelated-Zn is quite expensive compared to inorganic Zn sources including zinc sulfate. Moreover, in recent extensive field research in Pakistan and many other countries, chelated-Zn did not prove superior to zinc sulfate for correcting Zn deficiency in wheat, both by soil application as well as foliar feeding (Rashid et al., 2014; I. Cakmak, per. com.). Some soil amendments which are by-products of certain industries, like biosolids, are additional options as Zn sources in particular regions; however, their agronomic values are yet to be comprehensively assessed and may contain toxic quantities of other heavy metals. In calcareous soils of South Australia, liquid $ZnSO_4$ is more effective than granular and powdered $ZnSO_4$ (Bertrand et al., 2006). However, the role of liquid Zn in the South Australian research cannot be separated from the role of P and N as the solid and liquid fertilizers are a mix of P, N, and Zn (Holloway et al., 2001). Therefore, in consideration of availability, effectiveness, and cost, in Pakistan zinc sulfate is the best option for use as Zn fertilizer. Both $ZnSO_4.H_2O$ (33% Zn) and $ZnSO_4.7H_2O$ (21% Zn) are equally good sources of Zn. Therefore, depending upon their availability and comparative price, either source can be used to cater to the recommended dose per unit field area.

In the case of B, common fertilizer sources are borax (fertilizer borate), boric acid, *Solubor*, and *Granubor* (Table 7.1). Because of its relatively low solubility, borax is more suitable for soil application than boric acid. Boric acid is soluble and, thus, a better choice for foliar sprays, but its solubility is rather slow in cold water. *Solubor* is readily soluble but is quite expensive compared with boric acid. Also, *Solubor* is not as abundantly available in Pakistan as boric acid. Similarly,

Granubor is more soluble than borax; also being granular in nature, its field broadcast and placement are much more convenient compared with powder borax. Similar to *Solubor*, *Granubor* is a relatively more expensive B source and is less abundantly available, compared with borax. Thus, unless the availability of *Granubor* and *Solubor* is assured and their use is known to be cost-effective (which is true for high-value crops and plantations), in Pakistan borax is the considered choice for soil application and boric acid for foliar feeding.

For Fe, ferrous sulfate and Fe-*Sequestrene* are the predominant fertilizer sources. For curing Fe chlorosis, ferrous sulfate is relatively less effective than Fe-*Sequestrene*, especially for application in calcareous soils like ours. In case a fresh batch of spray solution of ferrous sulfate is prepared just prior to performing the spray, this source of Fe is also quite effective in managing Fe deficiency in plants. As Fe-*Sequestrene* is highly expensive compared with ferrous sulfate, the former source of Fe is suitable only for high-value crops and ornamental plants. And ferrous sulfate is used in all other situations.

14.6.2 RIGHT RATE

Unlike major nutrient fertilizers, the magnitude of variation in rates of micronutrient fertilizers for soil application is rather very small, if any. For instance, the generalized rate of Zn fertilizer for soil application varies from 2 to 5 kg Zn ha^{-1}. In the case of B, the recommended fertilizer rates for soil application vary from 0.75 to 1.0 kg ha^{-1}. However, in other countries like South Africa, B application dose also depends upon the clay contents of soils. For example, Chempac (Pty) Ltd. in South Africa advertises soil application of *Solubor* @ 4–5 kg ha^{-1} for maize, and sunflower if the soil contains up to 15% clay. In case clay contents are in the range of 16–30%, the recommended *Solubor* dose is much higher, i.e., 5–10 kg ha^{-1}. If clay contents are beyond 30%, the *Solubor* application rate is still higher, i.e., 11–15 kg ha^{-1}. Another aspect of B fertilization is that a relatively less fertilizer B rate is recommended for crop species which are more susceptible to B toxicity, like rice, compared with the crop species which are known to be relatively tolerant to high soil B levels. However, great care must be exercised for avoiding overdose of B fertilizer because the range between B deficiency and B toxicity is very small. As is true for major nutrient fertilizers, rates of micronutrient fertilizers are much smaller for band placement compared with the rates for field broadcast. Due to the lack of farm machinery equipment for band placement of fertilizers in Pakistan, soil application of micronutrient fertilizers is performed by broadcast method.

Quite obviously, the rates of micronutrient fertilizers for foliar fertilization are much smaller than for soil application. The generally recommended concentrations of B and Zn are 0.05–0.1% in the final/diluted spray solution. For wheat crop, foliar sprays of 0.5% $ZnSO_4.7H_2O$ solution are quite effective to correct the deficiency (Zou et al., 2012). For Fe, the recommended foliar spray solution concentrations are 0.5% of $FeSO_4$ and 1% of Fe-*Sequestrene*. Therefore, the R concerning the right rate within 4Rs is hardly variable for micronutrient fertilizers except for B, in case application dose is recommended in consideration of soil's clay content.

Summarily, the following calculation of fertilizer salts may be used at the farmers' end: For the preparation of 0.1% foliar Zn spray, dissolve 303 g of $ZnSO_4.H_2O$ 33% Zn *or* 476 g of $ZnSO_4.7H_2O$ 21% Zn in 100 liters of water (approx. five spray tanks, each of 20 liters) and spray over 1-acre crop area. For the preparation of 0.1% foliar B spray, dissolve 909 g of borax 11% B *or* 588 g of boric acid 17% B *or* 478 g of *Solubor* 20.9% B in 100 liters of water (approx. five spray tanks, each of 20 liters) and spray over 1-acre crop area. For the preparation of 0.05% foliar B spray, dissolve 455 g of borax 11% B *or* 294 g of boric acid 17% B *or* 239 g of *Solubor* 20.9% B in 100 liters of water (approx. five spray tanks, each of 20 liters) and spray over 1-acre crop area.

14.6.3   Right Time

In the case of micronutrients, the right time for soil application of fertilizers is just prior to seed sowing. In the case of flooded rice crop which is still at the seedling stage, however, Zn fertilizer can be broadcast-applied immediately on observing deficiency symptoms.

The timing of foliar sprays is quite important. Instead of waiting for the deficiency symptoms to appear (which results in yield loss or impairment of produce quality), a pre-emptive approach must be adopted in performing foliar feeding. An adequate supply of micronutrients, especially of B, is warranted at flower fertilization and fruit setting growth stages. Therefore, the first foliar spray must be performed prior to flowering and the second, a few days after flowering. For correcting Zn deficiency in wheat, two foliar sprays are needed; the first spray one week prior to heading and the second spray one week after heading. For managing B deficiency in cotton crop, three foliar sprays are needed; the first 45 days after sowing, the second after 60 days, and the third after 90 days. In short, foliar sprays must be performed at the recommended growth stages.

14.6.4   Right Place

In the case of micronutrients, placement of fertilizers refers to the method of application, i.e., soil application or foliar feeding, rather than broadcasting or band placement of fertilizers which is true for major nutrient/NPK fertilizers. However, after field broadcast, micronutrient fertilizer must be incorporated into the surface soil layer by light plowing. Whereas soil-based fertilization is a long-term strategy to increase or maintain micronutrient availability, foliar fertilization of micronutrients only resolves the immediate deficiency. Rates of micronutrient fertilizers for soil application are much higher than that for foliar sprays, and the fertilizer cost for field broadcast is much more compared with the fertilizer cost involved for foliar sprays. Labor cost is also involved in performing foliar sprays, and generally, 2–3 foliar sprays are needed to effectively correct micronutrient deficiencies.

Therefore, the decision about the right place (or method, i.e., soil or foliar application) in the case of micronutrient fertilizer use is dependent on the above-stated considerations rather than the effectiveness of the applied fertilizer *per se.*

14.7 4R PACKAGES FOR MANAGING MICRONUTRIENT DEFICIENCIES IN MAJOR CROPS

Now, micronutrient fertilizer use in the perspective of 4R nutrient stewardship is elaborated by citing salient examples of micronutrient deficiencies in major crops, i.e., wheat, rice, cotton, maize, and sugarcane. Summarized information regarding the 4R packages is given in Table 14.2. Details about operational aspects, various options, and precautionary information are given in the following paragraphs.

TABLE 14.2
Summary of 4R Packages for Managing Micronutrient Deficiencies in Major Crops of Pakistan

Crop	Micronutrient	Fertilizer Source, Rate, Method, and Place of Application*
Wheat	Zn	5 kg elemental Zn ha^{-1}, as zinc sulfate, broadcast-applied to the field area prior to crop sowing **AND** 2 foliar sprays of 0.5% $ZnSO_4.7H_2O$ (or 0.32% $ZnSO_4.H_2O$) solution; first foliar spray one-week prior to crop heading and the second one week after heading
	B	0.75 kg elemental B ha^{-1}, as borax, broadcast-applied to the field area prior to crop sowing
Rice	Zn	*Zn-enriched Rice Nursery*: 20 kg elemental Zn ha^{-1} nursery area, as zinc sulfate, broadcast-applied to nursery beds prior to seed sowing **OR** *Zn Fertilization of Entire Field Area*: 5 kg elemental Zn ha^{-1}, as zinc sulfate, broadcast-applied to the puddled field prior to nursery transplanting *The easier to practice and more economical method is the use of Zn-enriched nursery.*
	B	0.75 kg elemental B ha^{-1} field area, as borax, broadcast-applied to the puddled field prior to nursery transplanting
Cotton	B	1.0 kg elemental B ha^{-1}, as borax, broadcast-applied to the field area prior to raised bed formation and crop sowing **OR** 3 foliar sprays of 0.1% elemental B solution, as boric acid (or *Solubor*), 45, 60, and 90 days after cotton sowing
	Zn	5 kg elemental Zn ha^{-1}, as zinc sulfate, broadcast-applied to the field area prior to raised bed formation and crop sowing **OR** 3 foliar sprays of 0.1% elemental Zn solution, as zinc sulfate, 45, 60, and 90 days after cotton sowing
Maize	Zn	5 kg elemental Zn ha^{-1}, as zinc sulfate, broadcast-applied to the field area prior to ridge formation and crop sowing

TABLE 14.2   (CONTINUED)

Summary of 4R Packages for Managing Micronutrient Deficiencies in Major Crops of Pakistan

Crop	Micronutrient	Fertilizer Source, Rate, Method, and Place of Application*
	B	0.75 kg elemental B ha^{-1}, as borax, broadcast-applied to the field area prior to ridge formation and crop sowing
Sugarcane	Zn	5 kg elemental Zn ha^{-1}, as zinc sulfate, broadcast-applied to the field area prior to ridge formation and crop sowing
	B	0.75 kg elemental B ha^{-1}, as borax, broadcast-applied to the field area prior to ridge formation and crop sowing

* *Important Notes*:

A) **Soil Application of Micronutrients**:
   1) In rice–wheat and cotton–wheat systems, micronutrients may be applied to rice and cotton; wheat benefits from residual effect; in the maize–potato system, micronutrients may be applied to potato, and maize would benefit from residual effect; in the sugarcane-based system, micronutrient may be applied to sugarcane.
   2) Zinc fertilizer leaves a beneficial residual effect on the succeeding ~5 crops, except for the rice–wheat system in which each rice crop needs Zn application. Boron fertilizer leaves a beneficial residual effect on the succeeding ~3 crops. Therefore, Zn and B fertilizers may be applied accordingly. However, periodic *soil testing is suggested, after every two to three years, to ascertain the need for fertilizer use – especially for B*.
   3) For uniform field broadcast, the volume of micronutrient fertilizers must be enhanced by mixing the recommended doses with about five-time well-pulverized soil of the same field.

B) **Foliar Application of Micronutrients**:
   1) Add 0.05% surfactant (i.e., any detergent powder, like *Surf* or *Ariel*) to spray solutions.
   2) B and Zn fertilizer can be mixed safely with pesticide spray solutions.
   3) For the preparation of 0.1% foliar zinc spray, dissolve 303 g of $ZnSO_4.H_2O$ 33% Zn *or* 476 g of $ZnSO_4.7H_2O$ 21% Zn in 100 liters of water (approx. five spray tanks, each of 20 liters) and spray over 1-acre crop area. For the preparation of 0.1% foliar B spray, dissolve 909 g of borax 11% B *or* 588 g of boric acid 17% B *or* 478 g of *Solubor* 20.9% B in 100 liters of water (approx. five spray tanks, each of 20 liters) and spray over 1-acre crop area. For the preparation of 0.05% foliar B spray, dissolve 455 g of borax 11% B *or* 294 g of boric acid 17% B *or* 239 g of *Solubor* 20.9% B in 100 liters of water (approx. five spray tanks, each of 20 liters) and spray over 1-acre crop area.

14.7.1   4R FOR MANAGING ZINC AND BORON DEFICIENCIES IN WHEAT

Wheat crop in Pakistan suffers from deficiencies of Zn and B. The best option for Zn fertilizer source is zinc sulfate. Both $ZnSO_4.7H_2O$ (containing 21% Zn) and $ZnSO_4.H_2O$ (containing 33% Zn) are equally good. For optimizing wheat productivity and enriching its grains with Zn, soil applications, as well as foliar-feeding combinations, are required. Recent multi-year, multi-country (including Pakistan) field research has revealed that, in Zn-deficient situations, soil-applied Zn fertilizer is effective to optimize wheat grain yield and foliar sprays of Zn fertilizer are needed to enhance Zn density in grains (Zou et al., 2012). Deep placement of fertilizers (at least 3 cm below the seed) containing nutrients with low mobility, like

Zn, improves grain yields in environments where the topsoil is prone to drying out and subsequently, nutrients are immobilized (Lindsay, 1972; Ma et al., 2009). These situations include sandy soils and low-rainfall geographical regions, which is true for many wheat-growing areas across Pakistan. Based on the experimental evidence that band-placed phosphatic fertilizer is 100% more efficient in correcting P deficiency in wheat (Rashid et al., 2010); this is a formally recommended practice in the country. Because of logistic constraints, however, the adoption of this farmer-friendly technology is minimal. Such a recommendation and the requisite farm machinery for micronutrient fertilizer use are lacking. Therefore, micronutrient fertilizers are applied by broadcast and subsequent incorporation into the topsoil layer by light plowing. Micronutrient fertilizers so applied are quite effective in correcting the deficiencies. Therefore, for yield enhancement of wheat, the 4R strategy is to apply 2–5 kg Zn ha^{-1} as $ZnSO_4$ by field broadcast, and subsequent soil incorporation by light plowing, just prior to crop sowing. For enriching wheat grains with Zn, two foliar sprays of 0.5% $ZnSO_4.7H_2O$ (or 0.32% $ZnSO_4.H_2O$) solution are applied; the first foliar spray is applied one-week prior to crop heading and the second one a week after crop heading. Soil + foliar Zn fertilization in Zn-deficient soil situations is highly cost-effective (Zou et al., 2012), and foliar Zn mixed with pesticide sprays remains effective in enhancing grain Zn density (Ram et al., 2016). Both soil application of $ZnSO_4$ and foliar feeding of Zn are suggested by adopting the above-stated 4Rs.

For taking care of B deficiency, the best fertilizer source for soil application is borax (containing 11% B). The recommended fertilizer rate is 0.75 kg B ha^{-1}; B fertilizer is applied just prior to crop sowing, along with other fertilizers, and B fertilizer is broadcast-applied.

In the interest of uniform field broadcast, the volume of micronutrient fertilizers must be enhanced by mixing their recommended doses with about five-time well-pulverized soil of the same field. Another important consideration for soil application of Zn and B is that both of these micronutrients need not be applied if a previous crop in the rotation (like cotton or wheat) had received their application. In any case, soil testing is suggested, after every two to three years, to ascertain the need for fertilization – especially for B.

14.7.2   4R for Managing Zinc and Boron Deficiencies in Rice

Rice crop in Pakistan also suffers from deficiencies of Zn and B. Like wheat, the best fertilizer source for Zn is zinc sulfate. Depending upon availability and comparative price, either $ZnSO_4.H_2O$ (33% Zn) or $ZnSO_4.7H_2O$ can be used to cater to the recommended dose per unit field area. In rice, Zn deficiency can be managed by i) soil-based fertilization, i.e., Zn fertilizer application to the entire rice field; ii) use of Zn-enriched rice nursery; and iii) foliar sprays of Zn fertilizer.

Soil-based fertilization is a long-term strategy to increase or maintain Zn availability, while the use of Zn-enriched nursery and foliar Zn fertilization only resolves the immediate deficiency. The agronomic comparison of Zn fertilization of entire rice field and the use of Zn-enriched nursery revealed that both the approaches are equally effective in correcting Zn deficiency (Rashid et al., 2000). However, soil-based Zn fertilization requires a much higher Zn dose (i.e., ~5 kg Zn ha^{-1} rice field area) and broadcasting the small dose of Zn fertilizer *in puddled fields* is a highly cumbersome field operation. Contrarily, Zn-enriched rice nursery can be raised by using much less Zn fertilizer (i.e., @ 20 kg Zn ha^{-1} nursery bed area, which is equal to 0.8 kg Zn ha^{-1} rice field area), and broadcasting of Zn fertilizer in the small-sized (non-puddled) nursery beds is a much easier field operation compared with field broadcast of Zn fertilizer in puddled fields. Whereas Zn fertilization of the entire rice field leaves a beneficial residual effect for the subsequent wheat crop in the rotation, the use of Zn-enriched nursery is a farmer-friendly technology. As foliar fertilization of Zn in flooded rice fields is a labor-intensive and difficult-to-practice field operation, Zn-enriched nursery technology is the most suitable Zn application method for managing Zn deficiency in rice. Zinc fertilizer to nursery beds is applied just prior to seed sowing in the beds, and fertilizer is applied by broadcast at the time of broadcasting farmyard manure. Thus, in the 4R stewardship perspective, *the best strategy for managing Zn deficiency in rice crop is the use of Zn-enriched nursery prepared by applying ZnSO$_4$ to nursery beds @ 20 kg Zn ha^{-1} nursery area at the time of seed sowing.*

Soil application of borax is a suitable and economical strategy to manage B deficiency in rice. Borax is applied prior to nursery transplanting @ 0.75 kg B ha^{-1} field area, by surface broadcast along with other fertilizers. Thus, 4Rs for B deficiency management in rice are i) the use of borax (right source), ii) @ 0.75 kg B ha^{-1} field area (right rate), iii) prior to nursery transplanting (right time), and iv) by broadcast method (right place).

For uniform field broadcast of micronutrients, the volume of micronutrient fertilizers must be enhanced by mixing their recommended doses with about five-time well-pulverized soil of the same field. In the case of soil application of Zn, its use is needed for every rice crop in the rotation. However, B leaves a considerable residual effect; thus, it may be applied after two years, i.e., to every alternate rice crop. Periodic soil testing is suggested, after every two to three years, to ascertain the need for fertilization – especially for B.

14.7.3   4R FOR MANAGING ZINC AND BORON DEFICIENCIES IN COTTON

Cotton crop in Pakistan suffers from deficiencies of B and Zn. The deficiencies can be corrected effectively both by soil application by foliar sprays of these micronutrients. Suitable sources for soil application are zinc sulfate and borax. For foliar feeding of Zn, zinc sulfate is the suitable fertilizer source. However, for foliar sprays of B, boric acid or *Solubor* can be used. Depending upon availability and cost, the growers are the best judge to opt for either B source. For soil application, zinc sulfate is applied @ 5 kg Zn ha^{-1} and borax @ 1 kg B ha^{-1}. These fertilizers are applied just prior to raised bed formation and/or seed sowing, by the surface broadcast method (Rashid and Rafique, 1997). For attaining uniform field broadcast, the volume of micronutrient fertilizers must be enhanced by mixing their recommended doses with about five times well-pulverized soil of the same field.

For foliar feeding, foliar spray solution must contain 0.1% each of B and Zn, as boric acid (or *Solubor*) and zinc sulfate, respectively. The foliar solution must contain 0.05% surfactant (i.e., any detergent powder, like *Surf* or *Ariel*), and can be mixed with certain pesticide spray solutions. For effective management of deficiencies, foliar sprays of B and Zn must be applied three times, i.e., 45, 60, and 90 days after cotton sowing (Rashid and Rafique, 1997). In the case of soil application, micronutrient use is not needed for every cotton crop. Rather, Zn use will be needed for every sixth cotton crop and B use for every fourth cotton crop. For determining the need for applying these micronutrients, especially for B, *soil testing is suggested every two to three years.*

14.7.4   4R FOR MANAGING ZINC AND BORON DEFICIENCIES IN MAIZE

Micronutrient problems in maize crop are also Zn and B deficiencies. Theoretically, these deficiencies can be managed both by soil application and by foliar feeding. In consideration of logistic aspects and residual effect of soil-applied micronutrients, however, soil application appears to be a better option. Suitable fertilizer sources for soil application are zinc sulfate and borax. Zinc sulfate is applied @5 kg Zn ha^{-1} and borax @0.75 kg B ha^{-1}. Both micronutrient fertilizers are applied prior to ridge forming and/or seed sowing, by the surface broadcast method. For uniform field broadcast, micronutrient fertilizers must be pre-mixed with about five times the volume of well-pulverized soil. As hybrid maize is grown on ridges, the surface applied Zn and B will get accumulated in the ridges.

Soil-applied Zn and B leave considerable residual effects; therefore, they need not be applied to every crop in the rotation or even to every maize crop in the cropping sequence. The duration of residual effect may vary in different soil types, primarily because of variations in soil texture and calcareousness. In general, Zn fertilizer use will be needed for every third maize crop in the rotation and B use for every second maize crop. To determine the need for applying these micronutrients, especially for B, soil testing is suggested every two to three years.

14.7.5   4R FOR MANAGING ZINC AND BORON DEFICIENCIES IN SUGARCANE

It has been diagnosed that sugarcane crop in Pakistan suffers from deficiencies of Zn and B. Unlike wheat, rice, cotton, and maize, sugarcane is a longer-duration crop and produces more biomass per unit field area compared with other crops. Theoretically, micronutrient deficiencies in sugarcane can be managed both by soil application and by foliar feeding. However, soil application appears to be a better option. Suitable fertilizer sources for soil application are zinc sulfate and borax. Zinc sulfate is applied @ 5 kg elemental Zn ha^{-1} and borax @ 0.75 kg elemental B ha^{-1}. Both micronutrient fertilizers are applied prior to ridge forming and/or sugarcane sowing, by the surface broadcast method. For uniform field broadcast, micronutrient fertilizers must be pre-mixed with about five times the volume of well-pulverized soil.

Soil-applied Zn and B fertilizers leave residual effects on the subsequent crops; therefore, they need not be applied to every crop in the rotation. The duration of residual effect may vary in different soil types, primarily because of variations in soil texture and calcareousness. In general, Zn fertilizer use will be needed for every third sugarcane crop and B use for every second sugarcane crop. However, to determine the need for applying these micronutrients, especially for B, soil testing is suggested after every two to three years.

14.8 KEY MESSAGES FOR FARMING COMMUNITY

1. **What Are Micronutrients?** Plants are also living organisms, like humans, and, thus, need food for their growth and reproduction. Eight nutrients are essentially required by plants in very small amounts; these are called micronutrients. All the micronutrients are as essential for plants as nitrogen and phosphorus. The functions of micronutrients in plants are similar to those of vitamins in human beings. Deficiency of any one of the eight micronutrients results in reduction of crop growth, crop yield, produce quality, and farm income. Some well-known examples of micronutrient disorders in Pakistani agriculture are Zn and B deficiencies in rice, B deficiency in cotton, and Fe chlorosis in apple and citrus.

2. **Soil–Plant Micronutrient Deficiencies and Human Malnutrition**: Deficiencies of micronutrients in plants not only lead to reduced crop yields and farm profitability, but also result in deteriorating quality of the crop produce. Now it is well established that almost one-half (50%) of the Pakistani population suffers from Zinc malnutrition – because of low Zn content as well as its bioavailability in wheat grains produced in Zn-deficient soils. Zinc malnutrition in humans causes serious health problems including stunted growth, abnormal sexual growth, and acute diarrhea. Zinc malnutrition can be corrected by taking Zn tablets/ capsules and Zn-rich foods (like meat, seafood, and nuts). However, resource-poor segments of the society in Pakistan cannot afford such costly measures. Zinc-biofortified wheat (flour), recently introduced by the National Agriculture Research System and AARI Faisalabad, is probably the only cost-effective and sustainable strategy to ameliorate human Zn deficiency in Pakistan. The R&D contributions by Lowe et al. (2022) and Khan et al. (2017) have generated relevant scientific evidence in this regard.

3. **Why Micronutrients Are Deficient in Our Crops?** Soils in Pakistan are alkaline–calcareous and low in OM. These soil conditions are conducive to deficiencies of certain micronutrients. Well-established field-scale deficiencies of micronutrients exist in the case of zinc, boron, and iron in a number of field crops, vegetables, and fruits orchards. In micronutrient-deficient situations, fertilizers of the deficient micronutrients must be applied – along with fertilizers of nitrogen (like urea) and phosphorus (like DAP).

4. **Salient Symptoms of Micronutrient Deficiency in Plants**: Though it is not always easy to diagnose micronutrient deficiency problems accurately by looking at plant symptoms, some micronutrient deficiencies are exhibited by typical symptoms. For example, iron deficiency symptoms(known as iron chlorosis) always appear on young leaves and are characterized by interveinal *chlorosis*; in case of severe deficiency, entire leaves turn yellow and even may die; deciduous fruit plants and certain crops (like peanut and soybean) are highly sensitive to *iron chlorosis*. Zinc deficiency in maize crop results in broad chlorotic bands on both sides of leaf midrib, and in apple, it is exhibited as "little leaf" (small and narrow leaves), and buds fail to develop and tend to form rosettes at tips of shoots. Boron deficiency in cotton crop promotes premature flower abortion and in rice crop, results in empty panicles on the lower end of the ears.

5. **How Plant Micronutrient Deficiencies Can Be Corrected?** Micronutrient deficiencies in crops can be corrected by soil application of micronutrient fertilizers (by broadcasting, followed by mixing in the top-soil) as well as by foliar sprays of micronutrient fertilizers. Doses of micronutrient fertilizer for soil application are much higher than the doses in foliar sprays. Whereas one recommended dose of a micronutrient fertilizer is sufficient for more than one crop in the rotation, generally 2–3 foliar sprays are needed to correct micronutrient deficiencies in a single crop.

6. **Micronutrient Fertilizer Doses for Soil Application**: Average doses of zinc and boron fertilizers for most field crops and vegetables are:
   6 kg zinc sulfate (containing 30–33% Zn) per acre
   2.75–3.75 kg borax (containing 11% B) per acre

7. **Uniform Field Application of Micronutrient Fertilizers**: Doses of micronutrient fertilizers are much less than that of urea and DAP fertilizers; thus, uniform application of small quantities of micronutrient fertilizers in the fields is not easy. Therefore, the micronutrient fertilizer dose must be mixed with about five times the volume of well-pulverized soil of the same field or with urea and/or DAP fertilizer before broadcasting in the field.

8. **Beneficial Residual Effect of Soil-applied Micronutrient Fertilizers**: Soil-applied micronutrient fertilizers leave beneficial residual effects on subsequent crop(s) grown in the same field. Therefore, it is not necessary to apply micronutrient fertilizer to each and every crop. In certain cases, the beneficial residual effect can last for 3–6 crops. Therefore, soil testing after every two to three years is recommended to determine the need for micronutrient fertilizer use.

9. **Micronutrient Fertilizers Can Be Sprayed Effectively along with Pesticide/Herbicide Sprays:** Many of the micronutrient fertilizers can be mixed safely with foliar solutions of pesticides and/or herbicides–because these remain effective in correcting micronutrient deficiencies.

For example, zinc sulfate mixed with *Confidor* insecticide remains effective in ameliorating deficiency of zinc in wheat crop as well as in increasing zinc density in wheat grains. For cotton as well, boron fertilizer can be mixed safely with foliar sprays of certain pesticides/herbicides. For details about specific pesticide/herbicide compatibility with micronutrient fertilizers, readers are referred to section 15.3 of this book and **Annex 17**.

10. **Micronutrient Fertilizer Use in Cotton**: Cotton crop suffers from deficiencies of boron and zinc. Deficiencies of these micronutrients in cotton can be corrected effectively by soil application of 3.75 kg Borax per acre and 6 kg zinc sulfate per acre. These micronutrient deficiencies can also be managed by performing three foliar sprays of 1.0% boric acid solution + 0.5% zinc sulfate solution: first spray 45 days after sowing; second spray 60 days after sowing; and third spray 90 days after sowing.

11. **Micronutrient Fertilizer Use in Rice:** Rice crop suffers from deficiencies of Zn and B. These deficiencies can be managed by soil application of 2.75 kg Borax per acre and 6 kg $ZnSO_4$ per acre. The deficiency of Zn in rice can also be managed by using Zn-enriched nursery seedlings, i.e., by applying 3.4 kg $ZnSO_4$ per 506 square meters (~ 1 *Kanal* = 1/8th acre) area of nursery beds. For flooded rice, application of zinc is recommended every year, although in other crops Zn residual effect is observed for the next five crops.

12. **Zinc-Enriched Rice Nursery**: Zinc deficiency in rice crop can be avoided effectively by using a zinc-enriched rice nursery. For raising zinc-enriched nursery, 3.4 kg $ZnSO_4$ fertilizer is broadcast applied per 506 square meters (~ 1 *Kanal*=1/8th acre) area of nursery beds, just prior to seed sowing. The nursery so raised will contain adequate zinc in the seedlings, and rice crop grown in zinc-deficient fields will not suffer from zinc deficiency. Compared with field broadcast of zinc sulfate fertilizer, zinc-enriched nursery technology is much easier to practice and less expensive.

13. **Micronutrient Fertilizer Use in Wheat:** Wheat crop suffers from deficiencies of Zn and B. In case these micronutrients are soil-applied to the previous crop in the rotation (e.g., to cotton or rice), their residual effect will also benefit the wheat crop. If not, the recommended doses of these micronutrients are:
    6.00 kg zinc sulfate (containing 30–33% Zinc) per acre
    3.75 kg borax (containing 11% B) per acre

14. **Micronutrient Fertilizer Use in Maize:** The deficient micronutrients in maize crop are Zn and B. The recommended doses of these micronutrients are:
    6.00 kg zinc sulfate (containing 30–33% Zn) per acre
    3.75 kg borax (containing 11% B) per acre

15. **Wheat Grains Can Be Enriched with Zinc by Foliar Sprays of Zinc Sulfate:** Wheat grains in Pakistan contain around 25 mg Zn kg^{-1}, which is much less than 40–60 mg Zn kg^{-1} required for good human health under a resource poor setting. The concentration of zinc in wheat grains

can be increased effectively by applying two foliar sprays of 0.32% zinc sulfate (containing 33% zinc) solution. The first spray must be applied one week prior to heading and the second spray one week after heading. This foliar supplementation strategy is in addition to the soil-applied recommended dose of zinc.

16. **Effective Foliar Sprays of Micronutrient Fertilizers**: For more effective absorption of foliar-sprayed micronutrients (like zinc) by plant leaves, 50 g (approx. tea-spoons) of detergent powder (like *Surf* or *Ariel*) must be added in 100 liters of spray solutions of micronutrient fertilizers (like zinc sulfate). The 100-liter spray solution (approx. five shoulder-carried tanks) would be sufficient to cover spray over 1 acre of field crop like wheat. Also, foliar sprays of micronutrient fertilizers must be applied either very early in the morning or late in the evening.

   For the preparation of 0.1% foliar Zn spray, dissolve 303 g of $ZnSO_4$. H_2O 33% Zn *or* 476 g of $ZnSO_4.7H_2O$ 21% Zn in 100 liters of water (approx. five spray tanks, each of 20 liters) and spray over 1-acre crop area. For the preparation of 0.1% foliar B spray, dissolve 909 g of borax 11% B OR 588 g of boric acid 17% B *or* 478 g of *Solubor* 20.9% B in 100 liters of water (approx. five spray tanks, each of 20 liters) and spray over 1-acre crop area. For the preparation of 0.05% foliar B spray, dissolve 455 g of borax 11% B *or* 294 g of boric acid 17% B *or* 239 g of *Solubor* 20.9% B in 100 liters of water (approx. five spray tanks, each of 20 liters) and spray over 1-acre crop area.

17. **Micronutrient Deficiencies in Apple, Peach, Plum**: Most fruit orchards in Pakistan suffer from deficiencies of zinc and iron, which can be corrected by applying 2–3 foliar sprays of 0.32% solution of zinc sulfate (containing 33% Zn) and 0.5% ferrous sulfate, containing 50 g detergent powder in each batch of 100-liter foliar spray solution. Fresh micronutrient solution must be prepared for each foliar spray.

18. **Managing Micronutrient Deficiencies in Citrus**: Quite frequently, citrus orchards suffer from deficiencies of Zn and Fe, and in many cases, their deficiency symptoms are quite obvious. These micronutrient problems can be cured both by foliar sprays and by soil application of micronutrient fertilizers, as described below:

   Apply 2–3 foliar sprays of 0.32% solution of zinc sulfate (containing 33% Zn) and 0.5% ferrous sulfate, containing 50 g detergent powder in each batch of 100-liter foliar spray solution. Fresh micronutrient solution must be prepared for each foliar spray.

19. **Managing Iron Chlorosis in Chili and Other Vegetables**: Certain vegetable crops, like chili, grown in light-textured soils (sandy soils) are highly prone to iron chlorosis. Iron chlorosis in vegetables can be controlled very effectively by applying two to three sprays of 0.5% solution of ferrous sulfate, containing 50 g detergent powder in 100 liters of spray solution. A fresh solution of ferrous sulfate must be prepared for performing each spray.

20. **Why Use Expensive Chelated Micronutrients**? Theoretically, chelated micronutrient fertilizers are considered more effective for curing micronutrient deficiencies compared with inorganic micronutrient fertilizers. This is especially true in the case of soil-applied iron fertilizers. As chelated micronutrient fertilizers are much more expensive than inorganic fertilizers, worldwide chelated micronutrients are hardly used by the growers, except for high-value floriculture or fruit plantations. Research in Pakistan (and many other countries of the world) has revealed that chelated micronutrient fertilizers are not often superior in performance to inorganic micronutrient fertilizers, except for Fe. Therefore, farmers should remain cautious while spending extra money on using chelated micronutrient formulations.

21. **Always Use Good-Quality Micronutrient Fertilizers**: The quality of micronutrient fertilizer products is a serious issue in Pakistan. Unless micronutrient fertilizers are of good quality, the expense and effort to correct micronutrient problems cannot be remunerative. Many small vendors in the country are marketing dubious quality micronutrient products (in liquid and solid forms). Therefore, farmers are advised to buy good-quality micronutrient fertilizers marketed by major fertilizer companies, i.e., Fauji Fertilizer Company Limited, Engro Fertilizers, Jaffer Brothers (Pvt.) Limited, and Imperial Chemical Industries (ICI), Pakistan.

22. **Which Other Fertilizers Can Affect Zinc Requirement of Crops**: High levels of P in plants have been shown to restrict Zn movement within the plant, resulting in accumulation of Zn in the roots and, thus, inducing Zn deficiency in the tops. Therefore, large applications of P fertilizer may contribute to Zn deficiency in Zn-responsive crops, especially where soil Zn availability is marginal. Besides heavy use of P fertilizer, overuse of nitrogenous fertilizers may also increase Zn requirements of plants.

23. **Why Zinc Deficiency in Crops Affects Yields**: Zinc governs the production of natural hormones called auxins within the plant. Adequate auxins are needed for optimum leaf size, and the size of these solar panels, in turn, determines the photosynthetic potential of plants. As about 95% of crop production comes from photosynthesis, adequate Zn nutrition of plants is crucial for optimum photosynthetic activity.

REFERENCES

Ahmad, M., S. Alam, W. Ahmad, I. Jan, and A. Zia. 2018. Application of the 4R nutrient stewardship concept for growing off-season tomatoes in high tunnels. *Journal of Soil Science and Plant Nutrition* 18:989–1001.

Bertrand, I., M. McLaughlin, R. Holloway, R. Armstrong and T. McBeath. 2006. Changes in P bioavailability induced by the application of liquid and powder sources of P, N and Zn fertilizers in alkaline soils. *Nutrient Cycling in Agroecosystems* 74: 27–40.

Bruulsema, T. W., F. Garcia, and T. Satyanaryana. 2012. The 4R nutrient stewardship concept. In: *4R Plant Nutrition a Manual for Improving the Management of Plant Nutrition Metric Version*. International Plant Nutrition Institute, Norcross, USA. p. 2-1–2-7.

Chaudhry, F. M., M. Sharif, A. Latif, and R. H. Qureshi. 1973. Zinc-copper antagonism in the nutrition of rice (*Oryza sativa* L.). *Plant and Soil* 38:573–580.

Chaudhry, F. M., S. M. Alam, A. Rashid, and A. Latif. 1977. Micronutrient availability to cereals from calcareous soils. IV. Mechanism of differential susceptibility of two rice verities to Zn deficiency. *Plant and Soil* 46:637–642.

Dang, Y. P., R. C. Dalal, D. G. Edwards, K. G. Tiller. 1994. Kinetics of zinc desorption from vertisols. *Soil Science Society of America Journal* 58:1392–1399.

FAO (Food and Agriculture Organization). 2017a. *Soil Fertility Atlas of Pakistan: The Punjab Province*. Waqar, A., Y. Niino, M.H. Zia, et al. (Ed.), ISBN 978-969-8304-08-9. Islamabad, Pakistan, 115 pp.

FAO (Food and Agriculture Organization). 2017b. *Soil Fertility Atlas of Pakistan: The Sindh Province*. Waqar, A., Y. Niino, M. H. Zia, et al. (Ed.), ISBN 978-969-8304-09-6. Islamabad, Pakistan, 105 pp.

Genc, Y., and G. K. McDonald. 2004. The potential of synthetic hexaploid wheats to improve zinc efficiency in modern bread wheat. *Plant and Soil* 262:23–32.

Holloway, R. E., I. Bertrand, A. J. Frischke, et al. 2001. Improving fertiliser efficiency on calcareous and alkaline soils with fluid sources of P, N and Zn. *Plant and Soil* 236:209–219.

IFA. 2009. *The global "4R" nutrient stewardship framework–developing fertilizer best management practices for delivering economic, social and environmental benefits*. International Fertilizer Industry Association, Paris, France.

IPNI. 2012. *4R Plant Nutrition Manual: A Manual for Improving the Management of Plant Nutrition*, Metric Version, Bruulsema, T. W., P. E. Fixen, and G. D. Sulewski (Ed.). International Plant Nutrition Institute, Norcross, GA, USA.

Johnston, A. M., and T.W. Bruulsema. 2014. 4R Nutrient Stewardship for Improved Nutrient Use Efficiency. *Procedia Engineering* 83:365–370.

Khan, M. J., U. Ullah et al. 2017. Effect of agronomically biofortified zinc flour on zinc and selenium status in resource poor settings; a randomised control trial. *Proceedings of the Nutrition Society* 76 (OCE4). DOI: 10.1017/S0029665117003457.

Lindsay, W. L. 1972. Zinc in soil and plant nutrition. *Advances in Agronomy* 24:147–186.

Loneragan, J. F., and M. J. Webb. 1993. Interactions between zinc and other nutrients affecting the growth of plants. p. 119–134, In: Robson, A. D. (Ed.), *Zinc in Soil and Plants*. Kluwer Academic Publishers, Dordrecht.

Lowe, N. M., M. Z. Afridi, M. J. Khan, et al. 2022. Biofortified wheat increases dietary zinc intake: A randomised controlled efficacy study of Zincol-2016 in rural Pakistan. *Frontiers in Nutrition* 8:809783. DOI: 10.3389/fnut.2021.809783.

Ma, Q., Z. Rengel, and T. Rose. 2009. The effectiveness of deep placement of fertilizers is determined by crop species and edaphic conditions in Mediterranean-type environments: a review. *Australian Journal of Soil Research* 47:19–32.

Maqsood, M. A., Rahmatullah, A. M. Ranjha, and M. Hussain. 2009. Differential growth response and zinc utilization efficiency of wheat genotypes in chelator buffered nutrient solution. *Soil and Environment* 28:174–178.

Marschner, H. 1993. Zinc uptake from soils. In: Robson, A. D. (Ed.), *Zinc in Soil and Plants*. Kluwer Academic Publishers, Dordrecht. p. 59–77.

McDonald, G. K., R. D. Graham, J. Lloyd, et al. 2001. Breeding for improved zinc and manganese efficiency in wheat and barley. In: Rowe, B., D. Donaghy, N. Mendham (Ed.), *Science and Technology: Delivering Results for Agriculture?* Proceedings of the 10th Australian Agronomy Conference, January 2001, Hobart, Tasmania.

Nawaz, N., M. S. Nawaz, M. A. Khan, et al. 2014. Effect of boron on peanut genotypes under rainfed conditions. *Pakistan Journal of Agricultural Research* 27:110–117.

NFDC. 1998. *Micronutrients in Agriculture: Pakistan perspective*. National Fertilizer Development Center, Islamabad, 51 pp.

Norton, R. M., and T. Roberts. 2012. Nutrient management to nutrient stewardship. In: Yunusa, I. (Ed.), *Capturing Opportunities and Overcoming Obstacles in Australian Agronomy*. Proc. 16th Australian Agronomy Conference 2012.

Ram, H., A. Rashid, W. Zhang, et al. 2016. Biofortification of wheat, rice and common bean by applying foliar zinc fertilizer along with pesticides in seven countries. *Plant and Soil* 403:389–401.

Rashid, A. 2006. Incidence, diagnosis and management of micronutrient deficiencies in crops: Success stories and limitations in Pakistan. In: Optimizing Resource Use Efficiency for Sustainable Intensification of Agriculture. IFA Agriculture Conference, Kunming, PR China. 27 February–2 March, 2006.

Rashid, A. 2006b. Boron Deficiency in Soils and Crops of Pakistan: Diagnosis and Management. *Pakistan Agricultural Research Council, Islamabad, Pakistan*, viii+34 pp. ISBN: 969-409-184-5

Rashid, A., and E. Rafique. 1997. Annual Report "Micronutrients/nutrient management in cotton in relation to cotton leaf curl virus (CLCV)", National Agricultural Research Center, Islamabad.

Rashid, A., and E. Rafique. 1998. *Micronutrients in Pakistan Agriculture: Significance and Use*. Pakistan Agricultural Research Council, Islamabad, Pakistan. 8 pp.

Rashid, A., and E. Rafique. 2017. Boron deficiency diagnosis and management in field crops in calcareous soils of Pakistan: A mini review. *BORON* 2:142–152.

Rashid, A., and J. Din. 1992. Differential susceptibility of chickpea cultivars to iron chlorosis grown on calcareous soils of Pakistan. *Journal of the Indian Society of Soil Science* 40:488–492.

Rashid, A., E. Rafique, and J. Ryan. 2002b. Establishment and management of boron deficiency in field crops in Pakistan: A country report. p. 339–348, In: Goldbach, H. E., et al. (Ed.), *Boron in Plant and Animal Nutrition*. Kluwer Academic Publishers, New York.

Rashid, A., K. Mahmood, Mahmood-ul-Hassan, M. Rizwan, and Z. Iqbal. 2014. HarvestPlus Zinc Fertilizer Project, Phase-II Country Report, 2011–2014. Nuclear Institute for Agriculture and Biology, Faisalabad.

Rashid, A., M. A. Kausar, F. Hussain, and M. Tahir. 2000. Managing zinc deficiency in transplanted flooded rice grown in alkaline soils by nursery enrichment. *Tropical Agriculture* 77:156–162.

Rashid, A., M. M. Mahmud, and E. Rafique. 2007. Potato responses to boron and zinc application in a calcareous UdicUstochrept. p. 103–116, In: Xu, F., et al. (Ed.), *Advances in Plant and Animal Boron Nutrition*. Proc. 3rd *Int. Symp. on All Aspects of Plant and Animal Boron Nutrition (Boron 2005), Wuhan, China, 9–13 Sep 2005.* Springer, Dordrecht.

Rashid, A., S. Muhammad, and E. Rafique. 2002a. Genotypic variation in boron uptake and utilization by rice and wheat. p. 305–310, In: Goldbach, H.E. et al. (Ed.), *Boron in Plant and Animal Nutrition*. Kluwer Academic Publishers, New York.

Rashid, A., Z. I. Awan, J. Ryan, E. Rafique, and H. Ibrikci. 2010. Strategies for phosphorus nutrition of dryland wheat in Pakistan. *Communications in Soil Science and Plant Analysis* 41:2555–2567.

Rietra, R. P. J. J., M. Heinen, C. O. Dimkpa and P. S. Bindraban. 2017. Effects of Nutrient Antagonism and Synergism on Yield and Fertilizer Use Efficiency. *Communications in Soil Science and Plant Analysis* 48:1895–1920.

Sally, A. F. 2020. Data management and variability: Precision agriculture considerations for 4R management planning. *Crops and Soils* 53:14–17. doi:10.1002/crso.20051.

Zou, C. Q., Y. Q. Zhang, A. Rashid, et al. 2012. Biofortification of wheat with zinc through zinc fertilization in seven countries. *Plant and Soil* 361:43–55.

15 Recommendations for Improving Micronutrient Nutrition of Crops

15.1 INTRODUCTION

The overall objective of writing this book is to improve the micronutrient nutrition of crops by research-based farmer-friendly use of fertilizers to sustain crop productivity, farm income, and soil resource sustainability. Improving the micronutrient nutrition of crops will certainly have an impact on the socio-economic status of the farming community. At the national level, a functional mechanism is a prerequisite to ensure that micronutrient nutrition of crops is given sufficient prominence, as a "do-able package", to attract the necessary resources. It may still be necessary to muster support for each of the programs separately (Zn-, B-, and Fe-deficient areas) since external support (technical expertise and funding) may be only available for the specific programs. But it is through the formulation of a coherent national plan, covering the whole micronutrient nutrition of crops "initiative", that better integration may be achieved, even if external funding comes separately. The challenge of sustaining crop productivity, farm income, and resource-based sustainability through crop nutrition can be attained only with concerted and collaborative efforts of the concerned researchers, agricultural extension personnel, and policymakers (Messer, 1992; Ockwell et al., 2015). Here, we have provided a set of R&D, technology transfer, and policy recommendations to help formulate a national micronutrient strategy for Pakistan. A national soil policy for Pakistan is yet to be surfaced that may encompass a national micronutrient strategy with integrated and multi-sectoral approaches to overcome challenges and fill up the gaps in the implementation. Therefore, the following sections of this chapter pertain to recommendations concerning future R&D, technology transfer, and policy formulation.

15.2 R&D RECOMMENDATIONS

- Standard fertilizer, plant, and soil reference materials and inter-laboratory comparisons should be introduced in research laboratories to ensure quality laboratory analyses. Such quality ensuring practices may also be adopted by the soil, plant, and fertilizer advisory laboratories.

DOI: 10.1201/9781003314226-15

- Nutrient indexing of farmer-grown crops in the left-over cropping systems should be carried-out to diagnose the nature, extent, and severity of micronutrient disorders.
- Residual and cumulative effects of micronutrient fertilizer use should be studied in long-term permanent layout field experiments in major cropping systems and soil orders.
- Micronutrient efficient crop varieties should be developed. The relevant screening data should be a prerequisite for the approval of new crop varieties.
- As "*Seeing is believing*", large-scale field demonstrations and farmers' field days should be organized for promoting the use of micronutrient fertilizers in deficient field crops, fruit orchards, and vegetables.
- Media campaigns (print and electronic) should be organized to emphasize the need and benefits of using micronutrients.

15.3 TECHNOLOGY TRANSFER RECOMMENDATIONS

- The public and private R&D sectors should incorporate crop-specific 4R messages for micronutrient fertilizers in the published literature.
- The value and accessibility of soil testing need to be communicated more widely to farmers and advisors, and users need to be educated to encourage the adoption of these services.
- This book on micronutrients must be translated into other local languages for dissemination to farmers and other stakeholders at the country-scale, free of cost, through district-level agriculture offices.
- Local language versions of the 4R Nutrient Stewardship manual should be developed by the public and private R&D sectors, in collaboration with FAO and international groups such as IFA.

15.4 POLICY RECOMMENDATIONS

- Micronutrient fertilizers are as important as macronutrient fertilizers (like urea, DAP, and potash fertilizers). Therefore, micronutrient fertilizer products (like Zn sulfate, borax, boric acid, ferrous sulfate, copper sulfate, and manganese sulfate) should be considered for subsidy across the country.
- National Fertilizer Development Center (NFDC) should maintain an inventory (central database) of micronutrient fertilizer products, district-wise and year-wise, with the provision of online access to the database to policy makers, industry, academics, and researchers. This would help in assessing micronutrient fertilizer availability in various cropping zones and geographic areas of the country. The central database of micronutrient fertilizer products will also enable Fertilizer Control Order (FCO) implementing authorities to check the legitimacy and authenticity of

marketed products in real time. For example, at present, Agriculture Extension Officers in Punjab are unable to verify whether the PSQCA-registered fertilizer products marketed in Punjab are under valid license and what is their guaranteed analysis.

- Soil testing laboratories at the district level, throughout the country, should be equipped with instruments and other facilities required for analyzing micronutrient fertilizers. This would help in improved monitoring and regulating the quality of micronutrient fertilizers being marketed.
- Effective soil–plant–fertilizer micronutrient advisory services and requisite staff training should be ensured and maintained at the district level throughout the country.
- Major fertilizer players in the country should ensure the provision of micronutrient-fortified fertilizers suitable for soil and foliar applications.
- Investment in agricultural R&D is required at federal and provincial levels to match the levels of other developing Asian countries. Private sector stakeholders, especially the fertilizer industry, should be provided enabling environment to invest in fertilizer-related R&D with financial incentives like income tax rebates to the industry scientists at par with that of government-employed researchers.

REFERENCES

Ockwell, D., A. Sagar, and H. de Coninck. 2015. Collaborative research and development (R&D) for climate technology transfer and uptake in developing countries: towards a needs driven approach. *Climatic Change* 131:401–415. doi:10.1007/s10584-014-1123-2.

Messer, E. 1992. Conference Report - Ending Hidden Hunger: A Policy Conference on Micronutrient Malnutrition. *Food and Nutrition Bulletin* 14:1–3. doi:10.1177/156482659201400105.

Annex 1
Suggested Plant Tissue Sampling Procedures for Selected Crops

Crop	Growth Stage	Plant Part to Sample	Plants to Sample
Field crops			
Wheat and barley	Seedling stage (<30 cm tall)	All the above-ground portion	50–100
	Before head emergence	Flag leaf	25–50
Rice (and other small grain crops)	Seedling stage (<30 cm tall)	All the above-ground portion	50–100
	Prior to or at heading	4th upper-most leaves at top of the plant	50–100
Cotton	Flower initiation	4th leaf (without petiole) on the main stem	25–30
Maize	Seedling stage (<30 cm tall)	All the above-ground shoot	20–30
	From tasseling to silking	The entire fully developed leaf below the whorl	15–25
Sorghum	Prior to or at heading	2nd leaf from the top of the plant	15–25
Sugarcane	6–8 months after planting	Leaf sheath nos. 3, 4, 5, and 6, counting spindle as zero, from five representative cane stalks (each from a different hill)	4 leaf sheaths from 5 cane stalks
Rapeseed-mustard	Seedling stage (<30 cm tall)	Whole shoots	30–40
	Bloom initiation to 50% flowering	Recently matured leaf	30–40
Peanut	Maximum tillering	Recently matured leaflets	25–35
alfalfa, clover, and other legumes	Prior to or at 1/10th bloom stage	Recently matured leaf blades	40–50
Chickpea, lentil, mungbean, and mash	Vegetative growth stage	Whole shoots	40–50
	Bloom initiation	Recently matured leaf	50–200
Vegetable crops			
Potato	Prior to or during early bloom	3rd to 6th leaf from growing tip	20–30
Cabbage (and other head crops)	Prior to heading	First mature leaf from center of the whorl	10–20

(Continued)

Crop	Growth Stage	Plant Part to Sample	Plants to Sample
Tomato (field)	Prior to or during early fruit set	3rd or 4th leaf from growing tip	20–25
Soybean and other beans	Seedling stage (<30 cm tall)	All the above-ground portion	20–30
	Prior to or during initial flowering	Two or three fully developed leaves at top of the plant	20–30
Peas	Prior to or during initial flowering	Leaves from the 3rd node down	30–60
Carrot, onion, and beets (and other root crops)	Prior to root or bulb enlargement	Center mature leaves	20–30
Cucumber and lettuce	Early stages of growth to fruit set	Mature leaf near the base portion of plant on the main stem	20–30
Fruits and nuts			
Apple, apricot, peach, pear, plum, and almond	September to October	Leaves near the base of the current season's growth or from spurs	50–100
Citrus (orange)	September to October	Spring cycle leaves, 5–7 months old, from non-fruiting twigs	20–30
Lemon	September to October	Mature leaves from the last flush on non-bearing terminals	20–30
Mango	September to October	4–7-month-old leaves (+petiole) from the middle of the shoot	15–25
Grape	End-of-bloom period	Petioles from leaves adjacent to fruit clusters	60–100
Banana	Bud differentiation, 4 months after planting	Petiole of 3rd open leaf from the apex	10–15

Adapted from Ryan et al., 2001; Jones and Case, 1990; Jones et al., 1971; Clements, 1980; Plank, 1979; Bharghava and Raghupathi, 1993; and Embleton et al., 1973.

When specific guidelines are unknown, the general *rule of thumb* is to sample *upper mature leaves* at flower initiation.

REFERENCES

Bharghava, B. S., and H. B. Raghupathi. 1993. Analysis of plant materials for macro and micronutrients. p. 49–82, In: Tandon, H. L. S. (Ed.) *Methods of Analysis of Soil, Plants*, Waters and Fertilizers. Fertilizer Development and Consultation Organization, New Delhi, India.

Clements, H.F. 1980. *Sugarcane crop logging and crop control: Principles and practices*. The University Press of Hawaii, Honolulu, HI.

Embleton, T. W., W. W. Jones, C. K. Labanauskas, and W. Reuther. 1973. Leaf analysis as a diagnostic tool and guide to fertilization. In: *Citrus Industry of California Division of Agriculture Science*. California Department of Food and Agriculture: Sacramento, CA, USA; pp. 83–210.

Jones, J. B., and V. W. Case. 1990. Sampling, Handling, and Analyzing Plant Tissue Samples. Chapter 15 In: R. L. Westerman (Ed.) *Soil Testing and Plant Analysis*, Volume 3, Third Edition, Soil Science Society of America, doi:10.2136/sssabookser3.3ed.c15.

Jones. J. B., Jr., R. L. Large, D. P. Pfleiderer, and K. S. Klosky. 1971. How to properly sample for plant analysis. *Crops and Soils* 23:15–18.

Plank, C.O. 1979. Plant Analysis Handbook for Georgia. University of Georgia Cooperative Extension Bulletin, 739.

Ryan, J., G. Estefan, and A. Rashid. 2001. *Soil and Plant Analysis Laboratory Manual*, 2nd ed. International Center for Agricultural Research in the Dry Areas (ICARDA), Aleppo, Syria.

Annex 2
Some Examples of Poor-Quality Micronutrient Research Data

Crop	% Yield Increase Over Control	Reference
Zinc fertilization, either alone or in combination with other micronutrients		
Sweet orange	50	Rahman and Haq (2006)
Wheat	60–120	Zafar et al. (2016)
Wheat	21–109	Khan et al. (2009)
Rice	14–45	
Citrus	46	Khan et al. (2012)
Tomato	36	Ejaz et al. (2012)
Mustard	20–54	Nawaz et al. (2012)
Flax	9–91	Tahir et al. (2014)
Onion	27–45	Khan et al. (2007)
Wheat	48	Ali et al. (2009)
Sunflower	6–32	Siddiqui et al. (2009)
Sunflower	9–31	Khan et al. (2009)
Sweet orange	2–33	Sajid et al. (2010)
Rice	7–31	Yaseen et al. (1999)
Maize	3–55	Marwat et al. (2007)
Boron fertilization, either alone or in combination with other micronutrients		
Sweet orange	27	Rahman and Haq (2006)
Seed cotton	48–124	Ahmad et al. (2009)
Coarse paddy	10–31	Ahmad and Irshad (2011)
Peanut	14–34	Nawaz et al. (2014)
Wheat	42	Ali et al. (2009)
Wheat	15–70	Khan et al. (2011)
Rice	17–54	
Tomato	27–64	Naz et al. (2012)
Manganese fertilization		
Sweet orange	31	Rahman and Haq (2006)

REFERENCES

Ahmad, R., and M. Irshad. 2011. Effect of boron application time on yield of wheat, rice and cotton crop in Pakistan. *Soil Environment* 30:50–57.

Ahmad, W., A. Niaz, S. Kanwal, and M. K. Rasheed. 2009. Role of boron in plant growth: a review. *Journal of Agricultural Research* 47:329–338.

Ali, S., A. Shah, M. Arif, et al. 2009. Enhancement of wheat grain yield and yield components through foliar application of zinc and boron. *Sarhad Journal of Agriculture* 25:15–19.

Ejaz, M., R. Waqas, C. M. Ayyub, et al. 2012. Efficacy of zinc with nitrogen as foliar feeding on growth, yield and quality of tomato grown under poly tunnel. *Pakistan Journal of Agricultural Sciences* 49:331–333.

Khan, M. A., J. Din, S. Nasreen, et al. 2009. Response of sunflower to different levels of zinc and iron under irrigated conditions. *Sarhad Journal of Agriculture* 25:159–163.

Khan, R., A. R. Gurmani, M. S. Khan and A. H. Gurmani. 2009. Residual, direct and cumulative effect of zinc application on wheat and rice yield under rice-wheat system. *Soil and Environment* 28: 24–28.

Khan, R. U., A. R. Gurmani, M. S. Khan, et al. 2011. Residual, direct and cumulative effect of boron application on wheat and rice yield under rice-wheat system. *Sarhad Journal of Agriculture* 27:219–223.

Khan, A. S., W. Ullah, A. U. Malik, et al. 2012.Exogenous applications of boron and zinc influence leaf nutrient status, tree growth and fruit quality of feutrell's early (Citrus reticulata Blanco). *Pakistan Journal of Agricultural Sciences* 49:113–119.

Khan, A. A., M. Zubair, A. Bari, and F. Maula. 2007. Response of onion (*Allium cepa*) growth and yield to different levels of nitrogen and zinc in Swat valley. *Sarhad Journal of Agriculture* 23: 933–936.

Marwat, K. B., M. Arif, and M. A. Khan. 2007. Effect of tillage and zinc application methods on weeds and yield of maize.*Pakistan Journal of Botany* 39:1583–1591.

Nawaz, N., M. S. Nawaz, N. M. Cheema, and M. A. Khan. 2012. Zinc and iron application to optimize seed yield of mustard. *Pakistan Journal of Agricultural Research* 25: 28–33.

Nawaz, N., M. S. Nawaz, M. A. Khan, et al. 2014. Effect of boron on peanut genotypes under rainfed conditions. *Pakistan Journal of Agricultural Research* 27:110–117.

Naz, R. M. M., S. Muhammad, A. Hamid, and F. Bibi. 2012. Effect of boron on the flowering and fruiting of tomato. *Sarhad Journal of Agriculture* 28:37–40.

Rahman, H. and I. Haq. 2006. Diagnostic criteria of micronutrients for sweet orange. *Soil & Environment* 25:119–127.

Sajid, M., A. Rab, N. Ali, et al. 2010. Effect of foliar application of Zn and B on fruit production and physiological disorders in sweet orange cv. Blood orange. *Sarhad Journal of Agriculture* 26:355–360.

Siddiqui, M. H., F. C. Oad, M. K. Abbasi, and A. W. Gandahi. 2009. Zinc and boron fertility to optimize physiological parameters, nutrient uptake and seed yield of Sunflower. *Sarhad Journal of Agriculture* 25: 53–57.

Tahir, M., M. Irfan, and A. Rehman. 2014. Effect of foliar application of zinc on yield and oil contents of flax. *Pakistan Journal of Agricultural Research* 27:287–295.

Yaseen, M., H. Hussain, A. Hakeem, and N. Mahmood. 1999. Rice-wheat response to integrated nutrient management with special emphasis on zinc fertilization. *Pakistan Journal of Biological Sciences* 2:1533–1535.

Zafar, S., M. Y. Ashraf, S. Anwar, Q. Ali, and A. Noman. 2016. Yield enhancement in wheat by soil and foliar fertilization of K and Zn under saline environment. *Soil and Environment* 35: 46–55.

Annex 3
Extent of Zinc Deficiency in Various Cropping Systems and Regions of Pakistan

Region and Cropping System/Crop	No. of Sites/Samples		Reference
	Total	% Deficient	
All Pakistan: Orchard soils	329	64	Zia et al. (2006)
Punjab			
Wheat	*407*	*56*	
Wheat (Gujranwala, Hafizabad, and Sheikhupuradistts.)*	112	57	Rafique (2011)
Wheat (Vehariand R.Y. Khan distts.)*	170	54	Rafique (2010)
Wheat young shoots (rainfed Pothowar)*	61	80	Rafique et al. (2006)
Wheat (3 distts. in Punjab)**	50	30	PARC (1986)
Wheat (cv. Maxi-Pak)**	14	57	Tahir (1981)
Cotton leaves and soils (Khanewal, Multan, Lodhran, Vehari, Bahawalpur, and R.Y. Khan distts.)*	420	**41**	Rashid (1995, 1996); Rashid and Rafique (1997)
Rice	*401*	*76*	
Rice (Gujranwala, Hafizabad, and Sheikhupura distts.)*	112	69	Rafique (2011)
Rice young whole shoots and soils (Sheikhupura and Gujranwala distts.)*	110	72	Hussain (2006)
Rice (cv. Basmati-385)*	84	90	Zia (1993)
Rice soils (Gujrat, Sahiwal, and Okara)	40	82	NIAB (1987)
Rice (6 distts. in Punjab)**	39	80	PARC (1986)
Rice (cv. Basmati-370)**	16	57	Tahir (1981)
Maize (Sahiwal distt.)	*65*	*60*	Rafique (2011)
Sugarcane	*234*	*57*	
Sugarcane (Thatta and and Hyderabad distts.)	123	60	Rafique (2011)
Sugarcane (leaves) (Jhang distt.)*	63	51	Ahmad and Rafique (2008)
Sugarcane soils (Sargodha distt.)*	48	56	

(Continued)

Region and Cropping System/Crop	No. of Sites/Samples		Reference
	Total	% Deficient	
Sorghum young shoots (rainfed Pothohar)*	*255*	*67*	Rashid et al. (1997)
Rapeseed-mustard young shoots (rainfed Pothohar)*	*120*	*80*	Rashid (1993)
Peanut shoot terminals (rainfed Pothohar)*	*100*	*60*	Rashid (1995)
Potato (Sahiwal distt.)*	**80**	**40**	Rafique (2011)
Mango orchards	*1074*	*29*	
Mango leaves (Sargodha and T.T. Singh distts.)*	100	60	Rafique (2011)
Mango leaves*	800	21	Asif et al. (1998)
Mango leaves*	174	47	Siddique et al. (1998)
Citrus	*252*	*74*	
Citrus leaves (Sargodha and T.T. Singh distts.)*	100	60	Rafique (2011)
Citrus leaves (Sargodha distt.)*	43	62	Siddique et al. (1994)
Citrus leaves (Sargodha distt.)*	34	100	Rashid et al. (1991);
Citrus leaves (Sahiwal distt.)*	41	75	Rashid and Rashid (1994)
Citrus leaves (Faisalabad distt.)*	34	100	
Apple orchards (Murree distt.)	13	**100**	Ahmad et al. (2010)
Total soils	*755*	*55*	
Apple orchards soils (Murree distt.)	130	38	Ahmad et al. (2010)
Orchard soils	158	60	Zia et al. (2006)
Soils (Faisalabad distt.)	200	63	NIAB (1987)
Wheat soils (Jhang, Vehari, and M. Garh distts.)	50	24	
Soils (20 distts)	177	62	Sillanpää (1982)
Soils	40	50	Kausar et al. (1979)
KP			
Apple, Peach, Persimmon (Swat valley)	200	50	Rafique (2011)
Peach orchards (Swat Valley)	50	6	Samiullah et al. (2013)
Orchards in Abbottabad (Galliyat area)	50	80	Khattak and Hussain (2007)
Plum orchards (Peshawar, Nowshehra, Charsadda, and Mardan distts.)	40	95	Tariq et al. (2008)
Apple leaves*	52	37	Shah and Shahzad (2008)
Apple leaves*	*50*	*35*	Rehman (1990)
Citrus	*241*	*41*	
Citrus leaves (6 distts.)*	147	53	Khattak (1994)
Citrus leaves*	94	21	Rehman (1989)

Region and Cropping System/Crop	No. of Sites/Samples		Reference
	Total	% Deficient	
Total soils	*1131*	*37*	
Orchards' soils (Galliyat area – Abbottabad)	48	11	Khattak and Hussain (2007)
Orchard soils (Swat Valley and KP distts.) (0–15 and 15–30 cm)	52	6 – 27	Shah and Shahzad (2008)
Citrus orchards soils (Swat and Malakand distts.)	51	10	Shah et al. (2012)
Soils (Peshawar and Charsadda distts.)	167	63	Rashid et al. (2008)
Soils, *mostly acid soils* (Mansehra distt.)*	232	40	Rashid (1994)
Soils, *mostly acid soils* (Swat distt.)*	166	45	
Soils (10 distts.)*	270	21	Khattak and Perveen (1988)
Vegetable-growing sites in Peshawar distt.	18	33	Perveen et al. (2010)
Soils (4 distts.)	36	77	Sillanpää (1982)
Soils*	39	60	Kausar et al. (1979)
Sindh			
Cotton leaves and soils (Nawab Shah distt.)*	100	**48**	Rafique et al. (2002)
Sugarcanetopsoils (Thatta distt.)	123	**76**	Arain et al. (2017)
Sugarcane leaves(Thatta distt.)	123	**24**	
Banana (Hyderabad, Tando Allah Yar, Matiari, Tando Muhammad Khan, Sangarh, Sukhar, and Ghotki distts.)	65	**57**	Rafique (2011)
Total soils	*748*	*66*	
Orchard soils	94	90	Zia et al. (2006)
Soils (Sanghar distt.)*	100	94	Memon et al. (1989)
Soils (Thatta distt.)*	133	82	
Soils (Badin distt.)*	100	44	
Soils (Tharparkar distt.)*	100	31	
Soils (Hyderabad distt.)*	125	60	
Tomato fields' soils (Badin distt.)	32	65	Memon et al. (2012)
Soils	28	100	Sillanpää (1982)
Soils*	36	17	Kausar et al. (1979)
Balochistan			
Appleleaves (Quetta, Pashin, and Ziarat distts.)	70	**80**	Rafique (2011)
Total soils	*189*	*73*	

(*Continued*)

Region and Cropping System/Crop	No. of Sites/Samples		Reference
	Total	% Deficient	
Orchard soils	72	43	Zia et al. (2006)
Soils (Quetta, Kalat, Loralai, and Zhob distts.)*	60	92	Ismail et al. (1994)
Soils*	20	95	Veryal et al. (1984)
Soils*	37	90	Kausar et al. (1979)
Azad Jammu and Kashmir	*45*	*28*	
Soils (Muzaffarabaddistt.)*	15	25	Majeed (1987)
Soils (Kotlidistt.)*	30	30	Chaudhry (1990)

* Based on nutrient indexing of crop plants/soils/crop plants and soils
** Based on field experimental results

REFERENCES

Ahmad, S., and E. Rafique. 2008. Nutrient indexing and integrated nutrient management for sustaining sugarcane yields. Agricultural Linkages Program (ALP) Project, Final Technical Report. Sugarcane Crops Research Program, Crop Sciences Institute, National Agricultural Research Center, Islamabad. 104 pp.

Ahmad, H., M. T. Siddique, I. A. Hafiz, and Ehsan-ul-Haq. 2010. Zinc status of apple orchards and its relationship with selected physico-chemical properties in Murree tehsil. *Soil and Environment* 29:142–147.

Arain, M. Y., K. S. Memon, M. S. Akhtar, and M. Memon. 2017. Soil and plant nutrient status and spatial variability for sugarcane in lower Sindh. *Pakistan Journal of Botany* 49:531–540.

Asif, M., K. Daud, M. Ashraf, et al.1998. Nutrient status of mango orchards in Punjab. *Pakistan Journal of Soil Science* 15:1–6.

Chaudhry, B. A.1990. To Study the Micronutrient Status of Kotli district Soils. M.Sc. (Hons) thesis, NWFP Agricultural University, Peshawar.

Hussain, F. 2006. Soil Fertility Monitoring and Management in Rice-Wheat System. Agricultural Linkages Program (ALP) Project, Final Report, 2002–2006. Land Resources Research Program, National Agricultural Research Center, Islamabad. 83 pp.

Ismail, M., A. M. Noor-ud-Din, and R. Bashir. 1994. Micronutrients in Baluchistan soils. I. Concentrations in soils of mid-hill region. *Pakistan Journal of Soil Science* 9:91–95.

Kausar, M. A., S. M. Alam, M. Sharif, and M. I. Pervaiz. 1979. Micronutrient status of Pakistan soils. *Pakistan Journal of Scientific and Industrial Research* 22:156–161.

Khattak, J. K. 1994. Effect of foliar application of micronutrients in combination with urea on the yield and fruit quality of sweet oranges. Final Technical Report, 1991–1994. NWFP Agricultural University, Peshawar.

Khattak, R. A., and Z. Hussain. 2007. Evaluation of soil fertility status and nutrition of orchards. *Soil and Environment* 26:22–32.

Khattak, J. K., and S. Perveen. 1988. Micronutrient status of NWFP soils. p. 62–74, In: Proceedings of National Seminar on Micronutrients in Soils and Crops in Pakistan, 13–15 December 1987, NWFP Agricultural University, Peshawar.

Majeed, R. S. 1987. To Study the Micronutrient Status of Muzaffarabad, Azad Kashmir Soils.M.Sc. (Hons) thesis, NWFP Agricultural Univeristy, Peshawar.

Memon, M., G. M. Jamro, N. Memon, K. S. Memon, and M. S. Akhtar. 2012. Micronutrient availability assessment of tomato grown in Taluka Badin, Sindh. *Pakistan Journal of Botany* 44:649–654.

Memon, K. S., H. K. Puno, and S.M. Memon. 1989. Cooperative Research Program on Micronutrient Status of Pakistani Soils. 5th Annual Report, 1988–89. Sindh Agriculture University, Tandojam, 62 pp.

NIAB. 1987. Fifteen Years of NIAB. Nuclear Institute for Agriculture and Biology, Faisalabad.

PARC. 1986. Soil Status and Scope of Micronutrients in Pakistan Agriculture. Pakistan Agricultural Research Council, Islamabad, 36 pp.

Perveen, S., Z. Malik, and W. Nazif. 2010. Fertility status of vegetable growing areas of Peshawer, Pakistan. *Pakistan Journal of Botany* 42:1871–1880.

Rafique, E. 2010. Integrated Nutrient Management for Sustaining Irrigated Cotton-Wheat Productivity in Aridisols of Pakistan. PhD dissertation, Quaid-i-Azam University, Islamabad.

Rafique, E. 2011. Micronutrient Management for Sustaining Major Cropping Systems and Fruit Orchards. ASPL-II Coordinated Project, Final Report 2006–2011. Land Resources Research Institute, National Agricultural Research Center, Islamabad.

Rafique, E., A. Rashid, A. U. Bhatti, G. Rasool, and N. Bughio. 2002. Boron deficiency in cotton grown in calcareous soils of Pakistan. I. Distribution of B availability and comparison of soil testing methods. p. 349–356, In: Goldbach, H. E. et al. (Ed.), *Boron in Plant and Animal Nutrition*. Kluwer Academic Publishers, New York.

Rafique, E., A. Rashid, J. Ryan, and A. U. Bhatti. 2006. Zinc deficiency in rainfed wheat in Pakistan: Magnitude, spatial variability, management, and plant analysis diagnostic norms. *Communications in Soil Science and Plant Analysis* 37:181–197.

Rashid, A. 1993. Nutritional disorders of rapeseed-mustard and wheat grown in *Pothohar* area.Micronutrient Project, Annual Report, 1991–92. National Agricultural Research Center, Islamabad.

Rashid, A. 1994. Nutrient Indexing Surveys and Micronutrient Requirement of Crops. Micronutrient Project, Annual Report, 1992–93. National Agricultural Research Center, Islamabad.

Rashid, A. 1995. Nutrient Indexing of Cotton and Micronutrient Requirements of Cotton and Groundnut. Micronutrient Project, Annual Report, 1993–94. National Agricultural Research Center, Islamabad. 91 pp.

Rashid, A. 1996. Secondary and micronutrients. p. 341–385, Chapter 12, In: *Soil Science*, Rashid, A., and K. S. Memon (Managing Authors). National Book Foundation, Islamabad, Pakistan.

Rashid, M., A. U. Bhatti, F. Khan and Wasiullah. 2008. Physico-chemical properties and fertility status of soils of districts Peshawar and Charsadda. *Soil & Environment* 27: 228–235.

Rashid, A., F. Hussain, A. Rashid, and J. Din. 1991. Nutrient status of citrus orchards in Punjab. *Pakistan Journal of Soil Science* 6:25–28.

Rashid, A., and E. Rafique. 1997. Nutrient Indexing and Micronutrient Nutrition of Cotton and Genetic Variability in Rapeseed-Mustard and Rice to Boron Deficiency. Micronutrient Management in cotton in Relation to CLCV. Annual Project Report 1995–96. NARC, Islamabad. 111 pp.

Rashid, A., E. Rafique, N. Bughio, and M. Yasin. 1997. Micronutrient deficiencies in rainfed calcareous soils of Pakistan. IV. Zinc nutrition of sorghum. *Communications in Soil Science and Plant Analysis* 28:455–467.

Rashid, A., and A. Rashid. 1994. Nutritional problems of citrus on calcareous soils of Pakistan. p. 190–197, In: *Proc. First International Seminar on Citriculture in Pakistan.* Faisalabad, 2–5 Dec 1992. University of Agriculture, Faisalabad.

Rehman, H. 1989. Annual Report, Directorate of Soils and Plant Nutrition. 1988–89. Agricultural Research Institute, Tarnab, Peshawar.

Rehman, H. 1990. Annual Report, Directorate of Soils and Plant Nutrition. 1989–90. Agricultural Research Institute, Tarnab, Peshawar.

Samiullah, Z. Shah, M. Tariq, T. Shah, A. Latif, and A. Shah. 2013. Micronutrients status of peach orchards in Swat valley. *Sarhad Journal of Agriculture* 29:485–493.

Shah, Z., M. Z. Shah, M. Tariq, et al. 2012. Survey of citrus orchards for micronutrient deficiency in Swat valley of north western Pakistan. *Pakistan Journal of Botany* 44:705–710.

Shah, Z., and K. Shahzad. 2008. Micronutrients status of apple orchards in Swat valley of North West Frontier Province of Pakistan. *Soil and Environment* 27:123–130.

Siddique, M. T., M. Rashid, and M. Saeed. 1994. Micronutrient status of soils/plants of citrus growing areas in Punjab. p. 355–60, In: Efficient Use of Plant Nutrients. Proc. 4th National Congress of Soil Science, Islamabad, May 24–26, 1992. Soil Science Society of Pakistan.

Siddique, T., M. Rashid, and M. Saeed. 1998. Nutrient status of mango orchards in southern Punjab. *Pakistan Journal of Soil Science* 14: 78–84.

Sillanpää, M. 1982. Micronutrients and the Nutrient Status of Soils: A Global Study. FAO Soils Bulletin 48, FAO, Rome.

Tahir, M. 1981. Availability Status of Micronutrients in the Soils of West Pakistan and the Role and Behavior of Selected Micronutrients in the Nutrition of Crops. Final Technical Report of PL-480 Project. Nuclear Institute for Agriculture and Biology, Faisalabad. 124 pp.

Tariq, M., Z. Shah, and A. Ali. 2008. Micronutrients status of plum orchards in Peshawar valley. *Soil and Environment* 27:223–227.

Veryal, M. A., H. K. Puno, and A. S. Sheikh. 1984. Zinc status of some soil series of Baluchistan. *Pakistan Journal of Scientific and Industrial Research* 29: 38–39.

Zia, M. S. 1993. Fertilizer Use Efficiency Project and Soil Fertility ARP-II. Annual Report, 1992–93. National Agricultural Research Center, Islamabad.

Zia, M. H., R. Ahmad, I. Khaliq, A. Ahmad, and M. Irshad. 2006. Micronutrients status and management in orchards soils: applied aspects. *Soil and Environment* 25:6–16.

Annex 4
Extent of Boron Deficiency in Various Cropping Systems and Regions of Pakistan

Region and Cropping System/Crop	No. of Sites/Samples		Reference
	Total	% Deficient	
Punjab			
Wheat	*357*	*46*	
Wheat leaves (Gujranwala, Hafizabad, and Sheikhupura distts.)*	112	48	Rafique (2011)
Wheat young shoots (rainfed Pothohar)*	61	64	Rashid et al. (2011)
Wheat (Vehari and R.Y. Khan distts.)	170	37	Rafique (2010)
Wheat (cv. Maxi-Pak)**	14	50	Tahir (1981)
Cotton	*433*	*51*	
Cotton – younger leaves at flowering (Faisalabad, Jhang, and T.T. Singh distts.)	13	92	Niaz et al. (2002)
Cotton (Khanewal, Multan, Lodhran, Vehari, Bahawalpur, R.Y. Khan distts.)*	420	50	Rafique et al. (2002)
Rice	*212*	*35*	
Rice (Gujranwala, Hafizabad, and Sheikhupura distts.)*	112	40	Rafique (2011)
Rice (cv. Basmati-385)*	84	25	Zia (1993)
Rice (cv. Basmati-370)**	16	46	Tahir (1981)
Maize (hybrid; Sahiwal distt.)	**65**	**50**	Rafique (2011)
Sugarcane	*234*	*56*	
Sugarcane (Thatta and Hyderabad distts.)*	123	45	Rafique (2011)
Sugarcane soils (Jhang distt.)*	63	76	Ahmad and Rafique (2008)
Sugarcane soils (Sargodha distt.)*	48	60	
Sorghum young shoots (rainfed Pothohar)*	255	**67**	Rashid et al. (1997a)

(*Continued*)

Region and Cropping System/Crop	No. of Sites/Samples		Reference
	Total	% Deficient	
Rapeseed young shoots (rainfed Pothohar)*	56	**57**	Rashid et al. (2002)
Mustard young shoots (rainfed Pothohar)*	64	**78**	
Peanut shoot terminals (rainfed Pothohar)*	100	**50**	Rashid et al. (1997b)
Potato (Sahiwal distt.)*	80	**55**	Rafique (2011)
Citrus (Sargodha, T.T. Singh, Bahawalpur, and Multan distts.)*	100	45	
Mango (Sargodha, T.T. Singh, Bahawalpur, and Multan distts.)*	100	45	
Soils	*271*	*61*	
Soils (20 distts)*	177	49	Sillanpää (1982)
Cotton fields (Faisalabad, Jhang, and T.T. Singh distts.)	13	92	Niaz et al. (2002)
Soils (Faisalabad, Jhang, T.T. Singh, Sargodha, and Sahiwal distts.)	81	82	Niaz et al. (2007)
KP			
Citrus leaves (6 distts.)*	147	**51**	Khattak (1994)
Apple leaves	*102*	*30*	
Apple leaves	52	44	Shah and Shahzad (2008)
Apple leaves*	50	16	Rehman (1990)
Peach and plum leaves	90	46	
Peach leaves	50	6	Samiullah et al. (2013)
Plum leaves	40	95	Tariq et al. (2008)
Total soils	*1060*	*42*	
Soils (Peshawar distt.)	88	72	Rashid et al. (2008)
Soils (Charsadda distt.)	79	46	
Peach orchard soils (Swat Valley)	50	2	Samiullah et al. (2013)
Plum orchard soils (4 distts.)	40	90	Tariq et al. (2008)
Citrus leaves (Swat and Malakand distts.)	51	24	Shah et al. (2012)
Soils (Swat Valley)	52	0	Shah and Shahzad (2008)
Citrus Orchard soils (Swat and Malakand distts.)	51	0	Shah et al. (2012)
Soils, with pH ≥7.0 (Mansehra distt.)	139	0	Rashid (1994)
Soils with pH ≥7.0 (Swat distt.)	84	0	
Soils (Bannu distt.)*	56	98	Noor-Ur-Rehman et al. (2008)
Soils (10 distts.)*	270	60	Khattak and Perveen (1988)
Soils (Mardanand Peshawar distts.)	64	89	PARC (1986)
Soils (4 distts.)	36	55	Sillanpää (1982)

Region and Cropping System/Crop	No. of Sites/Samples		Reference
	Total	% Deficient	
Sindh			
Cotton leaves and Soils (Nawab Shah distt.)*	100	**56**	Rafique et al. (2002)
Sugarcane topsoils (Thatta distt.)*	123	**45**	Arain et al. (2017)
Sugarcane leaves (3rd leaf from spindle) (Thatta distt.)*	123	**21**	
Banana (Hyderabad, Tando Allah Yar, Matiari, Tando Muhammad Khan, Sangarh, Sukkur, and Ghotki distts.)	65	60	Rafique (2011)
Total Soils	*702*	*66*	
Banana Orchard soils***	48	94	Zia et al. (2006)
Tomato Fields (Badin Tehsil)	32	91	Memon et al. (2012)
Soils (6 distts)*	622	**62**	Sillanpää (1982); Memon et al. (1989)
Balochistan			
Apple leaves	*520*	*61*	
Apple leaves (Quetta, Pishin, and Ziarat distts.)	70	51	Rafique (2011)
Apple leaves (Pishin distt.)	450	62	Ziad et al. (2016)
Apple orchard soils	*146*	*60*	
Apple orchards' soils – Pishin distt.	90	58	Ziad et al. (2016)
Apple orchards' soils***	56	64	Zia et al. (2006)
Azad Jammu and Kashmir	*45*	*45*	
Soils (Muzaffarabad distt.)*	15	85	Majeed (1987)
Soils (Kotli distt.)*	30	25	Chaudhry (1990)

* Based on nutrient indexing of crop plants and soils.
** Based on field experimental results.
*** Samples depicting dilute-HCl-extractable B < 1.0 mg kg^{-1} were categorized as deficient.

REFERENCES

Ahmad, S., and E. Rafique. 2008. Nutrient indexing and integrated nutrient management for sustaining sugarcane yields. Agricultural Linkages Program (ALP) Project, Final Technical Report. Sugarcane Crops Research Program, Crop Sciences Institute, National Agricultural Research Center, Islamabad. 104 pp.

Arain, M. Y., K. S. Memon, M. S. Akhtar, and M. Memon. 2017. Soil and plant nutrient status and spatial variability for sugarcane in lower Sindh. *Pakistan Journal of Botany* 49:531–540.

Chaudhry, B. A. 1990. To Study the Micronutrient Status of Kotli district Soils. M.Sc. (Hons) thesis, NWFP Agricultural University, Peshawar.

Khattak, J. K. 1994. Effect of foliar application of micronutrients in combination with urea on the yield and fruit quality of sweet oranges. Final Technical Report, 1991–1994. NWFP Agricultural University, submitted to Pakistan Science Foundation, Islamabad.

Khattak, J. K., and S. Perveen. 1988. Micronutrient status of NWFP soils. p. 62–74, In: Proceedings of National Seminar on Micronutrients in Soils and Crops in Pakistan, 13–15 December 1987, NWFP Agricultural University, Peshawar.

Majeed, R. S. 1987. To Study the Micronutrient Status of Muzaffarabad, Azad Kashmir Soils. M.Sc. (Hons) thesis, NWFP Agricultural Univeristy, Peshawar.

Memon, M., G. M. Jamro, N. Memon, K. S. Memon, and M. S. Akhtar. 2012. Micronutrient availability assessment of tomato grown in taluka Badin, Sindh. *Pakistan Journal of Botany*, 44: 649–654.

Memon, K. S.,H. K. Puno, and S. M. Memon. 1989. Cooperative Research Program on Micronutrient Status of Pakistani Soils. 5th Annual Report, 1988–89. Sindh Agriculture University, Tandojam, 62 pp.

Niaz, A., M. Ibrahim, N. Ahmad, and S. A. Anwar. 2002. Boron contents of light and medium textured soils and cotton plants. *International Journal of Agriculture and Biology* 4:534–536.

Niaz, A., A. M. Ranjha, A. Hannan, and M. Waqas. 2007. Boron status of soils as affected by different soil characteristics – pH, $CaCO_3$, organic matter and clay contents. *Pakistan Journal of Agricultural Sciences* 44:428–435.

Noor Ur Rehman, M., Z. Hussain, and S. Hussain. 2008. Hot-water soluble boron contents of Bannu soils. *Pakistan Journal of Soil Science* 3:7–10.

PARC. 1986. *Soil Status and Scope of Micronutrients in Pakistan Agriculture.* Pakistan Agricultural Research Council, Islamabad, 36 pp.

Rafique, E. 2010. Integrated Nutrient Management for Sustaining Irrigated Cotton-Wheat Productivity in Aridisols of Pakistan. PhD dissertation, Quad-i-Azam University, Islamabad.

Rafique, E. 2011. Micronutrient Management for Sustaining Major Cropping Systems and Fruit Orchards. ASPL-II Coordinated Project, Final Report 2006–2011. Land Resources Research Institute, National Agricultural Research Center, Islamabad.

Rafique, E., A. Rashid, A. U. Bhatti, G. Rasool, and N. Bughio. 2002. Boron deficiency in cotton grown in calcareous soils of Pakistan. I. Distribution of B availability and comparison of soil testing methods. p. 349–356, In: Goldbach, H. E. et al. (Ed.), *Boron in Plant and Animal Nutrition.* Kluwer Academic Publishers, New York.

Rashid, A. 1994. Nutrient Indexing Surveys and Micronutrient Requirement of Crops. Micronutrient Project, Annual Report, 1992–93. National Agricultural Research Center, Islamabad.

Rashid, M., A. U. Bhatti, F. Khan and Wasiullah. 2008. Physico-chemical properties and fertility status of soils of districts Peshawar and Charsadda. *Soil and Environment* 27:228–235.

Rashid, A., E. Rafique, and N. Ali. 1997b. Micronutrient deficiencies in rainfed calcareous soils of Pakistan. II. Boron nutrition of the peanut plant. *Communications in Soil Science and Plant Analysis* 28:149–159.

Rashid, A., E. Rafique, A. U. Bhatti, et al. 2011. Boron deficiency in rainfed wheat in Pakistan: Incidence, spatial variability, and management strategies. *Journal of Plant Nutrition* 34:600–613.

Rashid, A., E. Rafique, and N. Bughio. 1997a. Micronutrient deficiencies in rainfed calcareous soils of Pakistan. III. Boron nutrition of sorghum. *Communications in Soil Science and Plant Analysis* 28:441–454.

Rashid, A., E. Rafique, S. Muhammed, and N. Bughio. 2002. Boron deficiency in rainfed alkaline soils of Pakistan: Incidence and genotypic variation in rapeseed-mustard. p. 363–370, In: Goldbach, H. E. et al. (Ed.), *Boron in Plant and Animal Nutrition.* Kluwer Academic/Plenum Publishers, New York.

Rehman, H. 1990. Annual Report, Directorate of Soils and Plant Nutrition. 1989–90. Agricultural Research Institute, Tarnab, Peshawar, Pakistan.

Samiullah, Z., M. Tariq, T. Shah, A. Latif, and A. Shah. 2013. Micronutrients status of peach orchards in Swat valley. *Sarhad Journal of Agriculture* 29:485–493.

Shah, Z., M. Z. Shah, M. Tariq, et al. 2012. Survey of citrus orchards for micronutrient deficiency in Swat valley of north western Pakistan. *Pakistan Journal of Botany* 44:705–710.

Shah, Z., and K. Shahzad. 2008. Micronutrients status of apple orchards in Swat valley of North West Frontier Province of Pakistan. *Soil and Environment* 27:123–130.

Sillanpää, M. 1982. Micronutrients and the Nutrient Status of Soils: A Global Study. FAO Soils Bulletin 48, FAO, Rome.

Tahir, M. 1981. Availability Status of Micronutrients in the Soils of West Pakistan and the Role and Behavior of Selected Micronutrients in the Nutrition of Crops. Final Technical Report of PL-480 Project. Nuclear Institute for Agriculture and Biology, Faisalabad. 124 pp.

Tariq, M., Z. Shah, and A. Ali. 2008. Micronutrients status of plum orchards in Peshawar valley. *Soil and Environment* 27:223–227.

Zia, M. S. 1993. Fertilizer Use Efficiency Project and Soil Fertility ARP-II. Annual Report, 192–93. National Agricultural Research Center, Islamabad.

Zia, M. H., R. Ahmad, I. Khaliq, A. Ahmad, and M. Irshad. 2006. Micronutrients status and management in orchards soils: applied aspects. *Soil and Environment* 25:6–16.

Ziad, T., M. T. Siddique, K. S. Khan, et al. 2016. Indexing bioavailable and foliage boron content in apple orchards of Pishin district, Baluchistan, using GIS and geo-statistics as diagnostic tools. *Soil and Environment* 35:35–45.

Annex 5
Extent of Iron Deficiency in Various Cropping Systems and Regions of Pakistan

Region and Cropping System/Crop	No. of Sites/Samples		Reference
	Total	% Deficient	
All **Pakistan**: Soils*	242	0	Sillanpää (1982)
All **Pakistan**: Orchards' soils	329	19	Zia et al. (2006)
Punjab			
Total vegetation	*1457*	*15*	
Cotton leaves (Khanewal, Multan, Lodhran, Vehari, Bahawalpur, and R.Y. Khan distts.)*	420	0	Rashid (1995, 1996); Rashid and Rafique (1997)
Rice young whole shoots and soils (Sheikhupura and Gujranwala distts.)*	110	0	Hussain (2006)
Peanut shoot terminals (rainfed Pothohar)***	**100**	**50**	Rashid et al. (1997)
Wheat young shoots (rainfed Pothohar)*	61	0	Rashid (1993)
Rapeseed-mustard young shoots (rainfed Pothohar)**	**120**	**20**	
Sorghum young shoots (rainfed Pothohar)*	**255**	**3**	Rashid and Qayyum (1991)
Mango leaves*	174	0	Siddique et al. (1998)
Citrus	*152*	*71*	
Citrus leaves (Sargodha)*	43	37	Siddique et al. (1994)
Citrus leaves (Sargodha, Sahiwal, and Faisalabad)*	109	85	Rashid et al. (1991)
Apple leaves from 13 orchards (Murree tehsil)	65	46	Ahmed et al. (2012)
Sugarcane	*111*	*17*	
Sugarcane soils (Jhang distt.)*	63	22	Ahmad and Rafique (2008)
Sugarcane soils (Sargodha distt.)*	48	10	

(Continued)

Region and Cropping System/Crop	No. of Sites/Samples		Reference
	Total	% Deficient	
Total soils	*828*	*4*	
Orchards' soils	158	22	Zia et al. (2006)
Apple orchards' soils – 30 orchards (Murree tehsil)	540	0	Ahmed et al. (2014)
Apple orchards' soils – 13 orchards (Murree tehsil)	130	0	Ahmed et al. (2012)
KP			
Total vegetation	*683*	*17*	
Peach leaves (Swat Valley)	50	0	Samiullah et al. (2013)
Orchards leaves – apple, peach, pear, citrus, and plum (Abbotabad Galliat)	50	0	Khattak and Hussain (2007)
Plum leaves (Peshawar, Nowshehra, Charsadda, and Mardan distts.)	40	0	Tariq et al. (2008)
Apple leaves	*302*	*38*	
Apple leaves(Swat Valley)*	200	45	Rafique (2011)
Apple leaves (Swat Valley)	52	0	Shah and Shahzad (2008)
Apple leaves****	50	0	Rehman (1990)
Citrus	*241*	*0*	
Citrus leaves (Swat and Malakand distts.)	51	0	Shah et al. (2012)
Citrus leaves (6 distts.)****	147	0	Khattak (1994)
Citrus leaves****	94	0	Rehman (1989)
Total Soils	*1245*	*8*	
Citrus orchards' soils (Swat and Malakand distts.)	51	12	Shah et al. (2012)
Vegetable fields (Peshawar distt.)	18	0	Perveen et al. (2010)
Peach orchards' soils (Swat Valley)	50	0	Samiullah et al. (2013)
Apple orchard soils (Swat Valley)	52	0	Shah and Shahzad (2008)
Plum orchards' soils (Peshawar, Nowshehra, Charsadda, and Mardan distts.)	40	0	Tariq et al. (2008)
Soils (Peshawar distt.)	88	0	Rashid et al. (2008)
Soils (Charsada distt.)	79	0	
Soils (Mansehra distt.)*	232	0	Rashid (1994)
Soils (Swat distt.)*	166	1	
Soils (10 distts.)	270	26	Khattak and Perveen (1988)

Region and Cropping System/Crop	No. of Sites/Samples		Reference
	Total	% Deficient	
Soils (DI Khan, Abbotabad, and Mansehra distts.)	125	20	PARC (1986)
Sindh			
Total Vegetation	*155*	*0*	
Sugarcane leaves (3rd leaf from spindle)	123	0	Arain et al. (2017)
Tomato leaves (Badin tehsil)	32	0	Memon et al. (2012)
Sugarcane topsoils (Thatta distt.)*	123	**0**	Arain et al. (2017)
Total soils	*684*	*3*	
Tomato fields (Badin tehsil)	32	0	Memon et al. (2012)
Orchards' soils	94	16	Zia et al. (2006)
Soils (Hyderabad, Badin, Sanghar, and Tharparkar distts.)*	425	0	Memon (1989)
Soils (Thatta distt.)*	133	3	
Balochistan			
Apple leaves (Quetta, Pashin, and Ziarat distts.)	70	80	Rafique (2011)
Total soils	*169*	*9*	
Apple orchard soils	72	19	Zia et al. (2006)
Soils (Quetta, Kalat, Loralai, and Zhob distts.)*****	60	2	Ismail et al. (1994)
Soils*****	37	0	Kausar et al. (1979)

* Based on nutrient indexing of crop plants/soils/crop plants and soils
** Based on field experimental results
*** Nutrient indexing, based on Fe²⁺ concentration in fresh shoot terminals of farmer-grown peanut
**** Based on leaf analysis for total Fe, which is not a reliable indicator; also, analytical results contradict widespread Fechlorosis in fruit orchards of KP
***** Based on soil analysis; not reliable, as the results contradict widespread Fechlorosis in fruit orchards of Balochistan

REFERENCES

Ahmad, S., and E. Rafique. 2008. Nutrient indexing and integrated nutrient management for sustaining sugarcane yields. Agricultural Linkages Program (ALP) Project, Final Technical Report. Sugarcane Crops Research Program, Crop Sciences Institute, National Agricultural Research Center, Islamabad. 104 pp.

Ahmed, H., M. T. Siddique, S. Ali, N. A. Abbasi, A. Khalid, and R. Khalid. 2014. Micronutrient indexing in the apple orchards of Northern Punjab, Pakistan using geostatistics and GIS as diagnostic tools. *Soil and Environment* 33:07–16.

Ahmed, H., M. T. Siddique, S. Ali, A. Khalid, and N. A. Abbasi. 2012. Mapping of Fe and impact of selected physico-chemical properties on its bioavailability in the apple orchards of Murree region. *Soil and Environment* 31:100–107.

Arain, M. Y., K. S. Memon, M. S. Akhtar, and M. Memon. 2017. Soil and plant nutrient status and spatial variability for sugarcane in lower Sindh. *Pakistan Journal of Botany* 49:531–540.

Hussain, F. 2006. Soil Fertility Monitoring and Management in Rice-Wheat System. Agricultural Linkages Program (ALP) Project, Final Report, 2002–2006. Land Resources Research Program, National Agricultural Research Center, Islamabad. 83 pp.

Ismail, M., A. M. Noor-ud-Din, and R. Bashir. 1994. Micronutrients in Baluchistan soils. I. Concentrations in soils of mid-hill region. *Pakistan Journalof Soil Science* 9:91–95.

Kausar, M. A., S. M. Alam, M. Sharif, and M. I. Pervaiz. 1979. Micronutrient status of Pakistan soils. *Pakistan Journal of Scientific and Industrial Research* 22:156–161.

Khattak, J. K. 1994. Effect of foliar application of micronutrients in combination with urea on the yield and fruit quality of sweet oranges. Final Technical Report, 1991–1994. NWFP Agricultural University, Peshawar.

Khattak, R. A., and Z. Hussain. 2007. Evaluation of soil fertility status and nutrition of orchards. *Soil and Environment* 26:22–32.

Khattak, J. K., and S. Perveen. 1988. Micronutrient status of NWFP soils. p. 62–74, In: Proceedings of National Seminar on Micronutrients in Soils and Crops in Pakistan, 13–15 December 1987, NWFP Agricultural University, Peshawar.

Memon, M., G. M. Jamro, N. Memon, K. S. Memon, and M. S. Akhtar. 2012. Micronutrient availability assessment of tomato grown in taluka Badin, Sindh. *Pakistan Journal of Botany* 44:649–654.

Memon, K. S., H. K. Puno, and S. M. Memon. 1989. Cooperative Research Program on Micronutrient Status of Pakistani Soils. 5th Annual Report, 1988–89. Sindh Agriculture University, Tandojam, 62 pp.

PARC. 1986. Soil Status and Scope of Micronutrients in Pakistan Agriculture.Pakistan Agricultural Research Council, Islamabad., 36 pp.

Perveen, S., Z. Malik, and W. Nazif.2010. Fertility status of vegetable growing areas of Peshawer. Pakistan. *Pakistan Journal of Botany* 42:1871–1880.

Rafique, E. 2011. Micronutrient Management for Sustaining Major Cropping Systems and Fruit Orchards. ASPL-II Coordinated Project, Final Report 2006–2011. Land Resources Research Institute, National Agricultural Research Center, Islamabad.

Rashid, A. 1993. Nutritional disorders of rapeseed-mustard and wheat grown in *Pothohar* area. Micronutrient Project, Annual Report, 1991–92. National Agricultural Research Center, Islamabad.

Rashid, A. 1994. Nutrient Indexing Surveys and Micronutrient Requirement of Crops. Micronutrient Project, Annual Report, 1992–93. National Agricultural Research Center, Islamabad.

Rashid, A. 1995. Nutrient Indexing of Cotton and Micronutrient Requirements of Cotton and Groundnut. Microenutrient Project, Annual Report, 1993–94. National Agricultural Research Center, Islamabad. 91 pp.

Rashid, A. 1996. Nutrient Indexing of Cotton in Multan District and Boron and Zinc Neutrition of Cotton. Micronutrients Project, Annual Report, 1994–95. National Agricultural Research Center, Islamabad.76 pp.

Rashid, M., A. U. Bhatti, F. Khan and Wasiullah. 2008. Physico-chemical properties and fertility status of soils of districts Peshawar and Charsadda. *Soil and Environment* 27:228–235.

Rashid, A., F. Hussain, A. Rashid, and J. Din. 1991. Nutrient status of citrus orchards in Punjab.*Pakistan Journal of Soil Science* 6:25–28.

Rashid, A., and F. Qayyum. 1991. Micronutrient Status of Pakistan Soils and Their Role in Crop Production, Cooperative Program. Final Report, 1983–1990. NARC, Islamabad. 84 pp.

Rashid, A.,and E. Rafique. 1997. Nutrient Indexing and Micronutrient Nutrition of Cotton and Genetic Variability in Rapeseed-Mustard and Rice to Boron Deficiency. Micronutrient/Nutrient Management in cotton in Relation to CLCV. Annual Project Report 1995–96. NARC, Islamabad. 111 pp.

Rashid, A., E. Rafique, J. Din, S. N. Malik, and M. Y. Arain. 1997. Micronutrient deficiencies in rainfed calcareous soils of Pakistan. I. Iron chlorosis in peanut. *Communications in Soil Science and Plant Analysis* 28: 135–148.

Rehman, H. 1989. Annual Report, Directorate of Soils and Plant Nutrition. 1988–89. Agricultural Research Institute, Tarnab, Peshawar.

Rehman, H. 1990. Annual Report, Directorate of Soils and Plant Nutrition. 1989–90. Agricultural Research Institute, Tarnab, Peshawar.

Samiullah, Z. S., M. Tariq, et al. 2013. Micronutrients status of peach orchards in Swat valley. *Sarhad Journal of Agriculture* 29:485–493.

Shah, Z., M. Z. Shah, M. Tariq, et al. 2012. Survey of citrus orchards for micronutrient deficiency in Swat valley of north western Pakistan. *Pakistan Journal of Botany* 44:705–710.

Shah, Z., and K. Shahzad. 2008. Micronutrients status of apple orchards in Swat valley of North West Frontier Province of Pakistan. *Soil and Environment* 27:123–130.

Siddique, M. T., M. Rashid, and M. Saeed. 1994. Micronutrient status of soils/plants of citrus growing areas in Punjab. p. 355–360, In: *Efficient Use of Plant Nutrients. Proc. 4th National Congress of Soil Science*, Islamabad, May 24–26, 1992. Soil Science Society of Pakistan.

Siddique, T., M. Rashid, and M. Saeed. 1998. Nutrient status of mango orchards in southern Punjab. *Pakistan Journal of Soil Science* 14:78–84.

Sillanpää, M. 1982. Micronutrients and the Nutrient Status of Soils: A Global Study. FAO Soils Bulletin 48, FAO, Rome.

Tariq, M., Z. Shah, and A. Ali. 2008. Micronutrients status of plum orchards in Peshawar valley. *Soil and Environment* 27:223–227.

Zia, M. H., R. Ahmad, I. Khaliq, A. Ahmad, and M. Irshad. 2006. Micronutrients status and management in orchards soils: applied aspects. *Soil and Environment* 25: 6–16.

Annex 6
Extent of Copper Deficiency in Various Cropping Systems and Regions of Pakistan

Region and Cropping System/Crop	No. of Sites/Samples		Reference
	Total	% Deficient	
Soils (All **Pakistan**; 20 distts.)*	242	0	Sillanpää (1982)
Soils (All **Pakistan**;orchard soils)	329	2	Zia et al. (2006)
Punjab			
Total	*2456*	*6*	
Cotton (Khanewal, Multan, Lodhran, Vehari, Bahawalpur, and R.Y. Khan distts.)*	420	1	Rashid (1995, 1996); Rashid and Rafique (1997)
Rice (young whole shoots and soils) (Sheikhupura and Gujranwala distts.)*	110	0	Hussain (2006)
Rice (cv. Basmati-385)*	84	2	Zia (1993)
Rice (Sialkot, Gujrat, and Sahiwal distts.)**	39	1	PARC (1986)
Rice (cv. Basmati-370)**	16	0	Tahir (1981)
Wheat (cv. Maxi-Pak)**	14	0	
Sugarcane (Jhang distt.)*	63	0	Ahmad and Rafique (2008)
Sugarcane (Sargodha distt.)*	48	0	
Wheat (young shoots) (rainfed Pothohar)*	61	2	Rashid (1993)
Sorghum (young shoots) (rainfed Pothohar)*	255	6	Rashid and Qayyum (1991)
Rapeseed-mustard (young shoots) (rainfed Pothohar)*	120	1	Rashid (1993)
Peanut (shoot terminals) (rainfed Pothohar)*	100	1	Rashid (1994, 1995)
Mango leaves*	800	10	Asif et al. (1998)
Mango leaves*	174	8	Siddique et al. (1998)
Citrus leaves (Sargodha distt.)*	43	50	Siddique et al. (1994)
Citrus leaves (Sargodha distt.)*	34	8	Rashid et al. (1991; Rashid and Rashid, 1994)
Citrus leaves (Sahiwal distt.)*	41	0	
Citrus leaves (Faisalabad distt.)*	34	8	
Total soils	*668*	*3*	
Soils*	40	5	Kausar et al. (1979)
Faisalabad soils	200	6	NIAB (1987)

(Continued)

| Region and Cropping System/Crop | No. of Sites/Samples | | Reference |
	Total	% Deficient	
Apple orchard soils (30 orchards; Murree tehsil)	180	0	Ahmed et al. (2014)
Orchard soils	158	3	Zia et al. (2006)
Wheat soils (Jhang, M. Garh, D.G. Khan, and Vehari distts.)	50	0	PARC (1986)
Rice soils (Gujrat, Sahiwal, and Okara distts.)	40	0	NIAB (1987)
KP			
Total vegetation	*534*	*19*	
Citrus leaves (Swat and Malakand distts.)	51	16	Shah et al. (2012)
Peach leaves (Swat Valley)	50	0	Samiullah et al. (2013)
Plum leaves (Peshawar, Nowshehra, Charsadda, and Mardan distts.)	40	13	Tariq et al. (2008)
Apple leaves (Swat Valley)	52	2	Shah and Shahzad (2008)
Apple, peach, pear, citrus, and plum leaves (Abbotabad Galliat)	50	8	Khattak and Hussain (2007)
Citrus leaves (6 distts.)*	147	58	Khattak (1994)
Apple leaves*	50	0	Rehman (1990)
Citrus leaves*	94	0	Rehman (1989)
Total soils	*1378*	*2*	
Citrus orchard soils (Swat and Malakand distts.)	51	0	Shah et al. (2012)
Vegetable fields (Peshawar distt.)	18	0	Perveen et al. (2010)
Peach orchard soils (Swat Valley)	50	0	Samiullah et al. (2013)
Soils (Peshawar distt.)	88	0	Rashid et al. (2008)
Soils (Charsada distt.)	79	11	
Plum orchard Soils (Peshawar, Nowshehra, Charsada, and Mardan distts.)	40	0	Tariq et al. (2008)
Apple orchard soils (Swat Valley)	52	0	Shah and Shahzad (2008)
Apple, peach, pear, citrus, and plum orchard soils (Abbotabad Galliat)	74	0	Khattak and Hussain (2007)
Soils (Mansehra distt.)	232	0	Rashid (1994)
Soils (Swat distt.)	166	0	
Soils (10 distts.)	270	6	Khattak and Perveen (1988)
Soils (Peshawar, Mardan, DI Khan, Mansehra, Abbotabad, and Haripur distts.)	199	0	PARC (1986)
Soils (Bannu distt.)	20	15	
Soils*	39	15	Kausar et al. (1979)
Sindh			
Total	*255*	*0*	
Sugarcane leaves (Thatta distt.)*	123	0	
Cotton leaves and soils*	100	0	Rashid and Rafique (1997)
Tomato leaves (Badin tehsil)	32	0	Memon et al. (2012)

| Region and Cropping System/Crop | No. of Sites/Samples | | Reference |
	Total	% Deficient	
Total soils	*843*	*1*	
Tomato Fields (Badin tehsil)	32	0	Memon et al. (2012)
Sugarcane soils (Thatta distt.)*	123	0	Arain et al. (2017)
Orchard soils	94	0	Zia et al. (2006)
Soils (Thatta distt.)	133	2	Memon et al. (1989)
Soils (Hyderabad, Badin, Sanghar, and Tharparkar distts.)	425	0	
Soils*	36	14	Kausar et al. (1979)
Balochistan			
Total soils	*169*	*1*	
Orchard soils	72	1	Zia et al. (2006)
Soils (Quetta, Kalat, Loralai, and Zhob distts.)*	60	3	Ismail et al. (1994)
Soils*	37	0	Kausar et al. (1979)

* Based on nutrient indexing of crop plants and soils.
** Based on field experimental results.

REFERENCES

Ahmad, S., and E. Rafique. 2008. Nutrient indexing and integrated nutrient management for sustaining sugarcane yields. Agricultural Linkages Program (ALP) Project, Final Technical Report. Sugarcane Crops Research Program, Crop Sciences Institute, National Agricultural Research Center, Islamabad. 104 pp.

Ahmed, N., M. Abid, A. Rashid, R. Abou-Shanab, and F. Ahmad. 2014. Influence of boron nutrition on membrane leakage, chlorophyll content and gas exchange characteristics in cotton (*Gossypiumhirsutum* L.). *Journal of Plant Nutrition* 37:2302–2315.

Arain, M. Y., K. S. Memon, M. S. Akhtar, and M. Memon. 2017. Soil and plant nutrient status and spatial variability for sugarcane in lower Sindh. *Pakistan Journal of Botany* 49:531–540.

Asif, M., K. Daud, M. Ashraf, et al. 1998. Nutrient status of mango orchards in Punjab. *Pakistan Journal of Soil Science* 15:1–6.

Hussain, F. 2006. Soil Fertility Monitoring and Management in Rice-Wheat System. Agricultural Linkages Program (ALP) Project, Final Report, 2002–2006. Land Resources Research Program, National Agricultural Research Center, Islamabad. 83 pp.

Ismail, M., A. M. Noor-ud-Din, and R. Bashir. 1994. Micronutrients in Baluchistan soils. I. Concentrations in soils of mid-hill region. *Pakistan Journal of Soil Science* 9:91–95.

Kausar, M. A., S. M. Alam, M. Sharif, and M. I. Pervaiz. 1979. Micronutrient status of Pakistan soils.*Pakistan Journal of Scientific and Industrial Research* 22:156–161.

Khattak, J. K. 1994. Effect of foliar application of micronutrients in combination with urea on the yield and fruit quality of sweet oranges. Final Technical Report, 1991–1994. NWFP Agricultural University, Peshawar.

Khattak, R. A., and Z. Hussain. 2007. Evaluation of soil fertility status and nutrition of orchards.*Soil and Environment* 26(1):22–32.

Khattak, J. K., and S. Perveen. 1988. Micronutrient status of NWFP soils. p. 62–74, In: Proceedings of National Seminar on Micronutrients in Soils and Crops in Pakistan, 13–15 December 1987, NWFP Agricultural University, Peshawar.

Memon, M., G. M. Jamro, N. Memon, K. S. Memon, and M. S. Akhtar. 2012. Micronutrient availability assessment of tomato grown in taluka Badin, Sindh. *Pakistan Journal of Btany* 44: 649–654.

Memon, K. S., H. K. Puno, and S. M. Memon. 1989. Cooperative Research Program on Micronutrient Status of Pakistani Soils. 5th Annual Report, 1988–89. Sindh Agriculture University, Tandojam, 62 pp.

NIAB. 1987. *Fifteen Years of NIAB*. Nuclear Institute for Agriculture and Biology, Faisalabad.

PARC. 1986. *Soil Status and Scope of Micronutrients in Pakistan Agriculture*. Pakistan Agricultural Research Council, Islamabad, 36 pp.

Perveen, S., Z. Malik, and W. Nazif. 2010. Fertility status of vegetable growing areas of Peshawar. *Pakistan Journal of Botany* 42:1871–1880.

Rashid, A. 1993. Nutritional disorders of rapeseed-mustard and wheat grown in *Pothohar* area. Micronutrient Project, Annual Report, 1991–92. National Agricultural Research Center, Islamabad.

Rashid, A. 1994. Nutrient Indexing Surveys and Micronutrient Requirement of Crops. Micronutrient Project Annual Report, 1992–93. National Agricultural Research Center, Islamabad. 138 pp.

Rashid, A.1995. Nutrient Indexing of Cotton and Micronutrient Requirements of Cotton and Groundnut.Micreonutrient Project, Annual Report, 1993–94. National Agricultural Research Center, Islamabad. 91 pp.

Rashid, A.1996. Nutrient Indexing of Cotton in Multan District and Boron and Zinc Nutrition of Cotton. Micronutrients Project, Annual Report, 1994–95. National Agricultural Research Center, Islamabad. 76 pp.

Rashid, M., A.U. Bhatti, F. Khan and Wasiullah. 2008. Physico-chemical properties and fertility status of soils of districts Peshawar and Charsadda. *Soil and Environment* 27: 228–235.

Rashid, A., F. Hussain, A. Rashid, and J. Din. 1991. Nutrient status of citrus orchards in Punjab. *Pakistan Journal of Soil Science* 6:25–28.

Rashid, A. and F. Qayyum. 1991. Micronutrient Status of Pakistan Soils and Their Role in Crop Production, Cooperative Program. Final Report, 1983–1990. National Agricultural Research Center, Islamabad. 84 pp.

Rashid, A. and E. Rafique. 1997. Nutrient Indexing and Micronutrient Nutrition of Cotton and Genetic Variability in Rapeseed-Mustard and Rice to Boron Deficiency. Micronutrient/Nutrient Management in cotton in Relation to CLCV. Annual Project Report 1995–96. National Agricultural Research Center, Islamabad. 111 pp.

Rashid, A., and A. Rashid. 1994. Nutritional problems of citrus on calcareous soils of Pakistan. p. 190–197, In: Proc. First International Seminar on Citriculture in Pakistan. Faisalabad, 2–5 Dec 1992. University of Agriculture, Faisalabad.

Rehman, H. 1989. Annual Report, Directorate of Soils and Plant Nutrition. 1988–89. Agricultural Research Institute, Tarnab, Peshawar, Pakistan.

Rehman, H. 1990. Annual Report, Directorate of Soils and Plant Nutrition. 1989–90. Agricultural Research Institute, Tarnab, Peshawar, Pakistan.

Samiullah, Z. Shah, M. Tariq, et al. 2013. Micronutrients status of peach orchards in Swat valley. *Sarhad Journal of Agriculture* 29:485–493.

Shah, Z., M. Z. Shah, M. Tariq, et al. 2012. Survey of citrus orchards for micronutrient deficiency in Swat valley of north western Pakistan.*Pakistan Journal of Botany*
44:705–710.

Shah, Z., and K. Shahzad. 2008. Micronutrients status of apple orchards in Swat valley of
North West Frontier Province of Pakistan. *Soil and Environment* 27:123–130.

Siddique, M. T., M. Rashid, and M. Saeed. 1994. Micronutrient status of soils/plants of
citrus growing areas in Punjab. p. 355–60, In: Efficient Use of Plant Nutrients. Proc.
4th National Congress of Soil Science, Islamabad, May 24–26, 1992. Soil Science
Society of Pakistan.

Siddique, T., M. Rashid, and M. Saeed. 1998. Nutrient status of mango orchards in southern Punjab. *Pakistan Journal of Soil Science* 14:78–84.

Sillanpää, M. 1982. Micronutrients and the Nutrient Status of Soils: A Global Study. FAO
Soils Bulletin 48, FAO, Rome.

Tahir, M. 1981. Availability Status of Micronutrients in the Soils of West Pakistan and
the Role and Behavior of Selected Micronutrients in the Nutrition of Crops.Final
Technical Report of PL-480 Project. Nuclear Institute for Agriculture and Biology,
Faisalabad. 124 pp.

Tariq, M., Z. Shah, and A. Ali. 2008. Micronutrients status of plum orchards in Peshawar
valley. *Soil and Environment* 27:223–227.

Zia, M. S. 1993. Fertilizer Use Efficiency Project and Soil Fertility ARP-II.Annual Report,
192–93. National Agricultural Research Center, Islamabad.

Zia, M. H., R. Ahmad, I. Khaliq, A. Ahmad, and M. Irshad. 2006. Micronutrients status
and management in orchards soils: applied aspects. *Soil and Environment* 25:6–16.

.

Annex 7
Extent of Manganese Deficiency in Various Cropping Systems and Regions of Pakistan

Region and Cropping System/Crop	No. of Sites/Samples		Reference
	Total	% Deficient	
Soils (All **Pakistan**)*	242	0	Sillanpää (1982)
Orchard soils (All **Pakistan**)	329	4	Zia et al. (2006)
Punjab			
Total	*2192*	*30*	
Cotton (Khanewal, Multan, Lodhran, Vehari, Bahawalpur, and R.Y. Khan distts.)*	420	2	Rashid (1995, 1996); Rashid and Rafique (1997)
Rice (young whole shoots and soils) (Sheikhupura and Gujranwala distts.)*	110	0	Hussain (2006)
Peanut (shoot terminals) (rainfed Pothohar)*	100	5	Rashid (1994, 1995)
Wheat (young shoots) (rainfed Pothohar)*	61	0	Rashid(1993)
Rapeseed-mustard (young shoots) (rainfed Pothohar)*	120	0	
Sorghum (young shoots) (rainfed Pothohar)*	255	2	Rashid and Qayyum (1991)
Mango leaves*	800	75[a]	Asif et al. (1998)
Mango leaves*	174	13[b]	Siddique et al. (1998)
Citrus leaves (Sargodha distt.)*	43	29	Siddique et al. (1994)
Citrus leaves (Sargodha distt.)*	34	4	Rashid et al. (1991; Rashid and Rashid (1994)
Citrus leaves (Sahiwal distt.)*	41	0	
Citrus leaves (Faisalabad distt.)*	34	8	
Total soils	*589*	*9*	
Apple orchard soils(30 orchards) (Murree tehsil)	180	0	Ahmed et al. (2014)
Orchard soils	158	5	Zia et al. (2006)

(Continued)

Region and Cropping System/Crop	No. of Sites/Samples		Reference
	Total	% Deficient	
Sugarcane soils (Jhang distt.)*	63	16	Ahmad and Rafique (2008)
Sugarcane soils (Sargodha distt.)*	48	8	
Wheat soils (Jhang, Vehari, and M. Garh distts.)	50	24	NIAB (1987)
Wheat soils (3 distts.)**	50	30	PARC (1986)
Soils*	40	5	Kausar et al. (1979)
KP			
Total vegetation	*534*	*32*	
Citrus orchards' leaves (Swat and Malakand distts.)	51	96	Shah et al. (2012)
Peach orchard soils (Swat Valley)	50	0	Samiullah et al. (2013)
Plum leaves (Peshawar, Nowshehra, Charsadda, and Mardan distts.)	40	3	Tariq et al. (2008)
Apple leaves (Swat Valley)	52	58	Shah and Shahzad (2008)
Apple, peach, pear, citrus, and plum leaves (Abbotabad Galliat)	50	2	Khattak and Hussain (2007)
Apple leaves*	50	6	Rehman (1990)
Citrus leaves (6 distts.)*	147	60[a]	Khattak (1994)
Citrus leaves*	94	0	Rehman (1989)
Total soils	*1159*	*2*	
Citrus orchard soils (Swat and Malakand distts.)	51	4	Shah et al. (2012)
Vegetable fields (Peshawar distt.)	18	0	Perveen et al. (2010)
Peach orchards' soils (Swat Valley)	50	0	Samiullah et al. (2013)
Soils (Peshawar distt.)	88	8	Rashid et al. (2008)
Soils (Charsada distt.)	79	7	
Plum orchard soils (Peshawar, Nowshehra, Charsadda, and Mardan distts.)	40	5	Tariq et al. (2008)
Apple orchard soils (Swat Valley)	52	4	Shah and Shahzad (2008)
Apple, peach, pear, citrus, and plum orchard soils (Abbotabad Galliat)	74	0	Khattak and Hussain (2007)
Soils (Mansehra and Swat distts.)*	398	0	Rashid (1994)
Soils (10 distts.)*	270	0	Khattak and Perveen (1988)
Soils*	39	15	Kausar et al. (1979)
Sindh			
Total	*255*	*6*	
Tomato leaves (Badin Tehsil)	32	6	Memon et al. (2012)
Sugarcane top soils (Thatta distt.)*	123	0	Arain et al. (2017)
Sugarcane leaves (Thatta distt.)*	123	8	

Region and Cropping System/Crop	No. of Sites/Samples		Reference
	Total	% Deficient	
Cotton leaves and soils (Nawab Shah distt.)*	100	3	Rashid and Rafique (1997)
Total soils	*720*	*0*	
Tomato fields (Badin tehsil)	32	0	Memon et al. (2012)
Orchard soils	94	2	Zia et al. (2006)
Soils (Thatta distt.)	133	0	Memon et al. (1989)
Soils (Hyderabad, Badin, Sanghar, and Tharparkar distts.)	425	0	
Soils*	36	0	Kausar et al. (1979)
Balochistan			
Total soils	*169*	*3*	
Orchard soils	72	4	Zia et al. (2006)
Soils (Quetta, Kalat, Loralai, and Zhob distts.)*	60	3	Ismail et al. (1994)
Soils*	37	0	Kausar et al. (1979)

* Based on nutrient indexing of crop plants/soils/plants and soils
[a] These kinds of widespread Mn deficiency (60% and 75%) warrant verification by way of yield responses to Mn fertilization
[b] Used 25 mg Mn kg^{-1} as the critical level for data interpretation (whereas literature lists 50 mg kg^{-1}); thus, this interpretation is questionable.

REFERENCES

Ahmad, S., and E. Rafique. 2008. Nutrient indexing and integrated nutrient management for sustaining sugarcane yields. Agricultural Linkages Program (ALP) Project, Final Technical Report. Sugarcane Crops Research Program, Crop Sciences Institute, National Agricultural Research Center, Islamabad. 104 pp.

Ahmed, N., M. Abid, A. Rashid, R. Abou-Shanab, and F. Ahmad. 2014. Influence of boron nutrition on membrane leakage, chlorophyll content and gas exchange characteristics in cotton (*Gossypium hirsutum* L.). *Journal of Plant Nutrition* 37:2302–2315.

Arain, M. Y., K. S. Memon, M. S. Akhtar, and M. Memon. 2017. Soil and plant nutrient status and spatial variability for sugarcane in lower Sindh. *Pakistan Journal of Botany* 49:531–540.

Asif, M., K. Daud, M. Ashraf, et al. 1998. Nutrient status of mango orchards in Punjab. *Pakistan Journal of Soil Science* 15:1–6.

Hussain, F. 2006. Soil Fertility Monitoring and Management in Rice-Wheat System. Agricultural Linkages Program (ALP) Project, Final Report, 2002–2006. Land Resources Research Program, National Agricultural Research Center, Islamabad. 83 pp.

Ismail, M., A. M. Noor-ud-Din, and R. Bashir. 1994. Micronutrients in Baluchistan soils. I. Concentrations in soils of mid-hill region. *Pakistan Journal of Soil Science* 9:91–95.

Kausar, M. A., S. M. Alam, M. Sharif, and M. I. Pervaiz. 1979. Micronutrient status of Pakistan soils. *Pakistan Journal of Scientific and Industrial Research* 22:156–161.

Khattak, J. K. 1994. Effect of Foliar Application of Micronutrients in Combination with Urea on the Yield and Fruit Quality of Sweet Oranges. Final Technical Report, 1991–1994. NWFP Agricultural University, Peshawar.

Khattak, R. A., and Z. Hussain. 2007. Evaluation of soil fertility status and nutrition of orchards. *Soil and Environment* 26:22–32.

Khattak, J. K., and S. Perveen. 1988. Micronutrient status of NWFP soils. p. 62–74, In: Proceedings of National Seminar on Micronutrients in Soils and Crops in Pakistan, 13–15 December 1987, NWFP Agricultural University, Peshawar.

Memon, M., G. M. Jamro, N. Memon, K. S. Memon, and M. S. Akhtar. 2012. Micronutrient availability assessment of tomato grown in taluka Badin, Sindh. *Pakistan Journal of Botany* 44: 649–654.

Memon, K. S., H. K. Puno, and S. M. Memon. 1989. Cooperative Research Program on Micronutrient Status of Pakistani Soils. 5th Annual Report, 1988–89. Sindh Agriculture University, Tandojam, 62 pp.

NIAB. 1987. *Fifteen Years of NIAB*. Nuclear Institute for Agriculture and Biology, Faisalabad.

PARC. 1986. *Soil Status and Scope of Micronutrients in Pakistan Agriculture*. Pakistan Agricultural Research Council, Islamabad, 36 pp.

Perveen, S., Z. Malik, and W. Nazif. 2010. Fertility status of vegetable growing areas of Peshawer, Pakistan *Pakistan Journal of Botany* 42:1871–1880.

Rashid, A. 1993. Nutritional disorders of rapeseed-mustard and wheat grown in *Pothohar* area. Micronutrient Project, Annual Report, 1991–92. National Agricultural Research Center, Islamabad.

Rashid, A. 1994. Nutrient Indexing Surveys and Micronutrient Requirement of Crops. Micronutrient Project, Annual Report, 1992–93. National Agricultural Research Center, Islamabad.

Rashid, A. 1995. Nutrient Indexing of Cotton and Micronutrient Requirements of Cotton and Groundnut.Micreonutrient Project, Annual Report, 1993–94. National Agricultural Research Center, Islamabad. 91 pp.

Rashid, A. 1996. Nutrient Indexing of Cotton in Multan District and Boron and Zinc Nutrition of Cotton. Micronutrients Project, Annual Report, 1994–95. National Agricultural Research Center, Islamabad.76 pp.

Rashid, M., A. U. Bhatti, F. Khan and Wasiullah. 2008. Physico-chemical properties and fertility status of soils of districts Peshawar and Charsadda. *Soil and Environment* 27:228–235.

Rashid, A., F. Hussain, A. Rashid, and J. Din. 1991. Nutrient status of citrus orchards in Punjab. *Pakistan Journal of Soil Science* 6:25–28.

Rashid, A., and F. Qayyum. 1991. Micronutrient Status of Pakistan Soils and Their Role in Crop Production, Cooperative Program. Final Report, 1983–1990. National Agricultural Research Center, Islamabad. 84 pp.

Rashid, A., and E. Rafique. 1997. Nutrient Indexing and Micronutrient Nutrition of Cotton and Genetic Variability in Rapeseed-Mustard and Rice to Boron Deficiency. Micronutrient/Nutrient Management in cotton in Relation to CLCV. Annual Project Report 1995–96. National Agricultural Research Center, Islamabad. 111 pp.

Rashid, A., and A. Rashid. 1994. Nutritional problems of citrus on calcareous soils of Pakistan. p. 190–197, In: Proc. First International Seminar on Citriculture in Pakistan. Faisalabad, 2–5 Dec 1992. University of Agriculture, Faisalabad.

Rehman, H. 1989. Annual Report, Directorate of Soils and Plant Nutrition. 1988–89. Agricultural Research Institute, Tarnab, Peshawar.

Rehman, H. 1990. Annual Report, Directorate of Soils and Plant Nutrition. 1989–90. Agricultural Research Institute, Peshawar.

Samiullah, Z. Shah, M. Tariq, et al. 2013. Micronutrients status of peach orchards in Swat valley. *Sarhad Journal of Agriculture* 29:485–493.

Shah, Z., M. Z. Shah, M. Tariq, et al. 2012. Survey of citrus orchards for micronutrient deficiency in Swat valley of north western Pakistan. *Pakistan Journal of Botany* 44:705–710.

Shah, Z., and K. Shahzad. 2008. Micronutrients status of apple orchards in Swat valley of North West Frontier Province of Pakistan. *Soil and Environment* 27:123–130.

Siddique, M. T., M. Rashid, and M. Saeed. 1994. Micronutrient status of soils/plants of citrus growing areas in Punjab. p. 355–60, In: Efficient Use of Plant Nutrients. Proc. 4th National Congress of Soil Science, Islamabad, May 24–26, 1992. Soil Science Society of Pakistan.

Siddique, T., M. Rashid, and M. Saeed. 1998. Nutrient status of mango orchards in southern Punjab. *Pakistan Journal of Soil Science* 14:78–84.

Sillanpää, M. 1982. Micronutrients and the Nutrient Status of Soils: A Global Study. FAO Soils Bulletin 48, FAO, Rome.

Tariq, M., Z. Shah, and A. Ali. 2008. Micronutrients status of plum orchards in Peshawar valley. *Soil and Environment* 27:223–227.

Zia, M. H., R. Ahmad, I. Khaliq, A. Ahmad, and M. Irshad. 2006. Micronutrients status and management in orchards soils: applied aspects. *Soil and Environment* 25:6–16.

Annex 8
Crop Yield Increases with Zinc Fertilizer and Profitability in Pakistan

Crop (cv.)	Region	Field Expts.	¹Soil Zn (mg kg⁻¹)	²Zn Applied (kg ha⁻¹)	**Cost of Applied Zn (PKR ha⁻¹)	Control Yield (t ha⁻¹)	Yield with Zn (t ha⁻¹)	Yield Increase (t ha⁻¹)	% Yield Increase	³Value of Increased Yield (PKR ha⁻¹)	VCR*	Data source
Wheat												
Soil-applied zinc fertilizer		**61**	**0.2–0.93**	**2–22.5**	**3670**	**3.5**	**3.8**	**0.45**	**13.2**	**14,625**	**4.2:1**	
	D.I. Khan distt., KP	2	NR	10	4300	3.53	4.02	0.48	14.0	15,763	3.7:1	Nadim et al. (2016)
cv. Faisalabad-2008	Punjab	3	0.26–0.46	10.5	4515	3.52	4.14	0.62	17.6	20,150	4.5:1	Rashid et al. (2016)
	Punjab	3	0.72–0.78	4.90	2107	5.59	5.85	0.26	4.7	8450	4.0:1	FFC (2014) – unpublished data
	Sukkur, Sindh	1	0.20	4.90	2107	5.33	6.00	0.67	12.6	21,775	10:1	
cv. Sehar-2006	Punjab	5	0.30–0.71	10	4300	4.16	4.80	0.64	15.4	20,800	5:1	Zou et al. (2012)
		5	0.30–0.71	Soil + Foliar Zn	4300 + 960 = 5260	4.16⁴	4.94	0.78	18.8	25,350	5:1	
cv. AS-2002	Thal Desert and Layyah distt.	2	0.93	22.5	9675	3.61	4.07	0.46	13	15,145	2:1	Abbas et al. (2010)
cvs. Chakwal-97, Inqlab-91, Chakwal-2000	Rawalpindi, Jehlum, and Chakwal	12		5	2150	2.85	3.40	0.56	19.6	18,103	8.4:1	Chaudry et al. (2007)
	Chakwal, Attock, and Jehlum distts)	9		4 -8	5160	2.67	3.02	0.35	13.20	11,516	2.2:1	Hussain (2006)
cv. Pak-81	Rainfed Pothohar	2	0.49–0.52	2	860	2.60	2.92	0.32	12.3	10,400	12:1	Rafique et al. (2006)
	Sheikhupura distt.	3	0.65–0.76	5	2150	3.40	3.83	0.43	13	13,975	6:1	Hussain (2006)
Maxi-Pak	Punjab (cotton–wheat area)	8	0.3	10	4300	3.93	4.43	0.50	13	16,250	4:1	NIAB (unpublished data)

Pak-81	Rawalpindi distt.	1	0.5	5	2150	2.67	2.95	0.28	10.5	9100	4.2:1	Butt et al. (1995)
Maxi-Pak	Faisalabad distt.	1	NR	10	4300	3.32	3.68	0.36	10.8	11,700	3:1	Gill et al. (1996)
Pak-81	Islamabad (rainfed)	1	0.5	5	2150	2.32	2.60	0.28	12.1	9100	4:1	Rashid and Qayyum (1991)
Pak-81	Rawalpindi (rainfed)	2	0.6	10	4300	2.18	2.51	0.33	15.1	10,725	2.5:1	
Sonalika	D.I. Khan	1	NR	2.5	1075	3.48	3.81	0.33	9.5	10,725	10:1	Khattak and Perveen (1988)
Foliar-applied zinc fertilizer[5]		**23**	**0.30–0.88**	**Foliar Zn**		**3.92**	**4.67**	**0.75**	**19.0**	**24,267**	**25:1**	
Various cvs.	Punjab	10	0.30–0.88	Foliar Zn	960[4]	4.06	4.84	0.78	19.2	25,350	26:1	Rashid et al. (2016)
cv. Faisalabad-2008	Punjab (rice and mixed cropping areas)	3	0.26–0.46	2 Foliar sprays	680 + 280 = 960[4]	3.52	3.88	0.36	10.2	11,700	12:1	
cv. Faisalabad-2008	Punjab	2	0.45 and 0.56	Foliar Zn	960[4]	3.72	4.73	1.01	27.2	32,825	34:1	Ram et al. (2016)
	Punjab	2	0.35 and 0.88	Foliar Zn	960[4]	4.26	5.22	0.96	22.5	31,200	32:1	
cv. Lasani-2008	Punjab	1	0.52	Foliar Zn	960[4]	3.82	4.59	0.77	20.2	25,025	26:1	Zou et al. (2012)
cv. Sehar-2006	Punjab	5	0.30–0.71	Foliar Zn	960[4]	4.16	4.76	0.60	14.4	19,500	20:1	
Rice												
Fine rice cvs.		**81**				**3.49**			**11.5**		**7:1**	
cv. Basmati-515	Faisalabad	1	0.37	10	4300	3.9	4.6	0.7	17.9	28,000	6.5:1	Imran et al. (2015)
cv. Basmati-515	Faisalabad	1	0.37	2 Foliar Sprays	960[4]	3.9	4.4	0.5	12.8	20,000	21:1	
cv. Basmati-515	Faisalabad	1	0.37	Soil + Foliar	4300 + 960 = 5260[4]	3.9	4.8	0.9	23.1	36,000 [Grain Zn 22, 29]	6.8:1	

(Continued)

Crop (cv.)	Region	Field Expts.	¹Soil Zn (mg kg⁻¹)	²Zn Applied (kg ha⁻¹)	**Cost of Applied Zn (PKR ha⁻¹)	Control Yield (t ha⁻¹)	Yield with Zn (t ha⁻¹)	Yield Increase (t ha⁻¹)	% Yield Increase	³Value of Increased Yield (PKR ha⁻¹)	VCR*	Data source
cv. Super Basmati	Sheikhupura distt.	2	0.65–0.76	5	2150	3.42	3.89	0.47	13.7	18,800	8.7:1	Hussain (2006)
cv. Bas-385	Punjab	24	NR	5	2150	3.64	3.95	0.31	8.5	12,400	5.7:1	Gill et al. (1996)
cv. Bas-385	Punjab	4	0.5–0.9	2.5	1075	2.72	3.17	0.45	16.5	18,000	16.7:1	Zia (1993)
cv. Bas-385	Punjab	22	NR	5	2150	4.00	4.49	0.49	12.2	19,600	9.1:1	Chaudhry et al. (1992)
cv. Bas cvs.	Punjab	15	NR	6	2580	3.22	3.55	0.33	10.2	13,200	5.1:1	Malik et al. (1988)
cv. Bas-370	Punjab	11	NR	10	4300	2.70	3.10	0.40	14.8	16,000	3.7:1	Chaudhry et al. (1977)
Coarse rice cvs.		**119**		**7.5–10**		**5.5**			**13.3**		**5:1**	
Coarse cvs.	Tandojam, Sindh	2	0.30–0.43	9	3870	6.8	7.45	0.65	9.55	16,250	4:1	Chaudhry et al. (1992); AARI (2013)
cv. IR-6	Sheikhupura	4	NR	10	4300	4.32	4.96	0.64	14.8	16,000	3.7:1	Chaudhry et al. (1992)
cv. JP-5	Malakand and Swat	8	0.4–0.9	7.5	3225	6.12	7.48	1.36	22.2	34,000	10.5:1	Ghani et al. (1990); Rehman (1991)
cv. IR-6	Sindh	2	0.8	10	4300	6.31	7.02	0.71	11.2	17,750	4.1:1	Memon et al. (1989)
cv. IR-6	Punjab	63	NR	8	3440	4.78	5.21	0.43	9.0	10,750	3.1:1	Malik et al. (1988)
cv. KS-282	Punjab	40	NR	8	3440	4.78	5.40	0.62	13.0	15,500	4.5:1	
Cotton		**19**	**0.6–1.0**	**5**		**2.21**			**6.7**		**6:1**	
cv. CIM-240	Punjab cotton belt	7	0.6–1.0	5	2150	1.87	2.00	0.13	7.0	10,400	4.8:1	Rashid and Rafique(1997)
		10	0.6–1.0	5	2150	2.56	2.77	0.21	8.2	16,800	7.8:1	Rashid (1996)
		2	0.8–1.0	5	2150	1.71	1.94	0.23	13.4	18,400	8.5:1	Rashid (1995)
Maize		**17**	**0.48–0.72**	**5-6**		**3.35**			**16.1**		**6:1**	
Synthetic cv. Golden	Swabi, KP	1	0.48	5	2150	3.89	4.42	0.53	13.6	12,587	5.8:1	Iqbal (2016)
	Faisalabad	1	0.72	6	2580	4.6	5.2	0.6	13.0	14,250	5.5:1	Kanwal et al. (2010)

Crop / cultivar	Location											Reference
	Rawalpindi (rainfed)	5	0.6	5	2150	2.08	2.40	0.32	15	7600	3.5:1	Rashid and Qayyum (1991); Malik et al. (1988)
cv. Azam	Charsadda	3	NR	5	2150	2.50	3.04	0.54	21.6	12,825	5.9:1	Khattak (1988)
cvs. Sarhad White and Azam	Peshawar, Mardan, and D.I. Khan	5	NR	5	2150	3.59	4.04	0.45	12.5	10,687	4.9:1	Khattakand Bhatti (1986)
cv. Sarhad White	Peshawar	1	NR	5	2150	3.42	4.13	0.71	20.8	16,862	7.8:1	Bhatti et al. (1986)
Sugarcane		**4**		**10–15**		**85.5**			**7.4**		**5.5**	
	Faisalabad distt.	2	0.81–1.19	15	6450	95.1	102.8	7.6	8.04	30,587	4.7:1	Ghaffar et al. 2011
	Peshawar distt.	2	NR	10	4300	75.84	80.83	4.99	6.7	26,800	6.2:1	Khattak (1980, 1981)
Mungbean												
	Faisalabaddistt.	1	NR	5	2150	**0.657**	0.973	0.316	**48.1**	60,040	**28:1**	AARI (2013)
Rapeseed (canola), Punjab		1	0.56	8	3440	1.52	1.75	0.23	15.1	13,800	4:1	AARI (2005)
Sunflower		**6**	**0.26–0.68**	**3–15**		**2.04**			**19.0**		**4.4:1**	
	Tandojam, Sindh	1	0.68	15	6450	1.86	2.23	0.37	20.07	34,408	5.3:1	Siddiqui et al. (2009a, 2009b)
	Tandojam, Sindh	1	0.68	15	6450	1.71	2.03	0.32	18.9	29,946	4.6:1	Siddiqui et al. (2009a, 2009b)
	Islamabad	1	0.26	5	2150	1.86	2.24	0.37	19.90	34,224	16:1	Khan et al. (2009)
	Faisalabad	1	0.44, 0.50	12	5160	1.05	1.32	0.27	25.0	16,875	3.2:1	AARI (2005, 2006)
	Faisalabad	1	0.68	3	1290	4.48	4.88	0.40	8.9	25,000	19:1	AARI (2004)
	Tandojam, Sindh	1	NR	4	1720	1.26	1.53	0.27	21.4	16,875	10:1	M. Salim Saif (per. comm.)

(Continued)

Crop (cv.)	Region	Field Expts.	Soil Zn (mg kg⁻¹)[1]	2Zn Applied (kg ha⁻¹)	**Cost of Applied Zn (PKR ha⁻¹)	Control Yield (t ha⁻¹)	Yield with Zn (t ha⁻¹)	Yield Increase (t ha⁻¹)	% Yield Increase	3Value of Increased Yield (PKR ha⁻¹)	VCR*	Data source
Potato		**6**		**5–10**		**16.50**			**20.3**		**13:1**	
	Buner, KP	1	0.48	6	2580	16.0	18.2	2.2	13.7	27,500	11:1	Iftikhar-ul-Haq (1996)
	Tamab, Peshawar	1	NR	5	2150	8.11	9.89	1.78	21.9	22,250	10:1	
	Kasur distt.	3	NR	10	4300	18.23	22.39	4.16	22.8	52,000	12:1	Gill et al. (1996)
	Nowshehra distt.	1	NR	5	2150	20.44	24.44	4.00	17.8	50,000	23.2:1	Khattak (1988)

Two sprays of 0.5% ZnSO$_4$·7H$_2$O solution (or of 0.32% ZnSO$_4$·H$_2$O solution)

NR = Not reported

* VCR = Value-Cost Ratio. One recommended dose soil use of Zn lasts 3–4 crop seasons; thus, actual VCR will be three to four times of the listed value.

** Cost of Zn fertilizer: PKR 430 per kg elemental Zn (based on FFC's PKR 425 per 3 kg zinc sulfate, containing 33% Zn); see **Annex 12**

For prices of crop produce, see **Annex 13**

1 DTPA or AB-DTPA-extractable soil Zn

2 Zn applied on top of NPK fertilizers at recommended doses

3 Considering grain yield increase only, increase in straw/cotton sticks yield is additional benefit

4 For foliar sprays, labor cost of two sprays included (PKR 280 per ha per spray) PKR 560 per day

5 Foliar sprays of Zn: 0.5% ZnSO$_4$·7H$_2$O solution (or 032. % ZnSO$_4$·H$_2$O solution; solution volume 250 L ha⁻¹ per spray – which contained 0.8 kg ZnSO$_4$.H$_2$O containing 33% Zn) costing Zn fertilizer of PKR 340 per spray or PKR 680 for two sprays

REFERENCES

AARI. 2004. Annual Report 2003–04. Ayub Agricultural Research Institute, The Government of Punjab, Pakistan.

AARI. 2005. Annual Report 2004–05. Ayub Agricultural Research Institute, The Government of Punjab, Pakistan.

AARI. 2006. Annual Report 2005–06. Ayub Agricultural Research Institute, The Government of Punjab, Pakistan.

AARI. 2013. Annual Report 2012–13. Ayub Agricultural Research Institute, The Government of Punjab, Pakistan.

Abbas, G., G. Hassan, M. A. Ali, et al. 2010. Response of wheat to different doses of $ZnSO_4$ under Thal desert environment.*Pakistan Journal of Botany* 42:4079–4085.

Bhatti, A. U., J. K. Khattak, and Z. Shah. 1986. Studies on the effect of trace elements (Zn, Cu, Fe, Mn) on the yield of maize crop. *Pakistan Journal of Soil Science* 1:33–36.

Butt, A. H., M. A. Khan, and M. Yousaf. 1995. Response of wheat to soil application of zinc and boron under rain-fed conditions. *Pakistan Journal of Soil Science* 10:66–68.

Chaudhry, F. M., S. M. Alam, A. Rashid, and A. Latif. 1977. Micronutrient availability to cereals from calcareous soils. IV. Mechanism of differential susceptibility of two rice verities to Zn deficiency. *Plant and Soil* 46:637–642.

Chaudhry, R. A., D. M. Malik, T. Amin, Ghafoor-ul-Haq, and S. Sabir. 1992. Rice response to micronutrients. p. 342–350. In: Proc. 3rd. National Congress of Soil Science, Lahore, March 20–22,1990. Soil Science Society of Pakistan.

Chaudry, E. H., V. Timmer, A. S. Javed, and M. T. Siddique. 2007. Wheat response to micronutrients in rain-fed areas of Punjab. *Soil and Environment* 26:97–101.

Ghaffar, A., Ehsanullah, N. Akbar, and S. H. Khan. 2011. Influence of zinc and iron on yield and quality of sugarcane planted under various trench spacing. *Pakistan Journal of Agricultural Sciences* 48:25–33.

Ghani, A., M. Shah, and D. R. Khan. 1990. Response of rice to elevated levels of zinc in mountainous areas of Swat. *Sarhad Journal Agriculture* 6:411–415.

Gill, K. H., G. Hassan, M. Y. Ahmad, and Ehsan-ul-Haq. 1996. Soil Fertility Investigations on Farmers Fields.Annual Report, 1995–96. Soil Fertility Survey and Soil Testing Institute, Lahore.

Hussain, F. 2006. Soil Fertility Monitoring and Management in Rice-Wheat System. Agricultural Linkages Program (ALP) Project, Final Report, 2002–2006. Land Resources Research Program, National Agricultural Research Center, Islamabad. 83 pp.

Iftkhar-ul-Haq. 1996. Annual Report, 1995–96. Directorate of Soils and Plant Nutrition, Agricultural Research Institute, Tarnab, Peshawar.

Imran, M., S. Kanwal, S. Hussain, T. Aziz, and M. A. Maqsood. 2015. Efficacy of zinc application methods for concentration and estimated bioavailability of zinc in grains of rice grown on a calcareous soil. *Pakistan Journal of Agricultural Sciences* 52:169–175.

Iqbal, J. 2016. Effect of nitrogen and zinc on maize productivity. MSc (Hons) thesis, Department of Soil and Environmental Sciences, The University of Agriculture, Peshawar.

Kanwal, S., Rahmatullah, A. R. and Ahmad. 2010. Zinc partitioning in maize grain after soil fertilization with zinc sulfate. *International Journal of Agriculture and Biology* 12:299–302.

Khan, M. A., Khan, M. A., J. Din, S. Nasreen, M. Y. Khan, S. U. Khan, and A. R. Gurmani. 2009. Response of sunflower to different levels of zinc and iron under irrigated conditions. *Sarhad Journalof Agriculture* 25(2): 159–163.

Khattak, J. K. 1980. Cooperative Research Program on Micronutrient Status of Pakistan Soils and Its Role in Crop Production. Annual Reports, 1979–80. NWFP Agricultural University, Peshawar.

Khattak, J. K. 1981. Cooperative Research Program on Micronutrient Status of Pakistan Soils and Its Role in Crop Production. Annual Reports,1980–81. NWFP Agricultural University, Peshawar.

Khattak, J. K. 1988. Cooperative Research Program on Micronutrient Status of Pakistan Soils and Its Role in Crop Production. Annual Reports, 1987–88. NWFP Agricultural University, Peshawar.

Khattak, J. K., and A. U. Bhatti. 1986. National outreach research project on soil fertility and fertilizer use in Pakistan. Final Technical Report, 1982–86. NWFP Agricultural University, Peshawar.

Khattak, J. K., and S. Perveen. 1988. Micronutrient status of NWFP soils. p. 62–74, In: Proceedings of National Seminar on Micronutrients in Soils and Crops in Pakistan, 13–15 December 1987, NWFP Agricultural University, Peshawar.

Malik, D. M., G. Hassan, and R. A. Chaudhry. 1988. Zinc requirement of rice crop in the Punjab. *Pakistan Journalof Soil Science* 3:39–42.

Memon, K. S., H. K. Puno, and S. M. Memon. 1989. Cooperative Research Program on Micronutrient Status of Pakistani Soils. 5th Annual Report, 1988–89. Sindh Agriculture University, Tandojam, 62 pp.

Nadim, M. A., M. S. Baloch, E. A. Khan, A. A. Khakwani and K. Waseem. 2016. Integration of organic, synthetic fertilizers and micronutrients for higher growth and yield of wheat. *Sarhad Journal of Agriculture* 32:9–16.

Rafique, E., A. Rashid, J. Ryan, and A. U. Bhatti. 2006. Zinc deficiency in rainfed wheat in Pakistan: Magnitude, spatial variability, management, and plant analysis diagnostic norms. *Communications in Soil Science and Plant Analysis* 37:181–197.

Ram, H., A. Rashid, W. Zhang, et al. 2016. Biofortification of wheat, rice and common bean by applying foliar zinc fertilizer along with pesticides in seven countries. *Plant and Soil* 403:389–401.

Rashid, A.1995. Nutrient Indexing of Cotton and Micronutrient Requirements of Cotton and Groundnut. Micronutrient Project, Annual Report, 1993–94. National Agricultural Research Center, Islamabad. 91 pp.

Rashid, A. 1996. Nutrient Indexing of Cotton in Multan District and Boron and Zinc Nutrition of Cotton. Micronutrients Project, Annual Report, 1994–95. National Agricultural Research Center, Islamabad. 76 pp.

Rashid, A., K. Mahmood, M. Rizwan, Z. Iqbal, and A. Naeem. 2016. Harvest Plus Zinc Fertilizer Project, Country Report, July 2016. Nuclear Institute for Agriculture and Biology, Faisalabad.

Rashid, A., and F. Qayyum. 1991. Micronutrient Status of Pakistan Soils and Their Role in Crop Production, Cooperative Program. Final Report, 1983–1990. National Agricultural Research Center, Islamabad. 84 pp.

Rashid, A. and E. Rafique. 1997. Nutrient Indexing and Micronutrient Nutrition of Cotton and Genetic Variability in Rapeseed-Mustard and Rice to Boron Deficiency. Micronutrient/Nutrient Management in cotton in Relation to CLCV. Annual Project Report 1995–96. National Agricultural Research Center, Islamabad. 111 pp.

Rehman, H. 1991. Annual Report, Directorate of Soils and Plant Nutrition. 1990–91. Agricultural Research Institute, Tarnab, Peshawar.

Siddiqui, M. H., F. C. Oad, M. K. Abbasi, and A. W. Gandahi. 2009a. Effect of NPK, micronutrients and N placement on the growth and yield of sunflower. *Sarhad Journal of Agriculture* 25(1): 45–52.

Siddiqui, M. H., Oad, F. C., Abbasi, M. K. and Gandahi, A. W. 2009b. Zinc and boron fertility to optimize physiological parameters, nutrient uptake and seed yield of sunflower. *Sarhad Journal of Agriculture* 25(1): 53–57.

Zia, M. S. 1993. Fertilizer Use Efficiency Project and Soil Fertility ARP-II.Annual Report, 192–93. National Agricultural Research Center, Islamabad.

Zou, C. Q., Y. Q. Zhang, A. Rashid, et al. 2012. Biofortification of wheat with zinc through zinc fertilization in seven countries. *Plant and Soil* 361:43–55.

Annex 9
Crop Yield Increases with Boron Fertilizer and Profitability in Pakistan

Crop (cv.)	Region	Field Expts.	Soil B (mg kg⁻¹)	¹B Applied (kg ha⁻¹)	*Cost of Applied B (PKR ha⁻¹)	Control Yield (t ha⁻¹)	Yield with B (t ha⁻¹)	Yield Increase (t ha⁻¹)	% Yield Increase	²Value of Increased Yield (PKR ha⁻¹)	**Value:Cost Ratio (VCR)	Data Source
Wheat		**48**	**0.1–1.2**	**1.0–2.0**	**1151 (930–1860)**	**3.87**	**4.3**	**0.4**	**10.8**	**12,575**	**9.7:1**	
3 cvs. (3-year trial)	Pothohar (rainfed)	13	0.1–1.2	1	930	2.85	3.16	0.31	10.9	10,075	10.83:1	Chaudry et al. (2007)
Faisalabad-2008 (3-year trial)	Faisalabad distt.	3	0.42–0.53	1	930	4.36	4.61	0.25	5.7	8125	8.74:1	Muhmood et al. (2014)
Faisalabad (3-year trial)	Faisalabad distt.	3	0.42–0.53	2	1860	4.36	4.85	0.49	11.2	15,925	8.56:1	Muhmood et al. (2014)
cv. Gomal-8 (2-year trial)	D.I. Khan distt.	2	NR	2	1860	3.54	4.14	0.60	17.1	19,500	10.48:1	Nadim et al. (2016)
cv. Pak-81	Pothohar (rainfed)	2	0.3–0.4	1.0	930	2.62	2.91	0.29	12.3	9425	10:1	Rashid et al. (2011)
cv. Pak-81	Rawalpindi distt. (rainfed)	4	0.5	1.0	930	2.21	2.44	0.23	10.4	7475	8:1	Rashid and Qayyum (1991); Butt et al. (1995)
cv. Pak-81	Islamabad (rainfed)	2	0.3–0.4	1.0	930	2.33	2.67	0.34	15	11,050	12:1	Rashid and Qayyum (1991)
Local cvs	Jhang distt.	1	0.70	1.0	930	3.45	3.80	0.35	10.2	11,375	12:1	FFC (2007–08), unpublished data
	Kasur distt.	1	0.80	1.0	930	3.91	4.37	0.46	11.8	14,950	16:1	
	Vehari distt.	1	0.82	1.0	930	3.96	4.07	0.11	2.8	3575	4:1	
	Mirpur Khas distt.	1	0.72	1.0	930	3.98	4.40	0.42	10.6	13,650	15:1	
	D.I. Khan, KP	1	0.26	1.0	930	4.28	4.69	0.41	9.6	13,325	15:1	
	Jhang distt.	1	0.40	1.0	930	4.72	4.97	0.25	5.3	8125	9:1	
	Kasur distt.	1	0.54	1.0	930	4.67	5.09	0.42	9.0	13,650	15:1	

	Vehari distt.	1	0.82	1.0	930	5.31	5.78	0.47	8.8	15,275	16:1	
	Mirpur Khas, Sindh	1	0.40	1.0	930	5.40	5.67	0.27	5.0	8775	9:1	Hussain (2006)
	D.I. Khan, KP	1	0.92	1.0	930	4.35	4.83	0.48	11.0	15,600	17:1	
	Punjab rice belt	2	0.39–0.42	1.0	930	3.40	3.77	0.37	10.9	12,025	13:1	
cv. Mexi-Pak	Punjab cotton belt	6	NR	2.0	1860	3.95	4.47	0.52	13.2	16,900	9:1	NIAB (unpublished data)
cv. Punjab-81	Faisalabad	2	0.3–0.4	2.0	1860	4.17	4.59	0.42	10.1	13,650	7:1	Kausar et al. (1988)
cv. Tarnab-73	Peshawar	2	NR	2.0	1860	3.53	4.45	0.92	26	21,620	12:1	Iqtidar et al. (1979)
Rice												
Basmati cvs.		**44**		**1.0–2.0**	**1163**	**3.63**	**4.3**	**0.61**	**17.4**	**24,286**	**23.6:1**	
cv. Shaheen Basmati	Punjab	2	0.46–0.53	1.0 kg + seed priming	930 + seed priming	2.40	2.81	0.41	17	16,400	18:1	Rehman et al. (2015)
Basmati cv.		1	0.48–0.56	1.0 kg	930	3.02	3.73	0.71	23.5	28,400	30:1	Rehman et al. (2014)
cv. Super Basmati	Multan	1	0.24	1.5 kg	1395	3.61	4.0	0.39	10.8	15,600	11:1	Rehman et al. (2012)
cv. Super Basmati	Gakhar, Punjab	1	0.56	1.0	930	4.47	5.23	0.76	17.0	30,400	33:1	FFC, 2008
cv. Super Basmati	Punjab rice belt	5	0.26–0.44	1.0	930	3.23	3.89	0.66	20.4	26,400	28:1	Rashid et al. (2007)
cv. Basmati-385	Punjab rice belt	5	0.26–0.44	1.0	930	3.77	4.72	0.95	25.2	38,000	41:1	

(Continued)

Crop (cv.)	Region	Field Expts.	Soil B (mg kg⁻¹)	¹B Applied (kg ha⁻¹)	*Cost of Applied B (PKR ha⁻¹)	Control Yield (t ha⁻¹)	Yield with B (t ha⁻¹)	Yield Increase (t ha⁻¹)	% Yield Increase	²Value of Increased Yield (PKR ha⁻¹)	**Value:Cost Ratio (VCR)	Data Source
cv. Super Basmati	Punjab rice belt	6	0.15–0.42	1.0	930	3.78	4.39	0.69	18.3	27,600	30:1	FFC (2008) – unpublished data
cv. Basmati-385	Punjab rice belt	6	0.15–0.42	1.0	930	3.69	4.54	0.85	23.0	34,000	36:1	FFC (2007) – unpublished data
cv. Basmati-2000	Bahawalnagar, Punjab	1	0.51	1.0	930	4.84	5.39	0.55	11.4	22,000	24:1	
cv. Super Basmati	Kasur and Jhang distts.	2	0.45, 0.70	1.0	930	5.45	5.91	0.46	8.4	18,400	20:1	
cv. Super Basmati	Rice–wheat area of Punjab	2	0.39–0.42	1.0	930	3.42	3.78	0.36	10.5	14,400	15:1	Hussain (2006)
cv. Super Basmati	Pindi Bhatian distt.	1	NR	0.5–2.0	465–1860	4.1	4.3–5.0	0.2–0.9	4.9–22	22,000 (8000–36,000)	17.2–19.4	Hyder et al. (2012)
cv. Bas-385	Punjab	2	0.4–0.5	2.0	1860	2.66	3.35	0.69	26	27,600	15:1	Zia (1993)
cv. Bas-370	Punjab	9	NR	2.0	1860	2.41	2.88	0.47	19	18,800	10:1	Chaudhry et al. (1977)
IRRI-type rices		**15**		**0.5–2.0**		**4.9**	**5.6**	**0.72**	**14.9**		**17:1**	
cvs. Khushboo-95 and Mehak	Tandojam	1	0.49 HWE	0.5–1.0	465–930	3.94–4.44	4.20–5.28	0.26–0.84	7–19	6500–21,000	13.9–22.6	Shah et al. (2011)
Cvs. Shandar and Sarshar	Tandojam	1	0.37	0.5–2.0	465–1860	5.31	5.73–5.89	0.42–0.68	7.9–10.9	10,500–17,000	9.12–22.6	Shah et al. (2016)
cv. IR-6	D.I. Khan, KP	1	0.56	1.0	930	5.43	6.37	0.94	17.3	23,500	25:1	FFC (2007) – unpublished data

Cultivar	Location	n										Reference
cv. KS-282	Punjab rice belt	6	0.15–0.42	1.0	930	4.82	5.48	0.66	13.6	16,500	18:1	Rashid et al. (2007)
cv. IR-6	Sindh rice belt	3	0.43–.51	1.0	930	4.30	5.38	1.08	25.0	27,000	29:1	FFC (2006) – unpublished data
cvs. KS-282 and KS-133	Kasur, Punjab	1	0.69	1.0	930	5.88	6.63	0.75	12.7	18,750	20:1	
cv. IR-6	D.I. Khan, KP	1	0.45	1.0	930	5.88	6.73	0.85	14.4	21,250	23:1	Chaudhry et al. (1977)
cv. IR-6	Punjab	3	NR	2.0	1860	3.33	3.73	0.40	12.0	10,000	5:1	
Cotton												
	Soil application	**36**	**0.2–0.66**	**0.5–2.5**		**2.3**	**2.6**	**0.3**	**15.7**		**19.6:1**	
cvs. FH-113, MNH-786, and CIM-496	Faisalabad (2-year trials)	2	0.45	0.5–2.5	465–1860	1.7–2.2	1.77–2.70	0.07–0.50	4.2–21.9	5600–40,000	12–21.5	Saleem et al. (2016)
cv. CIM-240	Punjab	5	0.4–0.6	1.5	1395	2.85	3.28	0.43	15	34,400	25:1	Rashid (1995)
	Punjab	12	0.3–0.6	1.5	1395	2.40	2.70	0.30	12.5	24,000	17:1	Rashid (1996)
	Punjab	8	0.2–0.5	1.5	1395	1.72	1.99	0.27	15.7	21,600	15:1	Rashid and Rafique (1997)
cv. B-557	Tandojam, Sindh	2	0.5	2.0	1860	2.36	2.80	0.44	18.6	35,200	15:1	Chaudhary and Hisiani (1970)
cv. B-557	DI Khan	1	NR	2.0	1860	0.94	1.23	0.29	30.8	23,200	10:1	Khattak (1988)
	Vehari distt.	3	0.41–0.60	1.0	930	2.69	2.98	0.29	10.8	23,200	25:1	FFC unpublished data (Munir Zia, per. Com.)
	Mirpur Khas, Sindh	3	0.56–0.66	1.0	930	3.53	3.86	0.33	9.44	26,400	28:1	

(Continued)

Crop (cv.)	Region	Field Expts.	Soil B (mg kg⁻¹)	¹B Applied (kg ha⁻¹)	*Cost of Applied B (PKR ha⁻¹)	Control Yield (t ha⁻¹)	Yield with B (t ha⁻¹)	Yield Increase (t ha⁻¹)	% Yield Increase	²Value of Increased Yield (PKR ha⁻¹)	**Value:Cost Ratio (VCR)	Data Source
Cotton	**Foliar sprays**	**13**		**0.55**	**175**	**2.29**	**2.52**	**0.23**	**11.4**	**18,400**	**33:1**	
cv. CIM-240	Punjab	5	0.4–0.6	2 Foliar sprays	80+(2 x 375) = 830	2.85	3.02	0.17	6.0	13,600	22:1	Rashid (1995)
	Punjab	8	0.2–0.5	2 Foliar sprays		1.72	2.01	0.29	16.8	23,200	37:1	Rashid and Rafique (1997)
Maize		**9**		**1.0–1.5**		**2.8**	**3.4**	**0.7**	**23.1**		**14.3:1**	
cv. Sarhad White	Rawalpindi (rainfed)	4	NR	1	930	2.20	2.46	0.26	11.8	6175	6.6:1	Malik et al. (1991)
	DI Khan	1	0.4	1.5	1395	2.84	3.84	1.00	35	23,750	17:1	Aslam (1990)
	Charsadda	3	NR	1.1	1023	2.44	3.04	0.60	24.5	14,250	13.9:1	Khattak et al. (1988)
cv. Changez	Peshawar	1	0.8	1.0	930	3.64	4.41	0.77	21.1	18,287	19.7:1	Ghani (1981)
Sugarcane		**4**		**0.25–1.0**		**69**	**84**	**15**	**23.4**	**59,600**	**33:1**	
	Khanewal, Punjab	1	0.58	1.0	930	83.98	85.07	1.09	1.30	4360	5:1	FFC (2001) – unpublished data
	Charsadda, KP	1	0.56	1.0	930	51.08	66.00	14.92	28.80	59,680	64:1	FFC (2011) – unpublished data
cv. CP 65/63	Mardan, KP	2	NR	0.25 (Foliar Sprays)	80+7503	71.73	100.69	28.96	40	114,760	30:1	SCRI (1989, 1990)
Sugarbeet	Peshawar	1	0.7	2	1860	75.84	81.27	5.43	7	95,025	51:1	Khattak (1991)

Crop	Location		Cost of B fertilizer*	B rate	Price of produce**							Reference
Potato		**3**		**1.5–2.0**		**13.44**	**15.95**	**2.50**	**18.2**		**20.4**	
	Tarnab, Peshawar	1	NR	2	1860	12.11	14.65	2.54	20.9	31,750	19:1	Iftikhar-ul-Haq (1996)
	Nowshehra distt.	1	NR	2.0	1860	7.77	8.97	1.20	15.4	15,000	8.1:1	Khattak (1988)
	Nowshehra distt.	1	NR	1.5	1395	20.44	24.22	3.78	18.4	47,250	34:1	
Mungbean 4 cvs.		1	0.14	1.0	930	0.71	0.89	0.18	25.3	34,200	**37:1**	
Peanut		**14**	**0.12–0.6**	**1.0**	**930**	**1.73**	**2.06**	**0.33**	**19.64**		**32:1**	
Golden	Pothohar (rainfed)	2	0.12–0.25	**1.0**	930	1.95	2.33	0.38	19.5	34,200	37:1	Rafique et al. (2014)
cv. BARD-479	(rainfed)	2	0.12–0.25	**1.0**	930	1.74	2.16	0.42	24.1	37,800	41:1	
cv. BARI 2000		2	0.12–0.25	**1.0**	930	1.61	2.06	0.45	30.0	40,500	43:1	
	Pothohar (rainfed)	3	0.3	1.0	930	2.10	2.32	0.22	10.4	19,800	21:1	Rashid et al. (1997)
	Chakwal (2-year trials)	5	<1.0	**1.0**	930	1.26	1.44	0.18	14.2	16,200	17:1	LRRI, NARC-ICARDA (Unpublished)
Tobacco	Mardan	1	NR	0.15 (2 foliar sprays)	48+750[3]	2.28[4]	2.71	0.43[4]	18.8	75,250	94:1	Khattak (1988)
Brassica	Peshawar	3	0.4	0.7–1 (3 foliar sprays)	1776–2055	1.76–2.64	1.80–3.84	0.04–1.20	2.1–23.8	2400–72,000	1.4–35:1	Ali et al. (2016)

NR = Not reported

* Cost of B fertilizer: See **Annex 12**

** Price of Crop Produce: See **Annex 13**

[1] B applied on top of basal dressing of NPK at recommended doses

[2] Counsidering grain yield increase only; increase in straw yield is additional benefit.

[3] Labor cost of two foliar sprays

[4] Cured tobacco leaves

REFERENCES

Ali, M., W. Muhammad, and I. Ali. 2016. Yield of oil seed Brassica (napus and juncea) advanced lines as influenced by boron. *Soil and Environment* 35(1):30–34.

Aslam, K. M. 1990. To study the effect of boron fertilization on the yield of maize. MSc (Hons) thesis. NWFP Agricultural University, Peshawar.

Butt, A. H., M. A. Khan, and M. Yousaf. 1995. Response of wheat to soil application of zinc and boron under rainfed conditions. *Pakistan Journal of Soil Science* 10:66–68.

Chaudhary, T. M., and G. R. Hisiani. 1970. Effect of B on the yield of seed cotton. *The Pakistan Cottons* 1:13–15.

Chaudhry, F. M., S. M. Alam, A. Rashid, and A. Latif. 1977. Micronutrient availability to cereals from calcareous soils. IV. Mechanism of differential susceptibility of two rice verities to Zn deficiency. *Plant and Soil* 46:637–642.

Chaudry, E. H., V. Timmer, A. S. Javed, and M. T. Siddique. 2007. Wheat response to micronutrients in rain-fed areas of Punjab. *Soil and Environment* 26:97–101.

Ghani, S. M. 1981. Efect of boron fertilization on the yield and chemical composition of maize. MSc (Hons) thesis, NWFP Agricultural University, Peshawar.

Hussain, F. 2006. Soil Fertility Monitoring and Management in Rice-Wheat System. Agricultural Linkages Program (ALP) Project, Final Report, 2002–2006. Land Resources Research Program, National Agricultural Research Center, Islamabad. 83 pp.

Hyder, S. I., M. Arshadullah, A. Ali, and I. A. Mahmood. 2012. Effect of boron nutrition on paddy yield under saline-sodicsoils. *Pakistan Journal of Agricultural Research* 25:266–271.

Iftkhar-ul-Haq. 1996. Annual Report, 1995–96. Directorate of Soils and Plant Nutrition, Agricultural Research Institute, Tarnab, Peshawar.

Iqtidar, A., J. K. Khattak, S. Perveen, and T. Jabeen. 1979. Effect of boron on the yield and crude protein content of wheat. *Pakistan Journal of Scientific and Industrial Research* 22:248–250.

Kausar, M. A., M. Tahir and M. Sharif. 1988. Wheat response to field application of boron and evaluation of various methods for the estimation of soil boron. p. 132–138. In: Proceedings of National Seminar on Micronutrients in Soils and Crops in Pakistan, December 13–15. NWFP Agriculture University, Peshawar.

Khattak, J. K. 1988. Cooperative Research Program on Micronutrient Status of Pakistan Soils and Its Role in Crop Production. Annual Reports,1987–88. NWFP Agricultural University, Peshawar.

Khattak, J. K. 1991. *Micronutrients in NWFP Agriculture.* Barani Research and Development Project, Pakistan Agricultural Research Council, Islamabad.

Khattak, J. K., S. Perveen, and Farmanullah. 1988. Proceedings of National Seminar on Micronutrients in Soils and Crops in Pakistan, 13–15 December 1987, NWFP Agricultural University, Peshawar.

Malik, D. M., S. Mahmood, and Ehsan-ul-Haq. 1991. Soil Fertility Investigations on Farmers Fields.Annual Report, 1990–91. Soil Fertility and Soil Testing Institute, Lahore.

Muhmood, A., S. Javid, A. Niaz, et al. 2014. Effect of boron on seed germination, seedling vigor and wheat yield. *Soil and Environment* 33:17–22.

Nadim, M. A., M. S. Baloch, E. A. Khan, A. A. Khakwani and K. Waseem.2016. Integration of organic, synthetic fertilizers and micronutrients for higher growth and yield of wheat. *Sarhad Journal of Agriculture* 32:9–16.

Rafique, E., M. Mahmood-ul-Hassan, M. Yousra, I. Ali, and F. Hussain. 2014. Boron nutrition of peanut grown in boron-deficient calcareous soils: genotypic variation and proposed diagnostic criteria. *Journal of Plant Nutrition* 37:172–183.

Rashid, A. 1995. Nutrient Indexing of Cotton and Micronutrient Requirements of Cotton and Groundnut.Micronutrient Project, Annual Report, 1993–94. National Agricultural Research Center, Islamabad. 91 pp.

Rashid, A. 1996. Nutrient Indexing of Cotton in Multan District and Boron and Zinc Nutrition of Cotton.Micronutrients Project, Annual Report, 1994–95. National Agricultural Research Center, Islamabad. 76 pp.

Rashid, A. and E. Rafique. 1997. Nutrient Indexing and Micronutrient Nutrition of Cotton and Genetic Variability in Rapeseed-Mustard and Rice to Boron Deficiency. Micronutrient/Nutrient Management in cotton in Relation to CLCV. Annual Project Report 1995–96. National Agricultural Research Center, Islamabad. 111 pp

Rashid, A., Rafique, E., and Ali, N. 1997. Micronutrient deficiencies in rainfed calcareous soils of Pakistan. II. Boron nutrition of the peanut plant. *Communication in Soil Science and Plant Analysis* 28: 149–159.

Rashid, A., E. Rafique, A. U. Bhatti, et al. 2011. Boron deficiency in rainfed wheat in Pakistan: Incidence, spatial variability, and management strategies. *Journal of Plant Nutrition* 34:600–613.

Rashid, A., and F. Qayyum. 1991. Micronutrient Status of Pakistan Soils and Their Role in Crop Production, Cooperative Program. Final Report, 1983–1990. National Agricultural Research Center, Islamabad. 84 pp.

Rashid, A., M. M. Mahmud, and E. Rafique. 2007. Potato responses to boron and zinc application in a calcareous UdicUstochrept. p. 103–116, In: F Xu (Ed) Advances in Plant and Animal Boron Nutrition. Proc. 3rdInt.Symp.on All Aspects of Plant and Animal Boron Nutrition, Wuhan, China, 9–13 Sep 2005. Springer, Dordrecht.

Rehman, A., M. Farooq, A. Nawaz, and R. Ahmad. 2014. Influence of boron nutrition on rice productivity, kernel quality and biofortification in different production systems. *Field Crops Research* 169:123–131.

Rehman, A., M. Farooq, A. Nawaz, and R. Ahmad. 2015. Improving the performance of short-duration basmati rice in water-saving production systems by boron nutrition. *Annals of Applied Biology* 168:19–28.

Rehman, H., T. Aziz, M. Farooq, A. Wakeel, and Z. Rengel. 2012. Zinc nutrition in rice production systems: a review. *Plant and Soil* 361: 203–226.

Saleem, M., M. A. Wahid, S. M. A. Basra, and A. M. Ranjha. 2016. Influence of soil applied boron on the boll retention, productivity and economic returns of different cotton genotypes. *International Journal of Agriculture and Biolology* 18:68–72.

SCRI. 1989. Annual Report, 1988-89. Sugar Crops Research Institute (SCRI), Mardan, Khyber Pakhtunkhwa

SCRI. 1990. Annual Report, 1989-90. Sugar Crops Research Institute (SCRI), Mardan, Khyber Pakhtunkhwa.

Shah, J. A., N. Rais, Z. Hassan, M. Abbas, and M. Y. Memon. 2016. Evaluating non-aromatic rice varieties for growth and yield under different rates of soil applied boron. *Pakistan Journal of Analytical and Environmental Chemistry* 17:1–7.

Shah, J. A., M. Y. Memon, M. Aslam, et al. 2011. Response of two rice varieties viz., Khushboo-95 and Mehak to different levels of boron. *Pakistan Journal of Botany* 43:1021–1031.

Zia, M. S. 1993. Fertilizer Use Efficiency Project and Soil Fertility ARP-II. Annual Report, 192–93. National Agricultural Research Center, Islamabad.

Annex 10
Crop Yield Increases with Iron Fertilizer and Profitability in Pakistan

Crop	Region	Field Expts.	¹Soil Fe (mg kg⁻¹)	²Fe Applied (kg ha⁻¹)	**Cost of Applied Fe (PKR ha⁻¹)	Control Yield (t ha⁻¹)	Yield with Fe (t ha⁻¹)	Yield increase over Control (t ha⁻¹)	% Yield Increase	***Value of Increased Yield (PKR ha⁻¹)	³Value: Cost Ratio	Data Source
Wheat		**21**			**970**	**3.24**	**3.57**	**0.33**	**10.4**	**10,725**	**11:1**	
cvs. Chakwal-97, Chakwal-2000, Inqlab-910	Pothohar (rainfed)	13	0.25–7.8	10	970	2.85	3.23	0.38	13.3	12,350	12.7:1	Chaudry et al. (2007)
	Rawalpindi	3	NR	10	970	3.14	3.41	0.27	9	8775	9:1	Rashid and Qayyum (1991)
cv. Mexi-Pak	Punjab (cotton–wheat area)	5	3.1	10	970	3.74	4.08	0.34	9	11,050	11:1	NIAB (unpublished data)
Maize		**20**		**2.5–15**	**243–1458**	**3.9**	**4.1**	**0.2**	**5.0**	**4671**	**11.7:1**	
cvs. Neelam, Akbar, Sadaf (3-year trials)	Central Punjab	6	NR	5–15	486–1458	4.1	4.3	0.2	4.9	4750	3–10:1	Gill et al. (2003)
	RainfedZone	9	NR	5–15	486–1458	3.78	3.94	0.16	4.2	3800	3–8:1	
cvs. Sarhad, white Azam	Peshawar, Mardanand DI Khan distts.	5	NR	2.5	243	3.83	4.06	0.23	6	5463	23:1	Khattak and Bhatti (1986)
Rice		**22**		**5**	**485**	**3.17**	**3.56**	**0.38**	**12**	**15,200**	**31:1**	
cv. Bas-385	Sheikhupura and Sialkot distts.	14	NR	10 (5*)	485	3.45	3.73	0.28	8	11,200	23:1	Chaudhry et al. (1992)
cv. IR-6	Eminabad	1	NR	5	485	3.70	4.20	0.50	14	12,500	26:1	Chaudhry et al. (1976)
cv. Bas-370	Punjab	7	NR	5	485	2.54	3.07	0.53	21	21,200	44:1	
Sunflower (cv. Parsun-2)	Islamabad	1	0.21	2.5–5.0	364	1.87	2.02–2.12	0.15–0.25	7.9–13.4	9375–15,625	25.8–42.9	Khan et al. (2009)

Mustard (cv. BARD-1)	Islamabad	1	1.76	1.5–3.0	219	0.57–1.17	0.64–1.16	0.07	11.2	4200	19.2	Nawaz et al. (2012)
Peanut		**2**	**6.5**		**3029**	**1.34**	**1.68**	**0.35**	**29.5**	**31,050**	**21:1**	
cv. BARD-92	Islamabad (rainfed)	1	6.5	*Sequestrene* (foliar sprays)	4534+750[4]	0.93	1.32	0.39	42	35,100	7:1	Rashid et al. (1997)
cv. BARD-699		1	6.5	0.5 FeSO$_4$ (foliar sprays)	24+750[4]	1.74	2.03	0.30	17	27,000	35:1	
Chickpea		**2**	**7**		**3029**	**3.2**	**3.49**	**0.29**	**9**	**28,500**	**10.5:1**	
cv. C-44	Pothohar (rainfed)	1	7	Sequestrene (foliar sprays)	4534+750[4]	3.20	3.69	0.48	15	48,000	9:1	Rashid and Din (1992)
cv. C-44		1	7	FeSO$_4$ foliar sprays	24+750[4]	3.20	3.29	0.09	3	9000	12:1	
Potato		**4**	**NR**	**2.5–15**	**849**	**1.93**	**2.23**	**2.96**	**16**	**37,000**	**59:1**	
Nowshehra distt.		1	NR	2.5	243	2.04	2.20	1.55	8	19,375	80:1	Khattak (1988)
Kasur distt.		3	NR	15	1455	1.82	2.26	4.37	24	54,625	38:1	Gill et al. (1996)

* Adjusted Fe dose.

** Cost of fertilizer Fe; see **Annex 12**.

*** Price of crop produce; see **Annex 13**; 60 L ha^{-1} spray^{-1}; for 240 L spray solution used 2.4 kg *Sequestrene* ha^{-1} and 1.2 kg FeSO$_4$ ha^{-1}

[1] DTPA- or AB-DTPA-extractable Fe

[2] Fe applied on top of the basal dressings of NPK at recommended doses.

[3] Considering grain yield increase only; increase in straw yield is additional benefit.

[4] Labor cost of two foliar sprays.

REFERENCES

Chaudhry, F. M., A. Latif, A. Rashid, and S. M. Alam. 1976. Response of the rice varieties to field application of micronutrient fertilizers. *Pakistan Journal of Scientific and Industrial Research* 19:134–139.

Chaudhry, R. A., D. M. Malik, T. Amin, Ghafoor-ul-Haq, and S. Sabir. 1992. Rice response to micronutrients. p. 342–350. In: Proc. 3rd. National Congress of Soil Science, Lahore, March 20–22, 1990. Soil Sciience Society of Pakistan.

Chaudry, E. H., V. Timmer, A. S. Javed, and M. T. Siddique. 2007. Wheat response to micronutrients in rain-fed areas of Punjab. *Soil and Environment* 26:97–101.

Gill, K. H., G. Hassan, M. Y. Ahmad, and Ehsan-ul-Haq. 1996. Soil Fertility Investigations on Farmers Fields. Annual Report, 1995–96. Soil Fertility Survey and Soil Testing Institute, Lahore.

Gill, K. H., S. J. A. Sherazi, J. Iqbal, et al. 2003. Maize response to Fe application in central and Barani zones of the Punjab. *Pakistan Journal of Soil. Science* 22:36–41.

Khan, R., A. R. Gurmani, M. S. Khan, and A. H. Gurmani. 2009. Residual, direct and cumulative effect of zinc application on wheat and rice yield under rice-wheat system. *Soil and Environment* 28:24–28.

Khattak, J. K. 1988. Cooperative Research Program on Micronutrient Status of Pakistan Soils and Its Role in Crop Production. Annual Reports, 1987–88. NWFP Agricultural University, Peshawar.

Khattak, J. K., and A. U. Bhatti. 1986. National outreach research project on soil fertility and fertilizer use in Pakistan. Final Technical Report, 1982–86. NWFP Agricultural University, Peshawar.

Nawaz, N., M. S. Nawaz, N. M. Cheema, and M. A. Khan. 2012. Zinc and iron application to optimize seed yield of mustard. *Pakistan Journal of Agricultural Research* 25: 28–33.

Rashid, A., and J. Din. 1992. Differential susceptibility of chickpea cultivars to iron chlorosis grown on calcareous soils of Pakistan. *Journal of the Indian Society of Soil Science* 40:488–492.

Rashid, A., and F. Qayyum. 1991. Micronutrient Status of Pakistan Soils and Their Role in Crop Production, Cooperative Program. Final Report, 1983–1990. National Agricultural Research Center, Islamabad. 84 pp.

Rashid, A., E. Rafique, and N. Ali. 1997. Micronutrient deficiencies in rainfed calcareous soils of Pakistan. II. Boron nutrition of the peanut plant. *Communications in Soil Science and Plant Analysis* 28:149–159.

Annex 11
Crop Yield Increases with Copper Fertilizer and Profitability in Pakistan

Crop	Region	Field Expts.	¹Soil Cu (mg kg⁻¹)	²Cu Applied (kg ha⁻¹)	**Cost of Applied Cu (PKR ha⁻¹)	Control Yield (t ha⁻¹)	Yield with Cu (t ha⁻¹)	Yield Increase over Control (t ha⁻¹)	% Yield Increase	***Value of Increased Yield (PKR ha⁻¹)³	****Value:Cost Ratio	Data Source
Wheat		**39**				**3.25**			**8.76**	**86363**	**5.5:1**	
Year 2003	Rice and cotton zones	2	NR	2*	2400	3.54	3.78	0.24	6.9	7800	3:1	Nadim et al. (2016)
2–3 years trials; cv. Inqilab-91	Central Punjab	10	1.5–3.7	0.25–1.25*	300–1500	4.3	4.4–4.6	0.1–0.3	2.6–6.1	3250–9750	7–11:1	Sherazi et al. (2003)
	Rice belt	12	0.7–2.3	0.25–1.25*	300–1500	4.1	4.3–4.4	0.2–0.3	5.7–7.8	6500–9750	7–22:1	
	Cotton	3	1.7–3.8	0.25–1.25*	300–1500	3.3	3.4	0.1	2.3–4.3	3250	2–11:1	
cv. Mexi-Pak	Punjab (cotton-wheat belt)	5	0.90	10(5*)	6000	3.76	4.18	0.42	11	13,650	2:1	NIAB (unpublished data)
	Rawalpindi (rainfed)	5	NR	5	6000	2.52	3.1	0.57	23	18,525	3:1	Rashid and Qayyum (1991)
	Tarnab, Peshawar	2	NR	5	6000	1.25	1.33	0.080	6	2600	0.4:1	Hamid (1982)
Rice		**8**		**5**	**6000**	**4.43**	**4.86**	**0.43**	**9.7**	**10,8333**	**1.8:1**	
cv. IR-6	Punjab	2	NR	5	6000	4.71	5.03	0.33	7	8250	1.4:1	Chaudhry et al. (1992)
cv. IR-6	Sheikhupura distt.	4	NR	5	6000	4.32	4.74	0.42	9	10,500	2:1	Chaudhry et al. (1992)
cv. IR-6	Punjab	2	NR	5	6000	4.25	4.80	0.55	13	13,750	2:1	Chaudhry et al. (1992)

										Reference
Fine rice	21		5	6000	2.97	3.46	0.47	16.5	18,6003 3.3:1	
cv. Bas-385 Shiekhupura and Sialkot	14	NR	5	6000	3.43	3.82	0.40	12	16,000 3:1	Chaudhry et al. (1992)
cv. Bas-370 Punjab	7	NR	5	6000	2.54	3.1	0.53	21	21,200 3.5:1	Chaudhry et al. (1976)
Maize	6		5	6000	2.83	3.22	0.39	13.5	9,2633 1.6:1	
cvs. Sarhad White, Azam — Peshawar, Mardan, and DI Khan	5	NR	5	6000	3.60	4.10	0.51	14	12,113 2:1	Khattak and Bhatti (1986)
Rawalpindi (rainfed)	1	NR	11.6 (5*)	6000	2.06	2.33	0.27	13	6413 1.1:1	Rashid and Qayyum (1991)
Potato Nowshehra	2	NR	2	2400	20.440	20.440	0	0		Khattak (1988)
Tobacco	2	NR	0.3–6	4155	2.21	2.61	0.40	18.5	70,000 12:1	
Mardan	1	NR	6	7200	2.13	2.63	0.50	24	87,500 15:1	Khattak (1988)
Mardan	1	NR	0.3 (2 foliar spray)	360+750[4]	2.28	2.58	0.30	13	52,500 9:1	Khattak (1988)

* Adjusted Cu dose.

** Cost of fertilizer Cu: PKR 1200 kg^{-1} Cu (commercial-grade imported $CuSO_4$, $5H_2O$ containing 25% Cu @ PKR300 kg^{-1}); see Annex 12.

*** Price of crop produce per 40 kg; see Annex 13.

**** One good soil application of Cu lasts for 3–4 crop seasons; thus fertilizer cost/crop could be 1/3–1/4 and actual VCR may be 3–4 times.

[1] DTPA- or ABDTPA-extractable Cu.

[2] Cu applied along with based dressings of NPK at recommended doses.

[3] Considering grain yield increase only; increase in straw yield is additional benefit.

[4] Labor cost of two foliar sprays.

REFERENCES

Chaudhry, F. M., A. Latif, A. Rashid, and S. M. Alam. 1976. Response of the rice varieties to field application of micronutrient fertilizers. *Pakistan Journal of Scientific and Industrial Research* 19:134–139.

Chaudhry, R. A., D. M. Malik, T. Amin, Ghafoor-ul-Haq, and S. Sabir. 1992. Rice response to micronutrients. p. 342–350. In: Proc. 3rd. National Congress of Soil Science, Lahore, March 20–22,1990. Soil Science Society of Pakistan.

Hamid, A. K. 1982. Cooperative Research Program on Micronutrients. Annual Report, 1981–82. Agricultural Research Institute, Tarnab, Peshawar.

Khattak, J. K. 1988. Cooperative Research Program on Micronutrient Status of Pakistan Soils and Its Role in Crop Production. Annual Reports, 1987–88. NWFP Agricultural University, Peshawar.

Khattak, J. K., and A. U. Bhatti. 1986. National outreach research project on soil fertility and fertilizer use in Pakistan. Final Technical Report, 1982–86. NWFP Agricultural University, Peshawar.

Nadim, M. A., M. S. Baloch, E. A. Khan, A. A. Khakwani and K. Waseem. 2016. Integration of organic, synthetic fertilizers and micronutrients for higher growth and yield of wheat. *Sarhad Journal of Agriculture* 32:9–16.

Rashid, A., and F. Qayyum. 1991. Micronutrient Status of Pakistan Soils and Their Role in Crop Production, Cooperative Program. Final Report, 1983–1990. National Agricultural Research Council, Islamabad. 84 pp.

Sherazi, S. J. A., J. Iqbal, and K. H. Gill, et al. 2003. Wheat response to copper application in various zones of the Punjab. *Pakistan Journalof Soil Science* 22:26–29.

Annex 12
Retail Sale Prices (Average) of Micronutrient Fertilizers in Pakistan (in January 2017)

Micronutrient	Fertilizer Product	Price of Fertilizer Product and Elemental Micronutrient	Remarks
Zn	FFC's zinc sulfate ($ZnSO_4.H_2O$), 33% Zn	PKR 425 per 3 kg = *PKR 430 per kg elemental Zn*	Used this price for Zn, as it is a cheaper option
	Engro's zinc sulfate (granular), ($ZnSO_4.H_2O$), 33% Zn	PKR 625 per 3 kg = PKR 63 per kg elemental Zn	
B	FFC borax, 10.5% B	3 kg: PKR 315 per 3 kg = *PKR 930 per kg elemental B*	Used for *soil application*
	Boric acid (H_3BO_3), 17% B	PKR 5400 per 25 kg = *PKR 1270 per kg elemental B*	Used for *foliar sprays*
Fe	Ferrous sulfate ($FeSO_4.H_2O$), 33% Fe	PKR 800 per 25 kg *PKR 97 per kg elemental Fe*	Used this price for Fe for VCR calculation
	Fe-*Sequestrene* (6% Fe)	PKR 3778 per kg Fe-*Sequestrene*	
Cu	Copper sulfate ($CuSO_4.5H_2O$) 25% Cu	PKR 7500 per 25 kg PKR 300 per 1 kg *PKR 1200 per kg elemental Cu*	Used this price for Fe for VCR calculation

Annex 13
Market Prices of Crop Produce (Average) in Pakistan (in January 2017)

Crop Produce	Market Price (Range) (at Which Farmers Sell Their Produce in Local Market/Mandi)		Remarks
	PKR per 40 kg (i.e., Maund)	PKR per Tonne	
Field crops			
Wheat grains	1300	32,500	
Rice paddy: cv. *Basmati*	1600	40,000	Based on quality of paddy
Rice paddy: *IRRI types*	1000	25,000	
Seed cotton	3200	80,000	
Maize (hybrid) grains	950	23,750	
Sugarcane (cane)	160	4000	In Punjab
Mungbean	7600	190,000	
Chickpea	4000	100,000	
Rapeseed (canola)	2400	60,000	
Sunflower	2500	62,500	
Sugarbeet	700	17,500	
Peanut	3600	90,000	
Tobacco	7000	175,000	
Vegetables			
Potato	500	12,500	
Tomato	1800	45,000	
Onion	700	17,500	
Chili	1500	37,500	
Fruits			
Citrus	1500	37,500	
Apple	4200	105,000	
Plum	3600	90,000	
Peach	3800	95,000	
Pear	3600	90,000	
Grapes	4500	112,500	
Mango	3200	80,000	
Banana	50–60/dozen	31,250–42,500	In Sindh
Pomegranate	1500	37,500	

Annex 14
Market Survey of Available Micronutrient Fertilizer Brands and Prices in Central Punjab, Pakistan (in January 2017)

Company	Brand	Active Ingredient	Composition (%)	Packing	Recommendation	Price (PKR)
EFL	Librel TMX AP02	Fe, Zn, Cu, Mn, B	(30 g/kg) + (40 g/kg) + (10 g/kg) + (10 g/kg) + (10 g/kg)	500 g	150 g/acre/foliar spray	565
	Librel Zinc	Zn	14% Zn-EDTA	500 g	150 g/acre/foliar spray	800
FMC	ZincStar, 6C	Zn	6%	400 mL	400 mL/acre/foliar spray	400
	ZincStar, 6C	Zn	6%	800 mL	400 mL/acre/foliar spray	775
ICI	Zingreen	Zn	33%	3 kg	3 kg/acre/Soil application	708
Target	Target Zinc 10%	Zn	10%	3 L	3 L/acre/flooding	365
	Target Zinc 10%	Zn	10%	20 L	4 L/acre/flooding	2225
	Target Zinc 10%	Zn	10%	200 L	5 L/acre/flooding	21,215
	Element Zinc	Zn	6% Chelated Zn	500 mL	500 mL/acre/foliar spray	385
	Bridge 5%	B	5%	500 mL	500 mL/acre/foliar spray	285
	Stroke	Zn + B	6% Zn, 4% B	500 mL	500 mL/acre/foliar spray	305
4B	Sixer	Zn	10%	3 L	3 L/acre/flooding	300
	Sixer	Zn	10%	20 L	3 L/acre/flooding	1800
	Sixer	Zn	10%	200 L	3 L/acre/flooding	18,000
	Sugar Mover/ Fruit Feed	B	5%	500 mL	500 mL–1 L/acre/ foliar	250
	Sugar Mover/ Fruit Feed	B	5%	1 L	500 mL–1 L/acre/ foliar spray	1000

(Continued)

Company	Brand	Active Ingredient	Composition (%)	Packing	Recommendation	Price (PKR)
Warble/ Welcon	*Aala Zinc 5, Wel Zinc 5,Premium Zinc*	Chelated Zn	5% Zn			
	Foliaral OK	B, K	20% B, 20% K	125 g	125 g/acre/foliar spray	300
	Hit Zinc, Shinining Zinc	Zn	10%	3 L	3 L/acre/flooding	375
	Kissan Zinc	Zn	21%	5 kg	5 kg/acre/soil application	550
	Ideal Zinc	Zn	33%	3 kg	3 kg/acre	700
	Tower	Chelated Zinc	6% Zn			
	Warble Zinc	Zn	33%			
	Wealth	B	5%	500 mL	500 mL/acre/foliar spray	225
Helb	*Zanden 14*	Chelated Zn	5%	1 kg	500 g/acre/foliar spray	600
	Vertac	Zn	33%	3 kg	3 kg/soil application	1200
SunCrop	*Boll Guard*	Micro mix		500 mL	500 mL/acre/foliar spray	145
	Bouncer/ Share	B	5%	500 mL	500 mL/acre/foliar spray	295
	Zinkron Plus	Zn	10%	3 L	3L/acre/flooding	335
	Zinkron Plus	Zn	10%	20 L	3L/acre/flooding	2115
	Zinkron Plus	Zn	10%	200 L	3L/acre/flooding	18,800
	Zinkron 21 (G)	Zn	21%	3 kg	6 kg/acre/soil application	717
	Zinkron 33 (G)	Zn	33%	3 kg	3 kg/acre/soil application	750
	Zinkron 33 (G)	Zn	33%	25 kg	4 kg/acre/soil application	5940
Kanzo	*Direx 5%*	B	5%	500 mL	500 mL/acre/foliar spray	209
	Select 5%	Chelated Zn	5%	2 kg	2 kg/acre/foliar spray	790
	Select 5%	Chelated Zn	5%	25 kg	2 kg/acre/foliar spray	9560
Auriga	*Zarcoon*	Zn	10%	3L	6L/acre/foliar spray	465
	Zarcoon	Zn	10%	27 L	6L/acre/foliar spray	3905
	Zarcoon	Zn	10%	200 L	6L/acre/foliar spray	26,400
	Microcure	Multi-micronutrients		1 L	4L/acre/flooding	350

Company	Brand	Active Ingredient	Composition (%)	Packing	Recommendation	Price (PKR)
Swat Agro	Zinc Force	Zn	33%	3 kg	3 kg/acre/soil application	680
	Mycrobor DF	B	0.17%	1 kg	800–1250 g/acre/ foliar spray; 3 kg/acre for soil application	600
	Complex	Multi-nutrient	N 20%, P 8%, K 14%, Mg 2%, S 9.7%, B 0.04%, Cu 0.2%, Fe 0.02%, Mn 0.26%, Mo 0.006%, Zn 0.14% as chelated EDTA	1 kg	1–2 kg/acre/foliar spray	630

Annex 15
Crop Species Requiring Micronutrient Fertilizer Use in Pakistan

Crops	Zn	B	Fe
I. Field crops		◆	
Wheat, rice, cotton, maize, rapeseed (canola), sorghum, sugarcane	◆	◆	
Peanut		◆	◆
Soybean	◆		
Alfalfa, sugarbeet, sunflower, tobacco		◆	
II. Vegetables			
Potato, tomato	◆	◆	
Turnip, cauliflower, cabbage, carrot, radish, spinach, lettuce, sweet potato		◆	
Onion	◆		
III. Fruits			
Citrus	◆		◆
Apple, peach, pear, plum, apricot, grape	◆	◆	◆
Strawberry			◆
IV. Ornamentals			
Ornamentals		◆	◆
Conifers, rose		◆	

Adapted from Rashid, 1996, 2006; NFDC, 1998; Rafique et al., 2006, 2014; Rashid et al., 2002a, 2002b, 2002c, 2002d; Zahid et al., 2000; Rashid and Rafique, 2017.

◆ Indicates the need for using specific micronutrient fertilizer for the given crop species.

REFERENCES

NFDC. 1998. *Micronutrients in Agriculture: Pakistan Perspective.* National Fertilizer Development Center, Islamabad, 51 pp.

Rafique, E., A. Rashid, J. Ryan, and A. U. Bhatti. 2006. Zinc deficiency in rainfed wheat in Pakistan: Magnitude, spatial variability, management, and plant analysis diagnostic norms. *Communications in Soil Science and Plant Analysis* 37:181–197.

Rafique, E., M. Mahmood-ul-Hassan, M. Yousra, I. Ali, and F. Hussain. 2014. Boron nutrition of peanut grown in boron-deficient calcareous soils: Genotypic variation and proposed diagnostic criteria. *Journal of Plant Nutrition* 37:172–183.

Rashid, A. 1996. Secondary and micronutrients. p. 341–385, Chapter 12, In: *Soil Science*, Rashid, A., and K. S. Memon (Managing Authors). National Book Foundation, Islamabad, Pakistan.

Rashid, A. 2006. Incidence, diagnosis and management of micronutrient deficiencies in crops: Success stories and limitations in Pakistan. In: Optimizing Resource Use Efficiency for Sustainable Intensification of Agriculture. IFA Agriculture Conference, Kunming, PR China. 27 February - 2 March, 2006.

Rashid, A., S. Muhammad, and E. Rafique. 2002a. Genotypic variation in boron uptake and utilization by rice and wheat. p. 305–310, In: Goldbach, H. E. et al. (Ed.), *Boron in Plant and Animal Nutrition*. Kluwer Academic Publishers, New York.

Rashid, A., E. Rafique, and N. Bughio. 2002b. Boron deficiency in rainfed alkaline soils of Pakistan: Incidence and boron requirement of wheat. p. 371–379, In: Goldbach, H. E. et al. (Ed.), *Boron in Plant and Animal Nutrition*. Kluwer Academic Publishers, New York.

Rashid, A., and E. Rafique. 2017. Boron deficiency diagnosis and management in field crops in calcareous soils of Pakistan: A mini review. *BORON* 2 (3): 142–152.

Rashid, A., E. Rafique, S. Muhammed, and N. Bughio. 2002c. Boron deficiency in rainfed alkaline soils of Pakistan: Incidence and genotypic variation in rapeseed-mustard. p. 363–370, In: Goldbach, H. E. et al. (Ed.), *Boron in Plant and Animal Nutrition*. Kluwer Academic/Plenum Publishers, New York.

Rashid, A., E. Rafique, and J. Ryan. 2002d. Establishment and management of boron deficiency in field crops in Pakistan: A country report. p. 339–348, In: Goldbach, H. E. et al. (Ed), *Boron in Plant and Animal Nutrition*. Kluwer Academic Publishers, New York.

Zahid, M. A., Rashid, A., and Din, J. 2000. Balanced nutrient management in chickpea. *International Chickpea and Pigeonpea Newsletter (ICRISAT)* 7: 24–26.

Annex 16
Crop-Specific Micronutrient Fertilizer Recommendations

Crop(s)	Micronutrient, Dose, and Method of Application*
	Zinc
Rice	20 kg Zn ha^{-1} to nursery bed
	OR
	5 kg Zn ha^{-1} by soil application during puddling
Cotton	0.1 % Zn foliar sprays 40, 60, 80 days after sowing OR
	5 kg Zn ha^{-1} by soil application
Wheat, maize, rapeseed, soybean/ bean	5 kg Zn ha^{-1} as soil application
	OR
	0.1% Zn foliar sprays 30, 40, 50 days after sowing
Sorghum	3 kg Zn ha^{-1} as soil application
Sugarcane, sugarbeet, potato, tomato, onion	5 kg Zn ha^{-1} by soil application
Citrus	0.1% Zn foliar sprays in February/March, June, and August.
	OR
	40 g Zn/tree for 6-year-old trees; increase by 10 g each year making 80 g Zn/tree for > 10 years age. Apply under the tree canopy.
Apple, peach, plum, apricot	0.1% Zn ha^{-1} foliar sprays twice a year
	OR
	50–100 g Zn/tree under the canopy
Grape	0.05 % Zn foliar sprays twice a year
	Boron
Cotton	0.1 % B foliar sprays at 40, 60, and 80 days after sowing
	OR
	1.0 kg B ha^{-1} by soil application
Rice, wheat, maize, alfalfa, rapeseed, sunflower, potato, tobacco	1.0 kg B ha^{-1} by soil application
Cauliflower, cabbage, sugarbeet	0.75 kg B ha^{-1} by soil application
	OR
	0.05 % B foliar spray twice: 2 weeks after planting and 2 weeks before heading/curd formation
Tomato, radish, spinach, lettuce, turnip, carrot, peanut	0.75 kg B ha^{-1} by soil application
	OR
	0.05 % B foliar sprays at flowering and one month after that 0.75 kg B ha^{-1} by soil application
	OR
	2 Foliar sprays of 0.05 % B solution

(*Continued*)

Crop(s)	Micronutrient, Dose, and Method of Application*
Rose	2 Foliar sprays of 0.05% B solution, in March and September
Apple, peach, plum, pear, apricot	Foliar sprays of 0.1 % B solution, twice a year
Grape	Foliar sprays of 0.05 % B solution, immediately before and after full bloom
	Iron
Sensitive crop and vegetables	Foliar sprays of 0.5% ferrous sulfate OR 1.0% Fe-Sequestrene, 3–4 times a year OR Soil application @ 5–10 kg Fe ha^{-1} as ferrous sulfate
Sensitive fruits	Foliar sprays of 0.05 % ferrous sulfate or 1.0 % Fe-Sequestrene, 3–4 times a year, in March/April, August, and October/November
	Copper
Sensitive crop and vegetables	5 kg Cu ha^{-1} as soil application OR 2–3 Foliar sprays of 0.5 % CuSO$_4$ solution, in March/April, August, and October/November
	Manganese
Sensitive crop and vegetables	5 kg Mn ha^{-1} as soil application OR 22–3 Foliar sprays of 0.5 % MnSO$_4$ solution
Sensitive fruits	Foliar sprays of 0.5 % MnSO$_4$ solution, 2–3 times a year in March/April, August, and October/November

Adapted from Rashid, 1996, 2006a, 2006b; Rashid and Qayyum, 1991; Tahir, 1981; Khattak, 1995; Tandon, 2013.

For uniform broadcast distribution, the micronutrient fertilizer dose should be thoroughly mixed with sieved soil (~5 times the fertilizer weight).

One good dose of Zn by soil application lasts for 3–4 crop seasons and of B for 2–3 crop seasons.

Neutralize zinc sulfate sprays solution with 0.25 % unslaked lime or area.

Add Surf detergent in sprays solution @ 0.05%.

For the preparation of 0.1% foliar Zn spray, dissolve 303 g of ZnSO$_4$.H$_2$O 33% Zn *or* 476 g of ZnSO$_4$.7H$_2$O 21% Zn in 100 liters of water (approx. five spray tanks, each of 20 liters) and spray over 1-acre crop area. For the preparation of 0.1% foliar B spray, dissolve 909 g of borax 11% B *or* 588 g of boric acid 17% B *or* 478 g of *Solubor* 20.9% B in 100 liters of water (approx. five spray tanks, each of 20 liters) and spray over 1-acre crop area. For the preparation of 0.05% foliar B spray, dissolve 455 g of borax 11% B *or* 294 g of boric acid 17% B *or* 239 g of *Solubor* 20.9% B in 100 liters of water (approx. five spray tanks, each of 20 liters) and spray over 1-acre crop area.

REFERENCES

Khattak, J. K. 1995. *Micronutrients in Pakistan Agriculture*. Pakistan Agricultural Research Council, Islamabad and NWFP Agricultural University, Peshawar. 135 pp.

Rashid, A. 1996. Secondary and micronutrients. p. 341–385, Chapter 12, In: *Soil Science*, Rashid, A., and Memon, K.S. (Managing Authors). National Book Foundation, Islamabad, Pakistan.

Rashid, A. 2006a. Incidence, diagnosis and management of micronutrient deficiencies in crops: Success stories and limitations in Pakistan. In: *Optimizing Resource Use Efficiency for Sustainable Intensification of Agriculture*. IFA Agriculture Conference, Kunming, PR China. 27 February - 2 March.

Rashid, A. 2006b. Boron Deficiency in Soils and Crops of Pakistan: Diagnosis and Management. Pakistan Agricultural Research Council, Islamabad, Pakistan, viii+34 pp. ISBN: 969-409-184-5.

Rashid, A., and F. Qayyum. 1991. Micronutrient Status of Pakistan Soils and Their Role in Crop Production, Cooperative Program. Final Report, 1983–1990. NARC, Islamabad. 84 pp.

Tahir, M. 1981. Availability Status of Micronutrients in the Soils of West Pakistan and the Role and Behavior of Selected Micronutrients in the Nutrition of Crops. Final Technical Report of PL-480 Project. Nuclear Institute for Agriculture and Biology, Faisalabad. 124 pp.

Tandon, H. L. S. 2013. Micronutrient Handbook – from research to application. Fertilizer Development and Consultation Organization, New Delhi, India. 234 pp.

Annex 17
Chemical Compatibility Chart: Micronutrient Fertilizers

Chemicals Name/ Category	Copper Sulfate	Iron Sulfate	Manganese Sulfate	Zinc Sulfate	Copper Chelate	Iron Chelate	Manganese Chelate	Zinc Chelate
Herbicides								
Diurion	√	n.a.	√	√	√	n.a.	√	√
Triallate	√	n.a.	√	√	√	n.a.	√	√
Paraquat and diquat	X	n.a.	X	X	X	n.a.	X	X
Glyphosate	X	n.a.	X	X	X	n.a.	X	X
Pendimethalin	√	n.a.	√	√	√	n.a.	√	√
Trifluralin	√	n.a.	√	√	√	n.a.	√	√
Metribuzin	√	n.a.	√	√	√	n.a.	√	√
MCPA sodium	√	n.a.	√	√	√	n.a.	√	√
Quizalofop	n.a.	n.a.	√	√	n.a	n.a.	n.a	n.a.
Fluazifop	√	n.a.	√	√	√	n.a.	√	√
Diclofopmethyl	√	√	√	√	√	√	√	√
Haloxyfop	n.a.	n.a.	n.a	√	n.a	n.a.	n.a	n.a.
Butroxydim	√	n.a.	√	√	√	n.a.	√	√
Insecticides								
Dimethoate	√	n.a.	√	√	√	n.a.	√	√
Chlorpyrifos	√	n.a.	√	√	√	√	√	√
Alphacypermethrin	n.a.	n.a.	√	n.a.	n.a	n.a.	n.a	n.a.
Cypermethrin	C	n.a.	√	√	n.a	n.a.	√	√
Deltamethrin	√	n.a.	√	√	√	n.a.	√	√
Thiometon	n.a.	n.a.	n.a	√	n.a	n.a.	n.a	√
Fungicide								
Mancozeb	n.a.	√	√	n.a.	n.a.	√	√	n.a.
Fertilizers/acids								
Magnesium sulfate	√	√	√	√	√	√	√	√
Phosphoric acid	√	√	√	√	R	R	R	R

(*Continued*)

Chemicals Name/ Category	Copper Sulfate	Iron Sulfate	Manganese Sulfate	Zinc Sulfate	Copper Chelate	Iron Chelate	Manganese Chelate	Zinc Chelate
Sulfuric acid	√	√	√	√	√	√	√	√
Nitric acid	√	√	√	√	X	X	X	X

Adapted/Modified from Lamb and Poddar, 2008, with permission; and Roddy, 2008.

x = Not compatible

√ = Compatible

C = Not all formulations of are completely compatible. Check with the company

R = Reduced compatibility

n.a. = Data not available

REFERENCES

Lamb, J., and A. Poddar. 2008. Chemical Compatibility Chart. Chapter 8, In: *Grain Legume Handbook: For the Pulse Industry. Grain Legume Handbook Committee*, Finsbury Press, Riverton, SA, Australia.

Roddy, E. 2008. Fertigation Fertilizer Sources. http://en.engormix.com/MA-agriculture/news/fertigation-fertilizer-sources-t13141/p0.htm (Accessed on Jan 25, 2017.)

Annex 18
Some Important Conversion Factors for Micronutrient Fertilizers

1 kg borax ($Na_2B_4O_7.10H_2O$) = 105 g B
1 kg boric acid (H_3BO_3) = 170 g B
1 kg *Solubor* = 210 g B
1 kg *Granubor Natur* = 150 g B
1 kg Zinc sulfate ($ZnSO_4.H_2O$) = 330 g Zn
1 kg Zinc sulfate ($ZnSO_4.7H_2O$) = 210 g Zn
1 kg Zinc chelate (Na_2Zn-EDTA) = 120 g Zn
1 kg Zinc oxide (ZnO) = 780 g Zn
1 kg Ferrous sulfate ($FeSO_4.7H_2O$) = 190 g Fe

1 kg *Sequestrene* (NaFe-EDDHA) = 60 g Fe
Iron chelate (Fe-EDTA) = 120 g Fe
Copper sulfate ($CuSO_4.5H_2O$) = 250 g Cu
Copper sulfate ($CuSO_4.H_2O$) = 350 g Cu
Copper chelate (Na_2Cu EDTA) = 130 g Cu
Copper chelate (NaCu EDTA) = 90 g Cu
Manganese sulfate ($MnSO_4.H_2O$) = 350 g Mn
Manganese chelate (Mn-EDTA) = 50–140 g Mn

Adapted from Tandon (2013).

REFERENCE

Tandon, H. L. S. 2013. *Micronutrient Handbook – From Research to Application*. Fertilizer Development and Consultation Organization, New Delhi, India. 234 pp.

Annex 19
Crude Estimate of Non-Chelated Zinc Sulfate Fertilizer Products Marketed in Pakistan in 2012

Company	Brand Name	Product Volume (Tonnes)
Granular/powder formulations		
Engro Fertilizer	*Zingro*	2100
Fauji Fertilizer Company	*Sona Zinc*[1]	1547
4-Brothers (Biologic)	*Activator*	1200
4-Brothers (Tarzan)	*Zinc*	900
FMC	*Zintox Plus*	690
Welcon	*Zinc Sulfate*	385
SunCrop	*Prime*	310
Welcon	*Kissan Zinc*	215
Solex	*Pasban*	160
ICI	*Zingreen*[2]	108
Totalcare	*Zinc Minc*	65
Jaffer Agro Services	*Grozin*	58
Hexon	*Nobel*	22
UAQ	*Confirm*	20
NH Chemical	*Confirm*	20
Total granular/powder formulations		*7800*

Company	Brand Name	Product Volume (m³)
Liquid formulations		
Auriga	*Zarcoon*	1275
Agri Farm	*Agri-Zoro*	550
Syngenta	*Nia Zinc*	529
Target	*Target zinc*	500
Syaban	*Zeneca*	500
Warble	*Shining Zinc*	460
Agri Top	*Top-Zoro*	460
SunCrop	*Zinkron+*	385
DJC	*Lift up*	385
Hexon	*Nobel*	280
Welcon	*Shining Zinc*	250
Auriga – Bulk	*Zinc*	250

(Continued)

Company	Brand Name	Product Volume (Tonnes)
Ali Akbar – Bulk	Zinc	250
NH Chemical	Negheban	230
Eon	Zinc	190
Sun Crop – Bulk	Zinc	190
Solex	Solo Zinc	190
Agri International	Zinc	190
Target	Sixer	130
Kanzo	Kanzo zinc	130
Welcon	Hit Zinc	130
HR Chemicals	Zinc	130
Abdul Haseeb	Momal	110
Total Care	Totalzar	90
Green Crop	Buraq	85
Agri Force	Agri Power	80
Adv. AgroTech	Zarghoon	65
UDL	Zinccash	45
Farm Eco	Zink	25
Nice	AqvaZincool	22
Nice	Power Zinc	20
Solex	Zinc Force	18
Wel Wisher	Shining Zinc	18
Life Tech	Aafreen Super	16
Alfa Chemicals	Zinco	9
4-B (Tarzan)	Sixer	
Others	Zinc	85
Total liquid formulations		*8272*
Total (granular + liquid products)		*15,684*

[1] *Sona Zinc* of FFC (i.e., Zinc Sulfate, 33% Zn) is being marketed since 2014

[2] Zingreen of ICI (i.e., zinc sulfate, 33% Zn (but being soldas 21% Zn product, like that of Syngenta's *Nia Zinc*) is marketed since 2014

1 Tonne = 1000 kg

1 m^3 = 1000 L

Annex 20
Crude Estimate of Chelated Zinc Sulfate Marketed in Pakistan in 2012

Company	Brand Name	Product Volume (Tonnes)*
Granular/powder formulations		
UDL	*Uttah Plus*	900
Kanzo	*Select*	85
Solex	*Five Times*	25
Helb	*Zenbeen*	25
Adv. Agro Tech	*Gema*	6
SunCrop	*Muhafiz*	5
Total Care	*Zarcare*	3
	Total volume	*1049*

Company	Brand name	Product volume (m³)**
Liquid formulations		
4-Brothers (Biologic)	*Zinc Chelate*	90
4-Brothers (Tarzan)	*Accelerate*	80
Target	*Element*	55
FMC	*Zinc Star*	25
DJC	*Adjust*	20
Agri Force	*Demand*	18
Agri Farm	*Invento*	4
Agri Top	*Speedo*	3
Life Tech	*Blossom Blast*	2
Life Tech	*Bloom Super*	1
Others		77
	Total volume	*375*
	Total (solid + liquid formulations)	**1424**

* 1 Tonne = 1000 kg
** 1 m³ = 1000 L

Annex 21
Crude Estimate of Boron Fertilizer Products Marketed in Pakistan in 2012

Company	Brand Name	Product Volume (m³ or Tonne)	Form
FFC	Sona Boron	212	Solid
Syngenta	Solubor	114	Solid
Ali Akbar –Bulk	Bounce	70	Solid
4B – (Tarzan)	Step (chelated)/Sugar Mover/Fruit Feed	55	Liquid
Eon Fertilizer	Biolizer	45	Liquid
Agri Farm	Boll Duo	40	Liquid
FMC/AQ Enterprises	Borostar Plus	35	Solid
4B – (Biologic)	Fruit Feed	32	n.a.
Ali Akbar Enterprises (Pvt.) Ltd – Target	Bridge	45	Liquid
Agri Top	Top Bol	35	Liquid
Dada Jee Corporation (Pvt.) Ltd. (DJC)	Drive	25	Liquid
Warble (Pvt.) Limited	Wealth	25	Liquid
SunCrop	Bouncer / Share	22	Liquid
Kanzo	Direx	20	Liquid
Agri Force Chemicals	Winner	18	n.a.
UDL	Indeplex	15	n.a.
Alfa Chemicals	Flowral	12	n.a.
Welcon	Wealth	12	Liquid
Weal AG Corporation	Choice	14	Liquid
Wel wisher	Biolizers	11	Liquid
HR Chemicals	Zebon	10	n.a.
Solex	Big boll	7	Liquid
Abdullah Haseeb Agro Chemicals	Plant feed	7	Liquid
Adv. Agro Tech	Softer	5	Liquid
Solex	Hisum	5	Liquid
Agri Village International	Quick	2.5	Liquid
HR Chemicals	Foster	3.5	n.a.

(Continued)

Company	Brand Name	Product Volume (m³ or Tonne)	Form
Hexon	*FOLIAR-Bk/Flowral*	3.5	Solid
Others		20	n.a.

Total B products: 921 tonnes
Assuming an average of 11% B in all B products,
Total elemental B = 101.31 tonnes

n.a. – data about product form not available

Annex 22

Crude Estimate of Micronutrient Complexes Marketed in Pakistan in 2012

Micronutrients Complex	Company	Brand Name	Consumption (m³)	Remarks
Micronutrients Complex	Solex	*Supreme*	7.15	Zn= 5%, Fe= 2%, Mn= 1%, B= 1%, Cu= 1%
Micronutrients Complex	Welcon Chemicals (Pvt.) Ltd.	*Well Plus*	3.72	Zn= 5%, Cu= 2%, Fe= 2%, Mn= 1%
Micronutrients Complex	SunCrop	*Boll Guard*	80	Zn= 5%, Fe= 1%, Mn= 1%, B= 2%, Cu= 1%
Micronutrients Complex	United Distributor Pakistan Limited (UDL)	*Cash Plus*	88	Zn= 4%, Fe= 1%, Mn= 2%, Cu= 3%
Micronutrients Complex	HR Chemicals	*Tolo*	28.60	n.a.
Micronutrients Complex	Kanzo	*Crown Prince*	14.30	Zn= 5%, Fe= 2%, Mn= 1%, B= 1%, Cu= 1%
Micronutrients Complex	Eon Fertilizer	*Trace*	28.60	Zn= 5%, Fe= 2%, Mn= 1%, Cu= 1%, B= 1%
Micronutrients Complex	Hexon Chemicals (Pvt.) Ltd.	*Growth +*	28.60	n.a.
Micronutrients Complex	Afla Chemicals	*Grow enrich*	8.58	n.a.
Micronutrients Complex	NH Chemicals	*Coupon*	14.30	n.a.
Micronutrients Complex	SunCrop	*Result+*	57.20	n.a.
Micronutrients Complex	Warble (Pvt.) Ltd.	*Crop+*	1.43	Zn= 5%, Cu= 2%, Fe= 2%, Mn= 1%
Micronutrients Complex	Arysta	*Maximo Gold*	0.34	n.a.

(Continued)

Micronutrients Complex	Company	Brand Name	Consumption (m³)	Remarks
Micronutrients Complex	Advance Agrotech	*Agri Plus*	4.93	Zn= 5%, Fe= 2%, Mn= 1%, B= 1%, Cu= 1%
Micronutrients Complex	Total Care	*Total power*	1.43	n.a.
Micronutrients Complex	Auriga	*Microcure*	55	Zn= 4%, Fe= 2%, Mn= 2%, Cu= 1%, B= 1%
		Total complex products	**422**	

n.a. – composition data not available

Annex 23

Zinc Concentration in Zinc Sulfate Fertilizers Sampled from the Rice-Growing Areas of Punjab and Sindh Provinces

		Zn Concentration (%)					
		1987		1991		1993	
S. No.	Trader/Manufacturer	Claimed	Actual	Claimed	Actual	Claimed	Actual
1	NFC Zinc Sulfate	21	20.7	--	--	--	--
2	Ittehad Chemicals, Kala Shah Kaku	27	20.5	35	13.3	35	28
3	FICO Zinc Sulfate	35	23.8	35	35	35	25
4	Pesticide Industries, Lahore	36	23.8	96	34.5	--	--
5	Javedan Agro-Chemical, Gujranwala	33	13.3	--	--	33	24
6	Rich Green Seed Industries, Lahore	35	22.7	--	--	--	--
7	Rice Green	27	13.2	--	--	--	--
8	Sunny Chemical Industries, Sialkot	35	0.65	20	0.5	--	--
9	Seiko Chemical Ltd., Lahore	--	--	35	21	--	--
10	Agro-Farm Services, Karachi	--	--	35	28	40	7
11	Micro-Zinc			52	15	--	--
12	Super-Active China	--	--	35	0.8	--	--

Adapted from NFDC, 1998; and NARC unpublished data.
-- = No data

REFERENCE

NFDC. 1998. Micronutrients in Agriculture: Pakistan perspective. National Fertilizer Development Center, Islamabad, 51 pp.

Annex 24
Residual and Cumulative Effect of Soil-Applied Boron on Crop Productivity and Soil Boron Status of the Rice–Wheat System in Punjab, Pakistan

	Paddy/Grain Yield (t ha⁻¹)*				
B applied (kg ha⁻¹)	2003 Rice	2003–04 Wheat	2004 Rice	2004–05 Wheat	2005 Rice
0	2.80	3.24	2.88	3.14	4.03
0.5	3.20	3.97	3.25	3.55	4.27
1.0	3.51 (25%)[a]	4.11 (27%)	3.51 (22%)	3.90 (24%)	4.55 (13%)
1.5	3.28	4.38	3.61	4.04	4.67

Adapted from Rashid et al. unpublished data.
* Mean data of four field sites.
[a] Yield increase over control yield

Annex 25

Apparent Soil Zinc Balances, at Termination of a Five-Year Cotton–Wheat Cropping System Experiment, as Affected by Nutrient and Crop Residue Management in Two Irrigated Aridisols of Pakistan

	Crop Residue	5-year Input[1]	5-year Crop Uptake	Apparent Balance	Crop Residues	5-year Input[1]	5-year Crop Uptake	Apparent Balance
	\multicolumn{8}{c}{Apparent zinc balance (kg Zn ha^{-1})}							
	\multicolumn{4}{c}{Awagat soil}		\multicolumn{4}{c}{Shahpur soil}				
Flat bed								
Without residue								
FFU[a]		1.26	0.61	0.65		1.41	0.85	0.56
RFU[b]		26.26	1.03	25.23		26.41	1.25	25.16
INM[c]		26.86	1.14	25.72		27.01	1.34	25.67
With residue								
FFU	0.28	1.54	0.68	0.86	0.38	1.79	0.91	0.88
RFU	0.48	26.74	1.15	25.59	0.55	26.96	1.36	25.60
INM	0.54	27.40	1.27	26.13	0.61	27.62	1.47	26.15
Raised beds								
Without residue								
FFU		1.12	0.68	0.44		1.27	0.92	0.35
RFU		26.12	1.14	24.98		26.27	1.35	24.92
INM		26.72	1.26	25.46		26.87	1.46	25.41
With residue								
FFU	0.31	1.43	0.75	0.68	0.40	1.67	0.97	0.70
RFU	0.53	26.65	1.26	25.39	0.60	26.87	1.47	25.40

(Continued)

	Crop Residue	5-year Input[1]	5-year Crop Uptake	Apparent Balance	Crop Residues	5-year Input[1]	5-year Crop Uptake	Apparent Balance
	Apparent zinc balance (kg Zn ha^{-1})							
	Awagat soil				Shahpur soil			
INM	0.59	27.31	1.40	25.91	0.65	27.52	1.58	25.94
LSD ($P \leq 0.05$) Nutrient management	0.02*	1.4*	0.03*	1.7*	0.02*	1.6*	0.02*	1.5*

Adapted/Modified from Source: Rafique et al., 2012, with permission.

[1] Zn input: Fertilizer (in RFU and INM treatments), 25.0 kg ha^{-1}; Farm yard manure, 0.60 kg ha^{-1}; irrigation water (all treatments) – flat bed, 1.12 and 1.32 kg ha^{-1}; raised beds, 0.98 and 1.18 kg ha^{-1}; rainfall (all treatments), 0.14 and 0.09 kg ha^{-1} in Awagat and Shahpur soil

[a] FFU, Farmers' fertilizer use

[b] RFU, recommended fertilizer use

[c] INM, integrated nutrient management

* = significant

REFERENCE

Rafique, E., A. Rashid, and M. Mahmood-ul-Hassan. 2012. Value of soil zinc balances in predicting fertilizer zinc requirement for cotton-wheat cropping system in irrigated Aridisols. *Plant and Soil* 361:43–55.

Annex 26
Micronutrient Determination Methods for Soils

26.1 PLANT-AVAILABLE MICRONUTRIENT CATIONS (FE, CU, MN, ZN) IN SOILS

26.1.1 DTPA METHOD

The diethylene triamine pentaacetic acid (DTPA) test of Lindsay and Norvell (1978) is commonly used for evaluating fertility status with respect to micronutrient cations, i.e., Fe, Zn, Mn, and Cu. DTPA is an important and widely used chelating agent, which combines with free metal ions in the soil solution to form soluble complexes with metal elements. The DTPA method has a capacity to complex each of the micronutrient cations as ten times its atomic weight. The capacity ranges from 550 to 650 mg kg^{-1} depending upon the micronutrient cation. The "universal soil test" for alkaline soils (i.e., AB-DTPA described in the next section) is also equally effective for determining micronutrient cations availability status in alkaline soils.

26.1.1.1 Apparatus

Atomic absorption spectrophotometer (AAS)
  Mechanical shaker, reciprocal

26.1.1.2 Reagents

A. DTPA Extraction Solution

- Weigh 1.97 g diethylene triamine pentaacetic acid (DTPA) and 1.1 g calcium chloride (CaCl$_2$), or 1.47 g CaCl$_2$.2H$_2$O, into a beaker. Dissolve with DI water and then transfer to a 1 L volume.
- Into another beaker, weigh 14.92 g (or add 13.38 mL) triethanolamine (TEA), transfer with DI water into the 1 L flask, and then bring it to about 900 mL volume.
- Adjust its pH to exactly 7.3 with 6 N HCl, and bring it to 1 L volume. This solution contains 0.005 M DTPA, 0.1 M TEA, 0.1 M CaCl$_2$.
- Carbon black

B. Standard Stock Solutions for Iron, Zinc, Copper, and Manganese

Prepare a series of standard solutions for micronutrients in DTPA extraction solution, as given below:

1. **Iron (Fe) standard solutions**
   - Pipette 10 mL Fe Stock Solution (1000 ppm) into 100mL flask and then dilute to volume with DTPA solution. This solution contains 100 ppm Fe (diluted stock solution).
   - Pipette 1, 2, 3, 4, and 5 mL of diluted stock solution into 100 mL numbered flasks and then dilute to volume in each flask with DTPA solution. These solutions contain 1, 2, 3, 4, and 5 ppm Fe, respectively.

2. **Zinc (Zn) standard solutions**
   - Pipette 10 mL Zn stock solution (1000 ppm) into 100 mL flask, and then dilute to volume with DTPA solution. This solution contains 100 ppm Zn (diluted stock solution).
   - Pipette 10 mL diluted stock solution into 100 mL flask, and then dilute to volume with DTPA solution. This solution contains 10 ppm Zn (second diluted stock solution).
   - Pipette 1, 2, 4, 6, 8, and 10 mL second diluted stock solution into 50 mL numbered flasks, and then dilute to volume in each flask with DTPA solution. These solutions contain 0.2, 0.4, 0.8, 1.2, 1.6, and 2.0 ppm Zn, respectively.

3. **Copper (Cu) standard solutions**
   - Pipette 10 mL Cu Stock Solution (1000 ppm) into 100 mL flask, and then dilute to volume with DTPA solution. This solution contains 100 ppm Cu (diluted stock solution).
   - Pipette 10 mL diluted stock solution into 100 mL flask, and then dilute to volume with DTPA solution. This solution contains 10 ppm Cu (second diluted stock solution).
   - Pipette 2, 3, 4, 5, 6, and 7 mL second diluted stock solution in 50 mL numbered flasks, and then dilute to volume in each flask with DTPA solution. These solutions contain 0.4, 0.6, 0.8, 1.0, 1.2, and 1.4 ppm Cu, respectively.

4. **Manganese (Mn) standard solutions**
   - Pipette 10 mL Mn stock solution (1000 ppm) into 100 mL flask and then dilute to volume with DTPA solution. This solution contains 100 ppm Mn (diluted stock solution).
   - Pipette 10 mL diluted stock solution to 100 mL flask, and then dilute to volume with DTPA solution. This solution contains 10 ppm Mn (second diluted stock solution).
   - Pipette 2, 3, 4, 5, 6, and 7 mL second diluted stock solution into 50 mL numbered flasks, and then dilute to volume in each flask with DTPA solution. These solutions contain 0.4, 0.6, 0.8, 1.0, 1.2, and 1.4 ppm Mn, respectively.

26.1.1.3 Procedure

Extraction

1. Weigh 10 g air-dry soil (2mm sieved) into a 125 mL Erlenmeyer flask.
2. Add 20 mL DTPA extraction solution.cover the flasks with stretchable parafilm.
3. Shake for two hours on a reciprocal shaker at 180 cycles per minute.
4. Filter the suspension through Whatman No. 42 filter paper.

Measurement

1. Operate AAS according to the equipment instructions.
2. Run a series of suitable standards for the micronutrient cation to be measured (i.e., Zn, Cu, Fe, or Mn), and draw a calibration curve.
3. Measure the micronutrient cation in the soil extracts by an AAS using hallow cathode lamp for the respective element.
4. Calculate concentrations of the micronutrient cation in samples using the calibration curve.

Calculation

$$\text{Micronutrient cation}\,(\text{ppm}) = \text{ppm MC}\,(\text{from calibration curve})\times(\text{V}\,/\,\text{Wt})$$

where
  MC = Micronutrient cation
  V = Total volume of the extract (mL)
  Wt = Weight of air-dry soil (g)

26.1.1.4 Technical Remarks

1. The theoretical basis for DTPA extraction is the equilibrium of metals in soil solution with the chelating agent. The pH of 7.3 enables DTPA to extract Fe and other metals.
2. The DTPA reagent should be of the acid form (not a di-sodium salt).
3. To avoid excessive dissolution of $CaCO_3$, which may release occluded micronutrients that are not available to crops in calcareous soils and may give erroneous results, the extractant is buffered at slightly alkaline pH.
4. TEA is used as the buffer because it burns clearly during the atomization of extractant solution while being measured on the AAS.
5. Extracting solution can be stored for two weeks under mineral oil, and then the pH adjusted to 7.3, if necessary.
6. The time of shakingis important because trace elements continue to dissolve (non-equilibrium extraction). Therefore, factors such as shaking time, speed, and shape of the vessel are critical and should be standardized in every laboratory.
7. The AB-DTPA soil test data can be interpreted using the generalized guidelines in **Table 2.5**.

REFERENCES

Ryan, J., G. Estefan, and A. Rashid. 2001. *Soil and plant analysis laboratory manual.* 2nd ed. International Center for Agricultural Research in the Dry Areas, Aleppo, Syria.

Lindsay, W. L., and W. A. Norvell. 1978. Development of DPTA soil test for zinc, iron, manganese, and copper. *Soil Science Society of America Journal* 42:422–428.

26.1.2 AMMONIUM BICARBONATE-DTPA (AB-DTPA) METHOD

The AB-DTPA is a multi-element soil test for alkaline soils (for simultaneous extraction of NO_3-N, P, K, Zn, Fe, Cu, and Mn), developed by Soltanpour and Schwab (1977) and later modified by Soltanpour and Workman (1979). The extracting solution is 1 M ammonium bicarbonate (NH_4HCO_3) and 0.005 M DTPA, adjusted to pH 7.6. This method is highly correlated with the $NaHCO_3$ method for P, NH_4OAc method for K, and DTPA method for Zn, Fe, Mn, and Cu. Its range and sensitivity are the same as that of the DTPA test for micronutrients, $NaHCO_3$ test for P, and NH_4OAc test for K.

26.1.2.1 Apparatus

Atomic absorption spectrophotometer (AAS)
Spectrophotometer
Flame photometer
Mechanical shaker, reciprocal
Accurate automatic dilutor

26.1.2.2 Reagents

A. Extracting Solution

- Add 1.97 g diethylene triamine pentaacetic acid (DTPA) to 800 mL DI water; and then add 2 mL 1:1 NH_4OH to facilitate dissolution and to prevent effervescence when bicarbonate is added. This solution contains 0.005 M DTPA.
- When most of the DTPA is dissolved, add 79.06 g NH_4HCO_3 and stir gently until dissolved.
- Adjust pH to 7.6 with NH_4OH, and bring it to 1 L volume with DI water.
- This extraction solution is used immediately or stored under mineral oil.

B. Standard Solutions for Iron, Copper, Manganese, and Zinc

After preparing standard stock solutions for these micronutrients (as described in the previous section for DTPA Soil Test Method), prepare the following working standard solutions:

Fe: 0, 1, 2, 3, 4, 5 ppm
Cu: 0, 1, 2, 3, 4 ppm
Mn: 0, 1, 1.5, 2, 2.5 ppm
Zn: 0, 0.2, 0.4, 0.6, 0.8, 1.0 ppm

26.1.2.3 Procedure

Extraction

1. Weigh 10 g air-dry soil (2-mm sieved) into a 125 mL conical flask.
2. Add 20 mL AB-DTPA extracting solution.
3. Shake on a reciprocal shaker for 15 minutes at 180 cycles/minute, with flasks kept open.
4. Filter the suspensions using Whatman No. 42 filter paper.

Measurement

1. Operate the AAS according to the instructions provided for the equipment.
2. Run a series of suitable standards for the respective micronutrient, and draw a calibration curve.
3. Measure the concentration of the micronutrient in the soil extracts at a suitable wavelength for each micronutrient.
4. Calculate the micronutrient concentrations using the calibration curve.

Calculation

$$\text{Micronutrient Cation}(\text{ppm}): \text{ppm MC}(\text{from calibration curve}) \times \text{dilution factor}$$

26.1.2.4 Technical Remarks

1. The DTPA reagent should be of the acid form.
2. The extracting solution can be stored for two weeks under mineral oil, and then the pH adjusted to 7.6, if necessary.
3. The AB-DTPA soil test data can be interpreted using the generalized guidelines in **Table 2.5**.

REFERENCE

Ryan, J., G. Estefan, and A. Rashid. 2001. *Soil and plant analysis laboratory manual.* 2nd ed. International Center for Agricultural Research in the Dry Areas, Aleppo, Syria.

26.1.3 PLANT-AVAILABLE BORON DETERMINATION IN SOILS

26.1.3.1 Hot-Water Extraction Method

The hot-water extraction (HWE) procedure, introduced by Berger and Truog (1940), and modified later, is the most popular method for measuring "available" soil boron (B) or the fraction of B related to plant growth in alkaline soils. After extraction with hot-water, B in soil extracts is measured calorimetrically using the reagent Azomethine-H (Bingham, 1982). Also, B can be analyzed by colorimetric methods using reagents such as Carmine, and most recently by inductively coupled plasma (ICP) and Atomic Emission Spectrometry (AES). The colorimetric method, using reagent Azomethine-H, and ICP method are preferable because the use of AAS poses some limitations as B is not a metal.

Apparatus
  Erlenmeyer flasks (Pyrex), 50 mL volume
  Spectrophotometer or colorimeter
  Polypropylene test tubes, 10 mL capacity

26.1.3.2 Reagents

A. **Buffer Solution**
Dissolve 250 g ammonium acetate (NH_4OAc) and 15 g ethylenediaminetet-raacetic acid disodium salt (EDTA disodium) in 400 mL DI water. Slowly add 125 mL glacial acetic acid (CH_3COOH), and mix well.

B. **Activated Charcoal (Boron-free)**
This is prepared by giving repeated washings (8–9 times) of DI water (by boiling charcoal with water in a 1:5 ratio), and subsequent filtering. Boron in the filtrate is checked by Azomethine-H color development. Continue washing until charcoal becomes B-free.

C. **Azomethine-H Solution ($C_{17}H_{12}NNaO_8S_2$)**
Dissolve 0.45 g Azomethine-H and 1 g L-ascorbic acid in 100 mL DI water. A fresh reagent should be prepared every week and stored in a refrigerator.

D. **Standard Stock Solution of Boron**
  • Dissolve 0.114 g boric acid (H_3BO_3) in DI water, and bring it to 1 L volume. This solution contains 20 ppm B (stock solution).
  • Prepare a series of standard solutions from the stock solution as follows:

Dilute 2.5, 5.0, 7.5, 10.0, 12.5, and 15.0 mL stock solution of B to 100 mL numbered flasks by adding DI water, and then bringing to volume. These solutions contain 0.5, 1.0, 1.5, 2.0, 2.5, and 3.0 ppm B, respectively.

26.1.3.3 Procedure

A. **Extraction**
  1. Weigh 10 g air-dry soil (2-mm sieved) into a 250 mL Erlenmeyer flask (Pyrex).
  2. Add about 0.2 g B-free activated charcoal.
  3. Add 20 mL DI water.
  4. Boil on a hot plate for five minutes with the flask covered by a watch glass.
  5. Then, filter the suspension immediately through Whatman No. 40 filter paper.

B. **Measurement**
  1. Pipette 1 mL aliquot of the extract into a 10 mL polypropylene tube.
  2. Add 2 mL buffer solution.
  3. Add 2 mL Azomethine-Hsolution, and mix well.
  4. Prepare a standard curve as follows:
      • Pipette 1 mL of each standard (0.5–3.0 ppm B), and proceed as for the samples.
      • Also make a blank, by pipetting 1 mL DI water, and proceeding as for the samples.

5. Read the absorbance of blank, standards, and samples after 30 minutes on a spectrophotometer at 420nm wavelength.
6. Prepare a calibration curve for standards, by plotting absorbance against the respective B concentration.
7. Read B concentration in the unknown samples from the calibration curve.

Calculation

$$B(ppm) : ppm\, B(\text{from calibration curve}) \times (V / Wt)$$

where
V = Total volume of the soil extract (mL)
Wt = Weight of air-dry soil (g)

Where HWE soil B levels are less than 0.5 ppm, deficiency is likely to occur for most crops. However, where HWE soil B levels are greater than about 5 ppm, toxicity to plants may occur. Generalized criteria for interpreting HWE soil B data are given in **Table 2.5**.

26.1.3.4 Technical Remarks

1. Use of glassware should be minimal; where absolutely essential, always use concentrated HCl-treated glassware (for a week).
2. Use of Azomethine-H is an improvement over that of carmine and curcumin since the procedure involving this chemical does not require the use of concentrated acid.
3. The amount of charcoal added may vary with OM content of the soil and should be just sufficient to produce a colorless soil extract after five minutes of boiling on a hot plate. Unnecessary excessive amounts of charcoal can reduce extractable B values.
4. In humid regions' soils, borate ions may tend to leach from soils. Therefore, soluble B concentrations are low in highly leached soils. Since highly leached soils usually exhibit low soil pH, acid soils are frequently deficient in B.
5. In arid region soils, borate ions are usually not affected by leaching. Therefore, strongly alkaline soils may contain excessive amounts of B for plant growth.

REFERENCES

Ryan, J., G. Estefan, and A. Rashid. 2001. *Soil and plant analysis laboratory manual.* 2nd ed.International Center for Agricultural Research in the Dry Areas, Aleppo, Syria.

Berger, K. C., and E. Truog. 1940. Boron deficiency as revealed by plant and soil test. *American Society of Agronomy* 32:297–301.

Bingham, F.T. 1982. Boron. In: Page, A.L. (Ed.), *Methods of Soil Analysis, Part 2: Chemical and Microbial Properties.* American Society of Agronomy, Madison, WI, USA., p. 431–448.

26.1.3.5 Dilute Hydrochloric Acid (HCl) Method

Though hot-water extraction method (HWE) is quite popular for predicting B fertility in alkaline soils, the procedure is tedious and prone to error (because of difficulty in maintaining uniform boiling time for all conical flasks on the hot plate). In an effort of having a convenient substitute soil test, researchers in Pakistan (Kausar et al., 1990; Rashid et al., 1994) have found that the dilute HCl method of Ponnamperuma et al. (1981), originally designed for acid soils, is equally effective in diagnosing B deficiency in alkaline and calcareous soils. The HCl method is simple, economical, and more efficient (Rashid et al., 2002; Rashid, 2006; Rashid and Rafique, 2017).

26.1.3.5.1 Reagents

A. Buffer Solution

Dissolve 250 g ammonium acetate (NH_4OAc) and 15 g ethylenediaminetetraacetic acid disodium salt (EDTA disodium) in 400 mL DI water. Slowly add 125 mL glacial acetic acid (CH_3COOH), and mix well.

B. Azomethine-H Solution ($C_{17}H_{12}NNaO_8S_2$)

Dissolve 0.45 g Azomethine-H and 1 g L-ascorbic acid in 100 mL DI water. A fresh reagent should be prepared every week and stored in a refrigerator.

C. Activated Charcoal (boron-free)

This is prepared by giving repeated washings (8–9 times) of DI water (by boiling charcoal with water in a 1:5 ratio), and subsequent filtering. Check B in the filtrate by Azomethine-H color development; continue washing till charcoal becomes B-free.

D. Hydrochloric Acid (HCl), 0.05 N

Dilute 4.14 mL concentrated HCl (37%, sp. gr. 1.19) in DI water, mix well, and bring it to 1 L volume.

E. Standard Stock Solution of B

- Dissolve 0.114 g boric acid (H_3BO_3) in DI water, and bring it to 1 L volume. This solution contains 20 ppm B (stock solution).
- Prepare a series of standard solutions from the stock solution as follows:

Dilute 2.5, 5.0, 7.5, 10.0, 12.5, and 15.0 mL stock solution to 100 mL numbered flask by adding DI water, and then bringing to volume. These solutions contain 0.5, 1.0, 1.5, 2.0, 2.5, and 3.0 ppm B, respectively.

26.1.3.5.2 Procedure

A. Extraction

1. Weigh 10 g air-dry soil (2 mm sieved) into a polypropylene tube or into a 50 mL Erlenmeyer flask (Pyrex).
2. Add about 0.2 g activated charcoal (B-free).
3. Add 20 mL 0.05 N HClsolution.
4. Shake for five minutes, and then filter the suspension using Whatman No. 40 filter paper.

B. Measurement (Azomethine-H method)
1. Pipette 1 mL aliquot of the extract into a 10 mL polypropylene tube.
2. Add 2 mL buffer solution.
3. Add 2 mL Azomethine-H solution, and mix well.
4. Prepare a standard curve as follows:
   - Pipette 1 mL of each standard (0.5–3.0 ppm), and proceed as for the samples.
   - Also make a blank, by pipetting 1 mL DI water, and proceeding as for the samples.
5. Read the absorbance of blank, standards, and samples after 30 minutes on a Spectrophotometer at 420 nm wavelength.
6. Prepare a calibration curve for standards, by plotting absorbance against the respective B concentrations.
7. Read B concentration in the unknown samples from the calibration curve.

Calculation

$$B(ppm): ppm\, B(from\ calibration\ curve) \times (V / Wt)$$

where
   V = Total volume of the soil extract (mL)
   Wt = Weight of air-dry soil (g)

Generalized criteria for interpreting HCl extractable soil B data are given in **Table 2.5**.

REFERENCES

Kausar, M. A., M. Tahir, and A. Hamid. 1990. Comparison of three methods for the estimation of available boron for maize. *Pakistan Journal of Scientific and Industrial Research* 33:221–224.

Ponnamperuma, F. N., M. T. Caytan, and R. S. Lantin. 1981. Dilute hydrochloric acid as an extractant for available zinc, copper and boron in rice soils. *Plant and Soil* 61:297–310.

Rashid, A. 2006. Boron deficiency in soils and crops of Pakistan: Diagnosis and management. Pakistan Agricultural Research Council, Islamabad, Pakistan, viii+34 pp. ISBN: 969-409-184-5

Rashid, A., and E. Rafique. 2017. Boron deficiency diagnosis and management in field crops in calcareous soils of Pakistan: A mini review. *BORON* 2 (3): 142–152.

Rashid, A., E. Rafique, and B. Bughio. 1994. Diagnosing boron deficiency in rapeseed and mustard by plant analysis and soil testing. *Communications in Soil Science and Plant Analysis* 25:2883–2897.

Rashid, A., E. Rafique, and J. Ryan. 2002. Establishment and management of boron deficiency in field crops in Pakistan: A country report. p. 339–348, In: Goldbach, H. E. et al (Ed), *Boron in Plant and Animal Nutrition.* Kluwer Academic Publishers, New York.

Ryan, J., G. Estefan, and A. Rashid. 2001. *Soil and plant analysis laboratory manual.* 2nd ed. International Center for Agricultural Research in the Dry Areas, Aleppo, Syria.

26.2   IODINE DETERMINATION IN SOILS

26.2.1   TETRAMETHYL AMMONIUM HYDROXIDE (TMAH) EXTRACTABLE IODINE IN SOILS

- Weigh 0.25 g soil sample (dry weight basis) directly into a 15 mL poly(tetrafluoroethene) HDPE Nalgene bottle.
- Add 5 mL of 5% TMAH, and shake to mix.
- Place sample bottles, with loose lids in a drying oven at 70 °C for 3 h.
- Shake the bottles manually after 1.5 h.
- After 3 h of heating, add 5 mL of deionized water and then centrifuge the bottles at 2500 rpm for 20 minutes.
- Remove the supernatant from the top of the sample solution and dilute to a final matrix of 0.5% TMAH.
- Use certified reference material (CRM), within each extraction batch, to monitor the performance of the TMAH extraction and subsequent analysis by ICP-MS.
- Iodine concentration in the extract is analyzed by ICP-MS, after dilution, if needed.

REFERENCE

Watts, M. J., and C. J. Mitchell. 2009. A pilot study on iodine in soils of Greater Kabul and Nangarhar provinces of Afghanistan. *Environmental Geochemistry and Health* 31:503–509.

26.2.2   WATER-SOLUBLE IODINE IN SOILS

Soluble iodine in soils can be determined by cold water extraction. 12.5 mL of deionized water and 1.25 g of soil (dry weight basis) are shaken for 15 minutes, centrifuged at 3000 rpm for 10 minutes and the extract is adjusted to a matrix of 0.5% TMAH for analysis. Use certified reference material (CRM), within each extraction batch, to monitor the performance of the TMAH extraction and subsequent analysis by ICP-MS.

Iodine concentration in the extract is analyzed by ICP-MS, after dilution, if needed.

REFERENCE

Watts, M. J., and C. J. Mitchell. 2009. A pilot study on iodine in soils of Greater Kabul and Nangarhar provinces of Afghanistan. *Environmental Geochemistry and Health* 31:503–509.

Annex 27

Micronutrients Determination Methods for Plant-Origin Materials

Microwave-assisted automatic digestion techniques are not available in agricultural labs in Pakistan. As full recovery of micronutrient cations (Zn, Fe, Mn, and Cu) in high-silica-containing plant tissues (such as wheat, barley, rice, sugarcane, etc.) is not possible by the Dry Ashing procedure. Wet Digestion procedure is used, across the country, to determine total nutrient contents (like P, K, Zn, Cu, Fe, Mn) in plant tissues. This kind of plant material must be wet-digested using HNO_3-$HClO_4$ mixture. It is recommended to always use a reference plant material (like NIST reference material of plant origin) to verify the recovery efficiency of the element of interest.

27.1   ZINC, IRON, COPPER, AND MANGANESE DETERMINATION IN PLANT-ORIGIN MATERIAL (WET DIGESTION)

Digestion

1. Weigh 1 g dry and ground plant material, and then transfer quantitatively into a 100-mL Pyrex digestion tube.
2. Add 10 mL (2:1 ratio) Nitric Acid (HNO_3)-Perchloric Acid($HClO_4$) mixture, and allow to stand overnight or until the vigorous reaction phase is over.
3. Place small and short-stemmed funnels in the mouth of the tubes to reflux acid.
4. After the preliminary digestion, place the tubes in a cold block digester, and then raise the temperature to 150 °C for 1 hour.
5. Place the U-shaped glass rods under each funnel to permit exit of volatile vapors.
6. Increase temperature slowly until all traces of HNO_3 disappear, and then remove U-shaped glass rods.
7. Raise temperature to 235 °C. When the dense white fumes of $HClO_4$ appear in the tubes, continue digestion for 30 minutes more.
8. Lift the tubes rack out of the block digester, cool for a few minutes, and add a few drops of DI water carefully through the funnel.
9. After vapors condense, add DI water in small increments for washing down walls of tubes and funnels.

10. Bring to volume, mix the solution of each tube, and then leave the tubes undisturbed for a few hours.
11. Clear supernatant digest is analyzed for determining concentrations of Zn, Cu, Fe, and Mn.
12. Each batch of samples for digestion should contain at least one reagent blank (without plant material).

Measurement
1. Operate Atomic Absorption Spectrophotometeraccording to the instructions provided for the equipment.
2. Run a series of suitable standards for the intended micronutrient, and draw a calibration curve.
3. Decant the supernatant liquid and analyze for Fe, Cu, Mn, and Zn in the aliquots by AtomicAbsorption Spectrophotometer.
4. Calculate the supernatant liquid concentrations according to the calibration curve.

Calculations

$$Fe, Cu, Mn, or\ Zn\,(ppm) = ppm\ Fe, Cu, Mn,$$

or

$$Zn\ from\ calibration\ curve \times (V\,/\,Wt)$$

where
V = Total volume of the plant digest (mL)
Wt = Weight of dry plant (g)

REFERENCE

Ryan, J., G. Estefan, and A. Rashid. 2001. *Soil and Plant Analysis Laboratory Manual*, 2 nd ed International Center for Agricultural Research in the Dry Areas (ICARDA), Aleppo, Syria.

27.2 BORON DETERMINATION IN PLANT-ORIGIN MATERIAL (DRY ASHING)

Boron in plant samples is measured by dry ashing (Chapman and Pratt, 1961) and subsequent measurement of B concentration by colorimetry using azomethine-H (Bingham, 1982).

Dry Ashing
1. Weigh 1 g dry and ground plant material in a porcelain crucible.
2. Ignite in a muffle furnace by slowly raising the temperature to 550 °C.

3. Continue ashing for 6 hours after attaining 550 °C.
4. Wet the ash with five drops DI water, and then add 10 mL 0.36 NH_2SO_4 solution into the porcelain crucibles.
5. Heat on a steam bath for 20 minutes.
6. Let stand at room temperature for 1 hour, stirring occasionally with a plastic rod to break up ash.
7. Filter through Whatman No. 1 filter paper into a 50 mL polypropylene flask and bring to volume. The filtrate is ready for B determination.

Measurement

1. Pipette 1 mL an aliquot of the extract into a 10 mL polypropylene tube.
2. Add 2 mL buffer solution.
3. Add 2 mL azomethine-Hsolution, and mix well.
4. Prepare a standard curve as follows:
   - Pipette 1 mL of each standard (0.5–3.0 ppm), and proceed as for the samples.
   - Make a blank with 1 mL DI water, and proceed as for the samples.
5. After 30 minutes, read the absorbance of blank, standards, and samples on the Spectrophotometer at 420 nm wavelength.
6. Prepare a calibration curve for standards, plotting absorbance against the respective B concentrations.
7. Read B concentration in the unknown samples from the calibration curve.

Calculation

$$B(ppm) = ppm\,B(from\ calibration\ curve) \times (V / Wt)$$

where
V = Total volume of the extract (mL)
Wt = Weight of dry plant (g)

Normal B concentration in mature leaves of most plants is reported to be between 20 and 100 mg kg^{-1} dry matter. Generalized criteria for interpreting plant analysis B data are given in **Table 2.6**.

REFERENCES

Ryan, J., G. Estefan, and A. Rashid. 2001. *Soil and Plant Analysis Laboratory Manual*, 2nd ed. International Center for Agricultural Research in the Dry Areas (ICARDA), Aleppo, Syria.

Bingham, F. T. 1982. Boron. In: Page, A. L. (Ed.), *Methods of Soil Analysis, Part 2: Chemical and Microbial Properties*. American Society of Agronomy, Madison, WI, USA., p. 431–448.

27.3 FERROUS IRON [FE(II)] ANALYSIS IN FRESH PLANT TISSUE

27.3.1 REAGENTS

27.3.1.1 Extraction Solution ($C_{12}H_8N_2$), 1.5% in HCl-buffer with pH 3.0

Add 15 g *1–10 o-phananthroline* to about 850 mL DI water. Dropwise, add 1 *N* HCl by continuously stirring the solution until last traces of *1–10 o-phananthroline* are solubilized. The final pH of the solution will be around 3.0. Make volume to 1 L with DI water.

27.3.1.2 Standard Solution

Prepare a working solution, using ferrous sulfate, containing 0, 1.0 1.5, 2.0, 2.5, and 3.0 ppm Fe(II) in the extraction solution.

Freshly prepared standards are run along with the plant extracts.

27.3.2 PROCEDURE

Extraction
1. Use carefully washed (and blotted) plant tissue for ferrous analysis.
2. Weigh 2 g fresh (chopped with a pair of stainless steel scissors) plant material into a 50 mL Erlenmeyer flask.
3. Add 20 mL extraction solution, and stir gently to ensure that all the plant tissue is completely immersed in the solution.
4. Close the flask using parafilm, and allow it to stand for about 16 hours at room temperature in dark (by keeping it inside a lab cabinet).
5. Filter the contents through Whatman No. 1 filter paper.

Measurement
1. Fe(II) content in the filtrate is determined by a spectrophotometer at 510 nm wavelength. Freshly prepared Fe(II) standards are run along with the extracts.
2. Ferrous content in plant tissue is expressed on an oven dry weight basis, after determining moisture content in a sub-sample of fresh plant tissue (or determining moisture content in the extracted plant tissue).

REFERENCE

Katyal, J.C., and B.D. Sharma. 1980. A new technique of plant analysis to resolve iron chlorosis. *Plant and Soil* 55:105–119.

27.4 IODINE DETERMINATION IN PLANT-ORIGIN MATERIALS

- Weigh 0.25 g of the plant sample directly into microwave vessels, add 5 mL of 5% TMAH, and shake to mix.

- Cap the vessels and place them in the microwave and heat at 1600W, to ramp up to 70°C over 10 minutes, and then hold at 70 °C for 60 minutes. This approach would produce a cleaner extract solution, especially for grain samples, compared to the heating method used for soil and leaf samples.
- After heating, add 5 mL of deionized water and then centrifuge the bottles at 2500 rpm for 20 minutes.
- Remove the supernatant from the top of the sample solution and dilute it to a final matrix of 0.5% TMAH.
- Use certified reference materials (CRMs), within each extraction batch, to monitor the performance of the TMAH extraction and subsequent analysis by ICP-MS.
- Then, analyze iodine concentration in the extract by ICP-MS, after needed dilution.

REFERENCE

Watts, M. J., and C. J. Mitchell. 2009. A pilot study on iodine in soils of Greater Kabul and Nangarhar provinces of Afghanistan. *Environmental Geochemistry and Health* 31:503–509.

Annex 28
Micronutrients Determination Methods for Fertilizers

28.1 WATER-SOLUBLE ZINC IN FERTILIZER

This test method is applicable to fertilizers that contain zinc contents as zinc sulfate. Extraction is done by boiling test sample in distilled water. Zinc contents are determined by flame atomic absorption spectrometry (FAAS).

28.1.1 REAGENTS

- Certified reference material (CRM) of zinc (1000 ± 5ppm Zn, stock solution). It should be NIST-traceable and manufactured by a ISO-certified company.
- Prepare a stock solution of 100ppm in Deionized water and can be stored for 15 days. The electrical conductivity (EC) of water should not be more than 20μS/cm.
- Prepare at least three working standards between 0 to 3ppm. Use graduated pipette for the preparation of working standards. Read the lower meniscus for colorless liquids keeping a mark on the pipette at eye-level.

28.1.2 PROCEDURE

Sample preparation: Grind almost 100 g of the sample as it is (do not dry or desiccate), and pass all the ground samples from Mesh No. 40 sieve. Place 1.00 g well-ground test portion (2mL for liquid sample) in a 150 mL glass beaker (tall form). Add 75 mL de-ionized water and boil for 30 minutes (count time after boiling begins) on a hot plate at 150 °C. Put watchglass on the beaker. Wash the watchglass into filtrate to avoid sample loss. Wash the beaker, watchglass into a 1 L volumetric flask. Make the volume up to the mark with distilled water. Shake well and filter through Whatman No. 42 filter paper. Re-dilute if necessary.

Measurements: Optimize Atomic Absorption Spectrophotometer (AAS) parameters, i.e., fuel flow, lamp energy current, lamp orientation, and burner height. Calibrate the AAS between 0 and 3 ppm standards. The R2 value of calibration curve should be 0.998 or higher otherwise recheck or repeat. Measure readings between 0.3 and 2.8 ppm. In case the reading is above or below this limit, dilute or concentrate the sample accordingly. Determine the concentration of the element in solution (ppm) from calibration curve or digital concentration readout following the standard operating parameters. Run the blank (D.I. water) between working standards and samples. Run the reference salt as a check sample. This is optional. Zinc will be determined using AAS at 213.8 nm wavelength within a range of 0–3.0 ppm.

28.1.3 CALCULATIONS

$$Zinc(\%) = Reading \times dilution / 10,000.$$

REFERENCES

Official Methods of Analysis of AOAC International, 20th ed. 2016. Volume-1. Method No. 2.6.01 (AOAC Official Method 965.09) Fertilizers, Chapter 2, p. 29–30.
Qazi, M. A., M. S. A. Khan, and F. Ahmed. 2021. *SFRI-Guide from Sample Receiving to Issuance of Test Results*. Directorate of Soil Fertility Research Institute, Department of Agriculture, Punjab, Lahore.

28.2 ACID-SOLUBLE ZN, CU, FE, AND MN IN INORGANIC FERTILIZER PRODUCTS AND MIXED FERTILIZERS

This test method is applicable to inorganic mixed fertilizers containing zinc, copper, iron, and manganese. The process involves wet digestion with acid to release and solubilize the nutrients, and flame atomic absorption spectrometry technique is used for determination.

28.2.1 REAGENTS

- Concentrated HCl (purity 37%)
- 2 M HCl solution: Dissolve 165.8 mL of 37% pure HCl in 1 L volumetric flask and makeup to 1 L.
- 0.5M HCl solution: Dissolve 82.89 mL of 37% pure HCl in 2 L volumetric flask and makeup to 2 L.
- Certified reference material (CRM) of zinc, copper, iron, and manganese (1000 ± 5ppm, stock solution). CRM should be NIST traceable and manufactured by ISO-certified company.

Prepare a stock solution of 100 ppm in D.I. water and store for 15 days. The electrical conductivity (EC) of water should not be more than 20 µS/cm.

Prepare at least three working standards considering the detection range of respective element. Use graduated pipette for preparing working standards. Read the lower meniscus for colorless liquids keeping a mark on the pipette at the eye-level.

28.2.2 PROCEDURE

Grind almost 100 g of the sample as it is (do not dry or desiccate), and pass all the ground samples from Mesh No. 40 sieve. Weigh 1g well-ground, homogenized test portion or 2 mL liquid sample into a 100 mL glass beaker. Add 10 mL concentrated 37% pure HCl. Boil and evaporate the solution nearly to dryness on a hot plate. Do not bake residue. Re-dissolve residue in 20 mL 2 M HCl, by boiling gently, if necessary. Wash the beaker, watchglass into a 100 mL volumetric flask. Make the volume up to the mark with distilled water. Shake well and filter through Whatman No. 41 filter paper. Measure the absorbance of respective micronutrient directly or dilute with 0.5 M HCl to obtain a solution within the range of the calibration curve by using instrument conditions for each element. Determine the concentration of the element in solution (mg/L) from the calibration curve or digital concentration readout following the standard operating parameters.

28.2.3 CALCULATIONS

$$\text{Element}\% = \text{Reading} \times \text{dilution} / 10,000.$$

REFERENCES

Official Methods of Analysis of AOAC International, 20th ed. 2016. Volume-1. Method No. 2.6.01-C (AOAC Official Method 965.09) Fertilizers, Chapter 2, Subchapter 6, p. 29–30.

Qazi, M. A., M. S. A. Khan, and F. Ahmed. 2021. *SFRI-Guide from Sample Receiving to Issuance of Test Results*. Directorate of Soil Fertility Research Institute, Department of Agriculture, Punjab, Lahore.

28.3 WATER-SOLUBLE MICRONUTRIENTS (ZN, CU, FE, AND MN) BY AAS IN INORGANIC (SYNTHETIC) FERTILIZERS

The process involves water extraction to solubilize the nutrients, and atomic absorption spectrophotometric method is applied for determination.

28.3.1 REAGENTS

- Certified reference material (CRM) of zinc, copper, iron, and manganese (1000 ± 5 ppm, stock solution). CRM should be NIST traceable and manufactured by ISO-certified company.
- Prepare a stock solution of 100ppm in D.I. water and store for 15 days. The electrical conductivity of water should not be more than 20 µS/cm.

28.3.2 PROCEDURE

Place 1.00 g test portion into a 100 mL glass beaker. Add 75 mL D.I. water and boil for 30 minutes.

Filter in 1 L volumetric flask, washing filter with D.I. water. Make the volume up to the mark with D.I. water. Re-dilute if necessary. Determine the concentration of the element in solution (mg/L) from calibration curve or digital concentration readout following the standard operating parameters.

28.3.3 CALCULATIONS

$$\%\text{element} = \left(\mu g / mL\right) \times \text{dilution} \times 10^{-4}.$$

REFERENCES

Official Methods of Analysis of AOAC International, 20th ed. 2016. Volume-1. Method No. 2.6.01-C (AOAC Official Method 965.09) Fertilizers, Chapter 2, Subchapter 6, p. 29–30.
Qazi, M. A., M. S. A. Khan, and F. Ahmed. 2021. SFRI-Guide from Sample Receiving to Issuance of Test Results. Directorate of Soil Fertility Research Institute, Department of Agriculture, Punjab, Lahore.

28.4 CHARRED/ASHED FRACTION OF MULTI-MICROS (ZN, CU, FE, AND MN) BY AAS IN ORGANIC FERTILIZERS

The method is applicable to organic fertilizers containing multi-micros nutrients (minors). The process involves furnace ashing of fertilizer to release/solubilize the nutrients and atomic absorption spectrophotometric method is applied for determination.

28.4.1 REAGENTS

*   Certified reference material (CRM) of zinc, copper, iron, and manganese (1000 ± 5 ppm, stock solution). CRM should be NIST traceable and manufactured by ISO-certified company.
*   Prepare a stock solution of 100ppm in D.I. water and store for 15 days. The electrical conductivity of water should not be more than 20 µS/cm.

28.4.2 PROCEDURE

Place 1.00 g sample in a 150 mL beaker (Pyrex, or equivalent). Char on a hot plate and ignite for 1 hour at 500 °C with the muffle door propped open to allow free access of air. Break up the cake with a stirring rod and dissolve in 10 mL HCl. Boil and evaporate solution nearly to dryness on a hot plate while covering the beaker with watchglass. Do not bake residue. Re-dissolve residue in 20 mL 2MHCl, boiling gently, if necessary. Wash the beaker, watchglass into a 100 mL volumetric flask. Make the volume up to the mark with distilled water. Shake well and filter through Whatman No. 41 filter paper. Measure absorption of solution directly or dilute with 0.5 M HC1 to obtain solutions within prescribed ranges of the AAS.

28.4.3 CALCULATIONS

$$\%element = (\mu g / mL) \times dilution \times 10^{-4}.$$

REFERENCES

Official Methods of Analysis of AOAC International, 20th ed. 2016. Volume-1. Method No. 2.6.01-C (AOAC Official Method 965.09) Fertilizers, Chapter 2, Subchapter 6, p. 29–30.

Qazi, M. A., M. S. A. Khan, and F. Ahmed. 2021. *SFRI-Guide from sample receiving to issuance of test results.* Directorate of Soil Fertility Research Institute, Department of Agriculture, Punjab, Lahore.

28.5 CHELATED ZN CONTENTS IN CHELATED FERTILIZERS

This method is applicable to fertilizers containing chelated zinc. Mineral-Zn fraction is masked by Na_2CO_3 solution and the filtrate is directly run for analysis of chelated fraction through AAS.

28.5.1 REAGENTS

- Certified reference material (CRM) of zinc (1000 ± 5 ppm Zn, stock solution). CRM should be NIST traceable and manufactured by ISO-certified company.
- Prepare stock solution of 100 ppm in D.I. water and store for 15 days. The electrical conductivity of water should not be more than 20 µS/cm.
- Prepare working standards at least 3 and between 0 to 3ppm. Use a graduated pipette for the preparation of working standards. Read the lower meniscus for colorless liquids keeping a mark on the pipette at the eye-level.
- Na_2CO_3 (0.1 M): Dissolve 21.2 g Na_2CO_3 in 2 L deionized water.
- H_2SO_4 AR grade

28.5.2 PROCEDURE

Take 1.25 g homogenized sample previously ground and sieved through Mesh No. 40 (or 2 mL filtered liquid sample) in a 250 mL flask. Add 100 mL Na_2CO_3 (0.1 M) solution and shake well. Make volume with Na_2CO_3 (0.1 M) solution up to the mark. Let the sample stay for 10 minutes. Non-chelated zinc will precipitate; filter the solution through Whatman filter paper No.42. Take 1 mL of filtrate in 100 mL volumetric flask and add 5 mL conc. H_2SO_4 and make a volume of 100 mL using distilled water. Take the reading on an AAS.

28.5.3 CALCULATIONS

$$Zinc(\%) = Reading \times dilution / 10,000.$$

For solid fertilizer, the dilution factor would be 2 and for liquid fertilizer, the dilution factor would be 1.25.

REFERENCES

Khan, M. S. A., M. A. Qazi, S. M. Mian, and M. Akram. 2013. Comparison of three analytical methods for separation of mineral and chelated fraction from an adulterated Zn-EDTA fertilizer. *Journal of the Chemical Society of Pakistan* 35:344–346.

Mendham, J., R. C. Denney, J. D. Barnes, M. Thomas. 2000. Vogel's textbook of quantitative chemical analysis. 6th ed. Prentice Hall, Harlow, England; New York. 806 pp.

Official Methods of Analysis of AOAC International, 20th ed. 2016. Volume-1. Method No. 2.6.01 (AOAC Official Method 965.09) Fertilizers, Chapter 2, p. 29–30.

Qazi, M. A., M. S. A. Khan, and F. Ahmed. 2021. *SFRI-Guide from Sample Receiving to Issuance of test Results*. Directorate of Soil Fertility Research Institute, Department of Agriculture, Punjab, Lahore.

28.6 CHELATED FE CONTENTS IN CHELATED FERTILIZERS

This method is applicable to fertilizers containing chelated Fe. The test portion is dissolved in water, and non-chelated Fe is precipitated as $Fe(OH)_3$ at pH 8.5, and removed. Chelated Fe is determined by using standard solutions containing Na_2H_2EDTA.

28.6.1 REAGENTS

- Sodium Hydroxide solution: 0.5 M. Dissolve 20 g NaOH in H_2O and dilute to 1 L.
- Disodium EDTA solution: 0.66%, dissolve 0.73 g $Na_2H_2EDTA.2H_2O$ in H_2O and dilute to100 mL.
- 0.5 M HCl solution: Dissolve 82.89 mL of 37% pure HCl in a 2 L volumetric flask and make it up to 2 L.
- Iron stock solution: Certified reference material (CRM) of iron (1000 ppm Fe). CRM should be manufactured by ISO-certified company and must be NIST traceable.
- Intermediate solution: 100 μg Fe/100 mL pipette 10 mL Fe stock solution and 10 M $Na_2H_2EDTA.2H_2O$ solution into 100 mL volumetric flask and dilute to volume with water.
- Working solutions: Dilute aliquots of intermediate solutions with 0.5 M HCl to make $\geq$ 4 standard solutions within the range of determination (2–20 μg Fe/mL)

28.6.2 PROCEDURE

Weigh 1 g well-ground, sieved through Mesh No. 40, and homogenized test portion or 2 mL liquid sample into 250 mL tall-form beaker. Wet with 2–3 drops of ethanol and dissolve in 100 mL H_2O. Add 4 drops of 30% H_2O_2; mix and adjust the pH of the solution to 8.5 with 0.5 M NaOH. If pH drifts above 8.8, discard the solution and repeat the analysis. Transfer solution to 250 mL volumetric flask. Dilute to volume with water and mix. Filter solution through quantitative paper. Pipette 1 mL filtrate into 100 mL volumetric flask and dilute to volume with 0.5 M HCl. Determine the concentration of iron in solution (mg/L) from the calibration curve or digital concentration readout following the standard operating parameters of AAS.

28.6.3 CALCULATIONS

$$\text{Chelated Fe}\,(\%) = \text{Reading} \times \text{dilution factor} / 10,000.$$

REFERENCES

Official Methods of Analysis of AOAC International, 20th ed. 2016. Volume-1. Method No. 2.6.01-C (AOAC Official Method 965.09) Fertilizers, Chapter 2, Subchapter 6, p. 29–30.

Qazi, M. A., M. S. A. Khan, and F. Ahmed. 2021. *SFRI-Guide from Sample Receiving to Issuance of Test Results.* Directorate of Soil Fertility Research Institute, Department of Agriculture, Punjab, Lahore.

28.7 CHELATED CU AND MN CONTENTS IN CHELATED FERTILIZERS

This is a modified method based on AOAC method 983.03. The sample is dissolved in water and pH of the solution is adjusted to 8.5–8.8. Non-chelated mineral fraction is precipitated and removed. Chelated fraction is detected by an AAS.

28.7.1 REAGENTS

- Sodium hydroxide solution: 0.5N. Dissolve 20g NaOH in H_2O and dilute to 1L.
- Hydrogen peroxide solution: 30%.

28.7.2 PROCEDURE

Weigh 0.5 g sample (ca 40 mg) into 200 mL tall-form beaker. Wet with 2–3 drops of alcohol and dissolve in 100 mL DW. Add 4 drops of 30% H_2O_2, mix and adjust pH of solution to 8.5–8.8 with 0.5 N NaOH. If pH drifts above 8.8, discard the solution and repeat the analysis. Transfer solution to 250 mL vol. flask, dilute to volume with DW, and mix. Filter solution using Whatman 42 filter paper. Dilute further, where necessary. Use air–acetylene flame and detect concentration from either calibration curve or digital readout. In the same manner, run blank on all reagents used and take the reading on an AAS following standard operating procedure.

28.7.3 CALCULATIONS

$$\%\text{element} = \left(\mu g \,/\, mL\right) \times \text{dilution} \times 10^{-4}$$

REFERENCES

Villén, M., J. J. Lucena, M. C. Cartagena, et al. 2007. Comparison of two analytical methods for the evaluation of the complexed metal in fertilizers and the complexing capacity of complexing agents. *Journal of Agricultural Food Chemistry* 55:5746–5753.

Qazi, M. A., M. S. A. Khan, and F. Ahmed. 2021. *SFRI-Guide from Sample Receiving to Issuance of Test Results*, Directorate of Soil Fertility Research Institute Department of Agriculture, Punjab, Lahore.

28.8   WATER-SOLUBLE BORON IN FERTILIZER (AZOMETHINE-H METHOD)

This test method is applicable to fertilizers containing borate fertilizers, etc.

- Weigh about 2.500 g of an analytical sample, and put it in a 300 mL beaker.
- Add about 200 mL of DI water, cover with a watch glass, and boil on a hot plate for about 15 minutes.
- Let it cool; then transfer to a 250 mL volumetric flask with DI water.
- Add DI water up to the marked line.
- Filter with Type 3 filter paper to make the sample solution.
- Transfer a predetermined volume of an analytical sample to a 100 mL volumetric flask.
- Add 25 mL of EDTA solution (to mask contamination of Cu, Fe, etc., if any), and add 10 mL of NH_4OAc solution and 10 mL of Azomethine-H solution successively, and further add DI water up to the marked line; then leave undisturbed for about 2 hours.
- Using a Spectrophotometer, measure absorbance at 415 nm to calculate the water-soluble boron (as B_2O_3).

REFERENCE

IAA (Incorporated Administrative Agency). 2020. Testing Methods for Fertilizers (2020). English Translated Version, Food and Agricultural Materials Inspection Center, Ministry of Agriculture, Forestry and Fisheries - Japan, p 316-317 (Accessible online, http://www.famic.go.jp/ffis/fert/obj/TestingMethodsForFertilizers2020.pdf).

Index

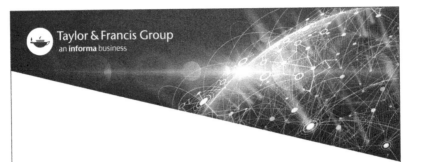

Printed in the United States
by Baker & Taylor Publisher Services

Printed in the United States
by Baker & Taylor Publisher Services